History of Nonlinear Oscillations Theory in France (1880–1940)

Archimedes

NEW STUDIES IN THE HISTORY AND PHILOSOPHY OF SCIENCE AND TECHNOLOGY

VOLUME 49

Archimedes has three fundamental goals; to further the integration of the histories of science and technology with one another: to investigate the technical, social and practical histories of specific developments in science and technology; and finally, where possible and desirable, to bring the histories of science and technology into closer contact with the philosophy of science. To these ends, each volume will have its own theme and title and will be planned by one or more members of the Advisory Board in consultation with the editor. Although the volumes have specific themes, the series itself will not be limited to one or even to a few particular areas. Its subjects include any of the sciences, ranging from biology through physics, all aspects of technology, broadly construed, as well as historically-engaged philosophy of science or technology. Taken as a whole, *Archimedes* will be of interest to historians, philosophers, and scientists, as well as to those in business and industry who seek to understand how science and industry have come to be so strongly linked.

More information about this series at http://www.springer.com/series/5644

Jean-Marc Ginoux

History of Nonlinear Oscillations Theory in France (1880–1940)

Springer

Jean-Marc Ginoux
Archives Henri Poincaré, CNRS, UMR 7117
Université de Nancy
France

Laboratoire des Sciences de l'Information
et des Systèmes, CNRS, UMR 7296
Université de Toulon
La Valette du Var
France

ISSN 1385-0180 ISSN 2215-0064 (electronic)
Archimedes
ISBN 978-3-319-55238-5 ISBN 978-3-319-55239-2 (eBook)
DOI 10.1007/978-3-319-55239-2

Library of Congress Control Number: 2017937654

Printed on acid-free paper

This Springer imprint is published by Springer Nature
The registered company is Springer International Publishing AG
The registered company address is: Gewerbestrasse 11, 6330 Cham, Switzerland

to Elisa...

If we wish to foresee the future of mathematics, our proper course is to study the history and present condition of the science.[1]

[1]Henri Poincaré, *Science and Method*, 1914, p. 25.

Foreword

I first heard of Jean-Marc Ginoux when he published his discovery of Poincaré's lectures on oscillations and limit cycles at the School of Posts and Telegraphs in 1908. As he made clear, their significance was considerable. Because they had been omitted from the eleven volumes of Œuvres de Poincare and as a result had been completely forgotten, there was a gap in our understanding of the contributions of the great French mathematician and physicist. What Ginoux went on to investigate was the even greater gap that this pointed to in our understanding of oscillation theory.

The household names in the field are those of the Dutchman Van der Pol and the Russian Aleksandr Andronov, who are remembered for their work in the 1920s, Van der Pol for the oscillator that bears his name and Andronov for among other things being apparently the first to connect the work of engineers and technologists with Poincaré's discoveries in the early 1880s. The first thing that Ginoux's discoveries made clear was that Poincaré had made this connection himself and had done so in lectures and in print to an audience of appropriate specialists. This was one of a number of contributions that he made to technology at the time, which makes its subsequent disappearance all the more striking.

What Ginoux then discovered was that there had been a considerable investigation into oscillation theory by many French investigators, and his account of this work forms the major part of this book. He introduces us to numerous mysterious devices, explains how experiments and theories of them evolved, and isolates their key novel feature, which was known at the time as the relaxation effect. His account sorts out numerous misconceptions and builds up to an important international conference, the Institut Henri Poincaré in 1933, that was dominated by Van der Pol and French and Russian speakers. Paradoxically, the meeting also lapsed from the community's memory, and Ginoux traces this to the unequal development of the subject in the two countries. The French engineers did not build an adequate theoretical framework with a substantial commitment of people and resources; the Russians did. As a result, he suggests the subjects passed for a time in the Soviet Union, before awakening again the international interest in dynamical systems.

It is evident that Ginoux's work is both interdisciplinary and international. It has much to say to both mathematicians and engineers; it establishes new features of the context for their work, both the interactions and the failures; it has a French core and a Russian dimension. It is based throughout on rich, often forgotten, archival material, and it establishes the historical record for the first time.

Emeritus Professor
Open University, Leeds, UK

Jeremy Gray

Preface

In the history of mathematics, from the nineteenth century, the extraordinary growth of the theory of dynamical systems is distinguished by a hitherto unknown development. This history resulted in several publications. The latest is the excellent article "Writing the history of dynamical systems and chaos: long life and revolution, disciplines and culture" by D. Aubin and A. Dahan Dalmedico (*Historia Mathematica*, 29 (2002), 273–339). Before the present general historical survey that offers us Jean-Marc Ginoux, as far as I know, the only book on this subject, written at the specialist level, is that of E.S. Boïko, published under the Russian title *Skola Akademika A.A. Andronova* (*The Academician A.A. Andronov* School, *Izdatelsvo Nauka*, Moscow, 1983). Although the introduction is devoted to important results achieved outside the USSR, among them the first rank key role played by what might be called the French School and its Poincaré leader, Boïko's book focuses primarily on the work carried on in Gorky (now Nizhny Novgorod), a city forbidden to foreigners at the time of the Soviet Union. Concerning this topic, it should be noted that some 60 years ago, American mathematicians J.P. La Salle and S. Lefschetz already noticed the lead taken by the researches in the USSR, in the framework of the Gorky School (*qualitative methods*) and the Kiev School (*analytical methods*), when they wrote:

> In USSR the study of differential equations has profound roots, and in this subject the USSR occupies incontestably the first place. One may also say that Soviet specialists, far from working in vacuum, are in intimate contact with applied mathematicians and front rank engineers. This has brought great benefits to the USSR and it is safe to say that USSR has no desire to relinquish these advantages.[2]

About the book *History of Nonlinear Oscillations Theory* by Jean-Marc Ginoux (Springer-Verlag, 2016), it is important to note that the germ of the *theory of nonlinear oscillations*, becoming after the *theory of dynamical systems*, occurred at a time when most mathematicians saw a source of inspiration for mathematical

[2]J.P. Lasalle and S. Lefschetz, "Recent soviet contributions to ordinary differential equations and nonlinear mechanics," *Journal of Mathematical Analysis and Applications*, 2, 1961, pp 467–499.

discovery in the study of phenomena of the "real world," that of physics in particular. This is perfectly illustrated by what Joseph Fourier wrote in the first half of the nineteenth century:

> The in-depth study of nature is the most productive source of mathematical discoveries. By offering a specific purpose, it has the advantage of excluding vague questions, and doomed calculations. It is also a sure way to train the Mathematical Analysis itself and discover the elements that matter most to know and maintain. These basic elements are those that breed in all natural effects.[3]

About that "germinal" time, we can talk of researches on "*concrete dynamical systems*" as opposed to "*abstract dynamic systems*" whose theory has developed subsequently in the field of pure mathematics, now considered the "noble" field of mathematical research, although most scientists have not been led to their discoveries by a process of deduction from general postulates, or principles, but rather by a thorough study of carefully selected individual cases. Generalizations came later, because it is much easier to generalize an established result than to discover a new line of argument. In this regard, already in the nineteenth century, the famous mathematician Halphen has often complained of nonessential generalizations overcrowding the publications media. Later Birkhoff, added:

> The systematic organization, or exposition, of a mathematical theory is always secondary in importance to its discovery, ... some of the current mathematical theories being no more than relatively obvious elaborations of concrete examples.[4]

After this introductory presentation, it is now possible to situate the context of Jean-Marc Ginoux's book: that of "*concrete dynamical systems* history." In an original form, combining many illustrations, publication extracts, author biographies, so far unknown documents, and correspondence copies, the whole accompanied by an extensive bibliography, the great merit of this contribution is to make us relive this pleasantly "germinal" time and its first phase of development with the birth of *analytical methods* and *qualitative methods* of nonlinear dynamics. More particularly, the researches on nonlinear oscillations and dynamical systems, published in France between 1880 and 1940, are meticulously recorded, both by the French scientific community and by foreign researchers who wanted to reveal their results at an international level (at that time the French language was widely spread in scientific exchanges). In particular, this is the case of the results of the American G. D. Birkoff, those of the Andronov School for qualitative methods, and those of the Kiev School for analytical methods.

The soundness of the above Fourier quotation, about which the mathematician Gaston Darboux said "*the development of modern Analysis has confirmed and revealed the penetrating Fourier's ideas*," is confirmed in Jean-Marc Ginoux's book throughout all the chapters. This is particularly the case for Poincaré.

[3]See *Éloge historique d'Henri Poincaré* by Gaston Darboux, "Œuvres d'Henri Poincaré", tome II, Gauthier Villars et Cie, Paris 1916, page XXXV.

[4]See M. Morse in *Bulletin of the American Mathematical Society*, May 1946, 52(5), 1, 357–391.

Studies of oscillatory phenomena generated by the electric arc, wireless telegraphy, electromagnetic waves, and other problems of physics, which shines his interest for experimental physics, led him to fundamental results. The introduction of Boïko's book already confirmed this point by enhancing the key role played by Poincaré, also quoting other results of the French school, including those of H. Léauté, Ch. Briot, and J.C. Bouquet, researchers also known for their interest in problems in the filed of physics. Boïko presents these authors as having inspired Poincaré.

Jean-Marc Ginoux also gives us all the elements of a discovery that makes clearer what was universally accepted. So far, many historical publications have attributed to Andronov (1928) the merit of having made the connection between *self-sustained oscillations* (also called *free oscillations*, independent of the initial conditions) of an *electronic oscillator* (i.e., a system of the "real world") and the notion of *limit cycle* introduced by Poincaré (1882) using the *phase plane* representation. A painstaking research of Jean-Marc Ginoux unveiled a series of Poincaré conferences, hitherto unknown to specialists in nonlinear dynamics. It was about lectures given in 1908 at the *École Supérieure des Postes et Télégraphes* (School of Post and Telegraph). To students of this school, Poincaré clearly established the correspondence between the periodic solution of the ODE model of the *singing electric arc* (another system of the "real world": here an *electric oscillator*) and *self-sustained oscillations*, without using the *phase plane* representation, that of *limit cycle*. Indeed the conferences' framework, that of engineering students, did not need to define the recent mathematical concept of *limit cycle*, which would have resulted in a longer development. The mere mention of *periodic solution*, familiar to the students, was sufficient.

In the last paragraph of the introduction to his book, Jean-Marc Ginoux regrets that the construction of the *theory of dynamical systems* has been practically discarded in France after Poincaré, a topic mainly carried out on a large scale in the USSR. He is surprised by the lack of interest by the French mathematicians to these questions, after the 1930s to the 1970s of the last century. It was in the last quarter of the twentieth century that the English-language publications on nonlinear dynamics, by authors from different countries (including France), had an exponential expansion, the majority of them ignoring the basic French contributions, followed by those of the Gorki School (*qualitative methods*) and the Kiev School (*analytical methods*).

Three questions then come naturally to mind: Is the vogue of the axiomatic method, and the renowned Bourbaki School, whose choices have long influenced the entire French research in mathematics, and even high school teaching, responsible for this situation? Is it the result of a kind of disdain of "pure" mathematicians toward "impure" mathematicians, i.e., those for whom "the in-depth study of nature is the most productive source of mathematical discoveries"? Or should we find the origin in the disappearance of a whole generation of young mathematicians of the French prestigious schools *École Normale Supérieure* and *École Polytechnique*, on the battlefields of the First World War? Mobilized from the very start, sent to the front line, many young mathematicians were killed.

Seriously wounded in the face, at the origin of his “leather nose,” Gaston Julia is one of the survivors of this tragedy. Issue 4751 (March 24, 1934) of *L'Illustration*, a famous weekly journal of the 1940 prewar period, devoted to him an article entitled “The youngest member of the Sciences Academy, great French mathematician.” Part of the paper underlined his heroic behavior as a young officer. So, Julia won the most famous military medals: *Chevalier de la Légion d'Honneur* (1915), *Officier de la Légion d'Honneur* (1925), and *Commandeur de la Légion d'Honneur* (1932). Having tenure of the *Chair of Analytical Mechanics and Celestial Mechanics* at the *Faculté des Sciences de Paris*, Gaston Julia belongs to that generation for which applied mathematics were still a “noble” activity. It is why the French *Académie des Sciences* had entrusted to him the publication of Volume X (632 pages) of the Œuvres de Poincaré, a volume devoted to articles dealing with problems of *Mathematical Physics* divided into two sections: *Hertzian Oscillations* and *Critics, Discussions Presentations on Physical Theories*.

Ginoux's important work on research and knowledge synthesis, implied for building the history of the era that saw the birth of the theory of nonlinear oscillations and the formation of the theory of “concrete” dynamical systems, is presented here in an English edition: *History of Nonlinear Oscillations Theory* (Archimedes Series, Springer, New Studies in the History and Philosophy of Science and Technology, 2016).

Professeur des Universités
Systèmes dynamiques non linéaires et Applications
Toulouse, France

Christian Mira

Translator's Preface

A harmonized combination of historical literature and mathematics constitutes Jean-Marc Ginoux's *History of Nonlinear Oscillations Theory*, making it a rewarding challenge to work on. Taking a year to complete, it took diligent research to find adequate ways to incorporate mathematical jargon and remain faithful to the original text. An effort has also been made to ensure that the author's literary style transpires as well in English as it does in French.

I would like to thank English mathematician Jeremy Gray, who has supported me throughout the whole process and ensured that I stayed on the right track. There is no doubt his tremendous contribution has had a beneficial impact on the final product's quality. I would also like to thank Mr Ginoux for trusting me with his work and for giving me this opportunity to broaden my translation experience.

Enjoy.

Glasgow
June 24, 2016

Laura Stenhouse

Acknowledgments

Here I want to show my gratitude to Springer Verlag – particularly to Professor Jeremy Gray – for the confidence he has given to this project and making possible the publication of this book.

I also wish to thank Professors Christian Gilain and David Aubin from Université Pierre and Marie Curie, Paris VI; Professor Bruno Rossetto; and Professor Christian Gerini from Université de Toulon, without which this work would not have been possible and would not be what it is.

First, I would like to thank Miss Laura Stenhouse who spent long hours translating the French version of this book and Miss Gaëlle Chapdelaine (librarian at the Université de Toulon) who provided me many original articles.

I also express my gratitude to Professor Christian Mira for our friendship and for giving me the honor to write the preface of this book.

Finally, I wish to thank very warmly all my loved ones who have helped, advised, and assisted me throughout the development of this manuscript. To my family and my wife who have supported me in this work, I would like to extend my gratitude and love.

Le Mourillon
July 14, 2016

Jean-Marc Ginoux

Contents

List of Figures

List of Tables

Introduction

From the end of the nineteenth century until the middle of the 1920s, the term "sustained oscillations" designated oscillations that are produced by systems moved by an external power such as maintained pendulum. It also referred to oscillations that are produced by self-sustaining systems such as the series dynamo machine, the singing arc, or the triode. The numerous researches conducted in the domain of oscillations in France and around the world during this time period have never been the subject of an in-depth study. Until now the historiography has primarily been focused on Balthasar Van der Pol's contribution entitled: "On relaxation-oscillations" (Van der Pol 1926d). In this publication, he introduced this terminology in order to distinguish a specific type of sustained oscillation, and the history of relaxation oscillations appears to establish itself with his work. In his essay titled *Mathématisation du Réel* (*Mathematisation of Reality*), Giorgio Israel announces as follows:

> Whilst searching for an explanation of how a triode assembled as an oscillator functioned, Van der Pol realized that the standard mathematical equations for oscillations were unusable. (Israel 1996, 39)

Similarly, David Aubin and Amy Dahan-Dalmedico offer the same considerations in an article titled "Writing the history of dynamical systems and chaos":

> By simplifying the equation for the amplitude of an oscillating current driven by a triode, van der Pol has exhibited an example of a dissipative equation without forcing, which exhibited sustained spontaneous oscillations:
>
> $$v'' - \varepsilon(1 - v^2)v' + v = 0$$
>
> In 1926, when he started to investigate its behavior for large values of ε (where in fact the original technical problem required it to be smaller than 1), van der Pol disclosed the theory of relaxation oscillations. (Aubin and Dahan Dalmedico 2002, 289)

In her article "Le difficile héritage de Henri Poincaré en systèmes dynamiques" ("The difficult legacy of Henri Poincaré regarding dynamical systems"), Amy Dahan-Dalmedico goes further stating that "Van der Pol had used Poincaré's concept of limit cycles"[5] (Dahan Dalmedico 2004a, 279). Indeed, based on a historiographical reconstruction, Van der Pol's contribution (1926d) seems to have been on several levels: Firstly, the discovery of the relaxation oscillations produced by a triode; secondly, the equation of the phenomenon; and, lastly, the theoretical formalization linked to Poincaré's work (1881–1886) most significantly: the "limit cycles theory."[6] However, this historical representation does not accurately reflect reality and provides yet another illustration of the "Matthew effect"[7]: By focusing almost exclusively on Van der Pol and a few of his publications[8] dealing with the *free oscillations of a triode*, this historical reconstruction resulted in the partial, or even complete, overshadowing of previous studies about sustained oscillations, as well as an overestimation of Van der Pol's contribution regarding the *forced oscillations of a triode*.[9] Moreover, the crystallization occurring around Van der Pol's article (1926d) has caused a misunderstanding about his own results. The term "relaxation oscillation" was indeed not introduced and defined in 1926, but rather in an article published in Dutch the previous year (Van der Pol 1925, 793). The analysis of Van der Pol's works (1927b,c, 1930) shows that the *relaxation oscillation* phenomenon was first observed in 1880 by Gérard-Lescuyer in a *series dynamo machine*, then in 1905 by Blondel in the *singing arc*. It will later be shown that the equation modeling for the triode oscillations was not carried out by Van der Pol in 1926, or even in 1922, as Giorgio Israel suggests:

> Of particular importance in our case is the 1922 publication in collaboration with Appleton, as it contained an embryonic form of the equation of the triode oscillator, now referred to as "van der Pol's equation." (Israel 2004, 4)

Van der Pol's article "A theory of the amplitude of free and forced triode vibrations," (Van der Pol 1920, 701), originally published in Dutch in 1920, points out that:

$$\ddot{x} - (\alpha - 3\gamma x^2)\dot{x} + \omega x = 0 \qquad (1)$$

> This equation has been previously considered * in connexion with the subject of triode oscillations.
>
> * Van der Pol, *Tijdsch . v. h. Ned. Radio Gen.* i. (1920); *Radio Review*, i. page 701 (1920). Appleton and Van der Pol, Phil. Mag. Xliii. page 177 (1922). Robb, Phil. Mag. Xliii. page 206 (1922). (Van der Pol 1926d, 979)

[5]See also Dahan Dalmedico (2004b, 237).

[6]See also Poincaré (1882, 261).

[7]See Robert Merton (1968).

[8]More specifically, "On relaxation-oscillations," (Van der Pol 1926d).

[9]This question has, however, been briefly addressed by Pechenkin (2002, 272) and Israel (2004, 9), who also uncovered Van der Pol's and Van der Mark's work (1928a, b) regarding the mathematical description of heartbeats. See Israel (1996, 34) and Israel (2004, 14). See Part III.

In fact, Blondel (1919b) found equation (1) symbolizing the free oscillations of a triode one year earlier, under a different form.

Lastly, the idea that Van der Pol developed relaxation oscillation theory, which originated in Van der Pol's (1922, 1926, 1934) and Philipe Le Corbeiller's articles (1931a, 1932, 1933) appears groundless. Indeed, Van der Pol did not use the concept of *limit cycle* in 1926 and did not cite[10] Henri Poincaré's works (1881–1886), as noted by Mary Lucy Cartwright (1960, 371). Van der Pol mentions it only after the publication of a note from Andronov to the C.R.A.S.[11] titled "Poincaré's limit cycles and the self-oscillation theory"[12] (Andronov 1929a) – in which he suggests that the periodic solution of a self-oscillator matches one of Poincarés' *limit cycles* – during a series of lectures at the École Supérieure d'Électricité, on March 10 and 11, 1930:

> On each of these three figures, we can see a closed integral curve, which is an example of what Poincaré called a limit cycle ([13]), because the neighboring integral curves approach it asymptotically. (Van der Pol 1930, 16)

Thus, in spite of an undeniable wish to develop a mathematical theory on the relaxation oscillation as early as 1926, Van der Pol did not manage to set its founding principles. By establishing a link between Poincaré's work (1881–1886, 1892) and this type of oscillation, Andronov (1929a) incorporated it into a broader viewpoint: self-oscillation theory or self-sustained oscillation theory.[14]

Consequently, it becomes clear that Van der Pol's contribution (1926d) did not consist in the modeling of the equations for free oscillations of a triode, which would have let him develop a relaxation oscillation theory, but rather in the discovery of a new type of oscillation and naming it *relaxation oscillation*. Van der Pol's crucial role (1926d) therefore resides, on the one hand, in the conceptualization[15] of an oscillatory phenomenon possessing two distinctive timescales when given specific parameter values or, in other words, two types of evolution: one slow, one fast. On the other hand, he describes a great number of seemingly differing processes using a single differential equation in an undimensionalized form. The duality of the "slow-fast" phenomenon became part of self-oscillation[16] theory, which Andronov (1928, 1929a,b,1930a,b, 1937) and Andronov and Witt (1935) developed based

[10]In his article, (Van der Pol 1926d, 981) names this closed curve "periodic solution."

[11]*Comptes Rendus de l'Académie des Sciences de Paris* or Proceedings of the Academy of Sciences.

[12]Contrary to Dahan Dalmedico's (2004a, 279; 2004b, 237) and Diner's (1992, 339) affirmations, this two-page note absolutely cannot be his "graduation work," i.e., his "senior thesis." See *infra* Part II.

[13]Andronov (1929a).

[14]In his original version, written in Russian, Andronov (1929b) did not use the phrase *self-sustained oscillations*, but rather the term *self-oscillation*, from the Greek *auto* and Russian *kolebania*: oscillations. See Pechenkin (2002, 288).

[15]Part II will demonstrate that Van der Pol was at the root of the concept of relaxation oscillation.

[16]Part II will establish the difference between *self-sustained oscillations* and *relaxation oscillations*.

on Poincaré's work (1881–1886, 1892). Therefore, among others, Andronov's note (1929a) is seen as a remarkable event in the history of nonlinear oscillation theory by scientists[17] as well as historians of science:[18]

> And it is only in 1929 that Russian researcher A.A. Andronov suggested that self-oscillations are expressed as limit cycles from Poincaré's theory. This date leads to a new era in this field of study. (Minorsky 1967, 2)

This milestone had not been questioned until now. Indeed, doing so would have meant finding a publication predating Andronov's where the same type of link between an equation similar to Van der Pol's – i.e., defining the self-oscillation phenomenon – and Poincaré's *limit cycles* was discussed. A study of most of the articles and books published during the period before Andronov's note has been carried out. The works of French researchers such as Alfred Liénard (1928, 1931), Henri and Élie Cartan (1925), Paul Janet (1919, 1925), and André Blondel (1919a,b,c, 1920, 1926) have proven to be of great interest by raising once again the question of Poincaré's scientific legacy. While these works did not lead to the discovery of a link similar to the one Andronov had introduced, they showed that it was another device, older than the *triode*, that should have been the subject of research: the *singing arc*. Thus far, this device had played an essential part during the rise of the wireless telegraphy,[19] approximately between 1900 and 1914. During this period, Poincaré published several crucial studies on *wireless telegraphy* (Poincaré 1902a,b, 1907a, 1908, 1909a,b, 1911). At a series of lectures at the *École Supérieure des Postes et Télégraphes* in 1908, twenty years before Andronov (1928, 1929a,b), Poincaré (1908, 1909a,b) set the connection between his own work on the *limit cycles* and the differential equation symbolizing the *self-oscillations* occurring in the *singing arc*. Thus, from the end of the nineteenth century to the middle of the twentieth century, three[20] devices – the *series dynamo machine*, the *singing arc*, and the *triode* – were the seat of a new type of oscillatory phenomenon, then considered as sustained oscillations, before Van der Pol (1925, 1926d) named it *relaxation oscillation* and before Andronov (1929a,b, 1930a,b, 1937) and Andronov and Witt (1935) incorporated it in the *self-oscillation theory*.

During this period, these three devices became the subject of much research in France and all over the world, the aim being:

- to isolate the cause of this phenomenon,
- to schematize the device's *current-voltage characteristic*[21]

---

[17]See Mandel'shtam et al. (1935, 83), Abelé (1943, 18), Rytov (1957, 170).

[18]See Diner (1992, 340), Dahan Dalmedico (1996, 21), Aubin and Dahan Dalmedico (2002, 13), Pechenkin (2002, 274), Dahan Dalmedico (2004a, 279), and Dahan Dalmedico (2004b, 237).

[19]*Télégraphie Sans Fil*, or *T.S.F.* in France.

[20]Around the same time, Léauté (1885) also noticed this type of oscillation in a hydraulic sifting control device.

[21]Thompson (1893, 247) stated that the term *characteristic* was introduced by Marcel Deprez in 1881. See Deprez (1881, 893). The *current-voltage characteristic* of an electric dipole is the function relating the tension at its terminals with the intensity of the current flowing through it.

- to model the equation symbolizing this new type of oscillation in order to determine its amplitude and period.

The schematization of the *current-voltage characteristic*, i.e., the *e.m.f.* of each of these three devices, and the determination of the oscillation period were the two main obstacles to overcome in order to define the relaxation oscillation phenomenon.

With direct current, the characteristic of a resistance (R) is a linear function of the intensity: $u = f(i) = Ri$. The inclination of this straight line provides the value of the resistance, seen in this case as a constant. The characteristic of an electromotive force (*e.m.f.*) generator and an internal resistance (r) is a linear function of the intensity: $u = f(i) = E - ri$. The y-intercept represents the *e.m.f.* (E), also assumed to be constant. These relations are the mathematical expression of Georg Ohm's law.

Part I
From Sustained Oscillations to Relaxation Oscillations

Chapter 1
From the Series-Dynamo Machine to the Singing Arc: Gérard-Lescuyer, Blondel, Poincaré

1.1 The Series Dynamo Machine: *The Expression of Nonlinearity*

At the end of the nineteenth century, *magneto-* or *dynamo-electric* machines were used in order to turn mechanical work into electrical work and vice versa. With the former type of machine, the magnetic field is induced by a permanent magnet, whereas the latter uses an electromagnet. These machines produced either *alternating* or *direct current* indifferently. They were therefore the most economical of all appliances where powerful currents are required, such as supplying lighthouses with power using electrical arcs.[1] A *dynamo-electric machine* where the electromagnet is integrated into the circuit is called a series *dynamo machine*, or *self-exciting dynamo*.[2]

1.1.1 Jean-Marie-Anatole Gérard-Lescuyer's Paradoxical Experiment

There are few biographical elements regarding the man who conducted this experiment. His family name, Gérard-Lescuyer, probably originated from his father Jean-Baptiste Gérard marrying a woman called Marie-Anne Lescuyer in Paris. Jean-Marie-Anatole Gérard-Lescuyer was an engineer[3] and the director of a public liability electric company established in Courbevoie. From 1895 to 1902, he was a

[1]For more details see Hospitalier's (1881, 68–115) or Lemoine's work (1890, 19).

[2]See Hospitalier (1881, 86).

[3]On 10 September 1879, he invented an electric arc lamp and an automatic light, as shown in an article published in *La Nature* (Hospitalier 1881, 220–222) and signed E. H. who is in fact Édouard

J.-M. Ginoux, *History of Nonlinear Oscillations Theory in France (1880–1940)*, Archimedes 49, DOI 10.1007/978-3-319-55239-2_1

member of the French society of physics. Similarly, baron Paul Arnould Edmond Thénard (1819–1884), physicist and chemist, presented on July 26, 1880 his first and only scientific contribution: a note in the *Comptes rendus de l'Académie des Sciences de Paris* (*Proceedings of the Academy of Sciences*). This is where he detailed an experiment, describing it as an "electrodynamical paradox". However as time passed he lost its paternity.

Gérard-Lescuyer's research on electrical generators led to his invention of a machine named after him.[4] In all likelihood it also brought about his experiment associating a *dynamo-electric machine* used as a generator with a *magneto-electric machine*, which in this case can be considered as the motor. He reports on the found effects in the following way:

> (…) If the current produced by a dynamo-electrical machine is sent into a magneto-electrical machine, a strange phenomenon is witnessed. As soon as the circuit is closed the magneto-electrical machine begins to move; it tends to take a regulated velocity in accordance with the intensity of the current by which it is excited; but suddenly it slackens its speed, stops, and starts again in the opposite direction, to stop again and rotate in the same direction as before. In short, it receives a regular reciprocating motion, which lasts as long as the current that produces it. (Gérard-Lescuyer 1880a, 226, 1880b, 215)

He also observed the periodical reversal of the magneto-electric machine's circular motion, despite the direct current, and wondered about the causes of this oscillatory phenomenon. He estimated that the change in the motor rotation can only happen if the current running through it also changes direction. He then researched how this inversion occurred:

> Some extraneous cause, then, must arise to reverse the polarities of the inductors of the generating dynamo-electrical machine, so that this machine may immediately give rise to a current of an opposite direction, which reverses the rotation direction of the receiving machine. (Gérard-Lescuyer 1880a, 226, 1880b, 215)

He indeed notices this polarity reversal by placing small compasses near the inductors, and notes that "movements of the compass-needle coincide with those of the galvanometer". This enables the establishment of a cause and effect relationship between the polarity reversal and the appearance of a reverted current. Thus, in order to try and provide an explanation, he hypothesizes that "the receiving magnetoelectrical machine can, for some unknown reason, receive periodically an increasing of its velocity". The experiment's trial confirms this:

> (…) if our hypothesis proves right, this phenomenon will no longer occur when, by any means whatsoever, we prevent the receiving magnetoelectrical machine from increasing its velocity: applying a brake suffices to do so. However, as soon as the brake intervenes the preceding effects disappear. (Lescuyer 1880a, 227, 1880b, 215)

Hospitalier, engineer at the *Arts et Manufactures* and chief editor for the periodic publication l'*Électricien*. He also invented a new incandescent light bulb in 1885 (see Rodet 1907, 67) and a chainless bicycle (see Picture n° 1, 7).

[4]In his book, Hospitalier (1881, 106–107) described A. Gérard's machine. Patent n° 336 636, 23 February 1886 (U.S. patent). See also Boulanger (1885, 111 and 120).

According to him this increase in the velocity of the magneto-electric machine, by inducing a reverse electrical current, caused the inductors' polarity reversal and inverted the rotation. It was actually proven by du Moncel (1880) a few weeks later, then by Witz (1889a,b), and by Janet (1893), that the gap situated between the brushes of the dynamo is the source of an electromotive force (*e.m.f.*), i.e. a potential difference at its terminals symbolized by a nonlinear function of the intensity that flows through there (see above note 18, 4). Therefore this "cause not investigated" by Gérard-Lescuyer, the essence of his paradox, is the presence of an *e.m.f*, which has a *nonlinear current-voltage characteristic* leading to *sustained oscillations*.[5] This reservation he emits is also notable:

> What are we to conclude from this? Nothing, except that we are confronted with a scientific paradox, the explanation which will come, but that does not cease to be interesting. (Gérard-Lescuyer 1880a, 227, 1880b, 215)

Gérard-Lescuyer's article was then published in the *Philosophical Magazine* with the title "On an electrodynamical paradox" (Gérard-Lescuyer 1880b), which allowed him to get some response in the United States particularly thanks to a paragraph in the *New York Times* of 22 August 1880:

> Gérard-Lescuyer finds that when the current from a dynamo-electric machine is sent into a magnetic electric machine the latter moves with increasing speed, then it slackens, stops, and turns in the opposite direction, and so on. The polarity of the inductors is reversed.

The New York Times which was founded in 1851 created very early on a column called *Scientific Gossip*, which aimed at retelling the most notable scientific events at the time. Later on this column would feature the announcement for the first *Congrès International d'Électricité* in Paris on the 11th of October 1882 (*New York Times* of 10 September 1882), as well as an article titled "Music in Electric Arcs", about the discovery of the *singing arc* (see *infra*) by William Du Bois Duddell (*New York Times* of 28 April 1901). As for Gérard-Lescuyer's note to the *C.R.A.S.*, it seems its publication in the *Philosophical Magazine* on one hand and the use of the word "paradox" on the other are the reason it caught this reporter's attention. Aside from this type of journalistic echo, the "paradox" uncovered by Gérard-Lescuyer did not seem to garner immediate reaction from the physicists and engineers researching electrical phenomena.[6] However the oscillatory phenomenon noted and detailed for the first time here later caused curiosity due to its paradoxical nature (Fig. 1.1).

[5]It was later established that they were actually *relaxation oscillations*. See *infra*.

[6]Except for Théodose du Moncel.

Fig. 1.1 J.M.A. Gérard-Lescuyer (*left*) and his daughter Marguerite in 1884. Document uploaded online by his great-grandson

1.1.2 Théodose du Moncel's Electrokinetic Interpretation of the Paradox

Being interested in science and more specifically in electricity Viscount Théodose du Moncel became an Electrician-Engineer for the *Administration des lignes télégraphiques françaises* (French telegraph lines administration) during the second half[7] of the nineteenth century. He submitted several notes to the *French Academy of Science* and became a member on the 21st of December 1874. He wrote numerous books[8] and articles, the latter published in the periodicals l'*Électricité* and *La Lumière Électrique*. He published on the 1st of September 1880 an analysis of Gérard-Lescuyer's experiment titled "Réactions réciproques des machines dynamo-électriques et magnéto-électriques" ("Reciprocal reactions of dynamo-electric and magneto-electric machines") (Moncel 1880). He then used the concept of counter-electromotive force in order to explain the inductors' polarity reversal found by Gérard-Lescuyer:

> The author of these experiments attributes this effect to a periodical increase in the magneto-electric machine's speed, which would generate a counter-electromotive force greater than the electromotive force developed by the dynamo-electric machine, able to

[7]In his youth, he went to visit the Greek archeological sites, and sold his sketches to earn money. See the *Annales Archéologiques*, 1848, volume eight, page 179 and page 236, *Bureau des Annales Archéologique*, Paris, *Librairie Archéologique de Victor Didron*. See Cornelius Herz's obituary (1884).

[8]See Moncel (1858, 1872–1878, 1878, 1879, 1882).

Fig. 1.2 Count Th. du Moncel, from Herz (1884, 383)

> reverse the inductor's polarity. A new reversed current would therefore be produced by the dynamo-electric machine, which would then turn off the current produced by the counter-electromotive force and as a consequence cause it to stop. This would be followed by movement in the opposite direction and so on. In order to verify this explanation, he forced the magneto-electric machine to move steadily by using a brake. In these conditions the machine's rotation became regular. (Moncel 1880, 352)

Gérard-Lescuyer's description enabled du Moncel to prove the existence of an electromotive force (*e.m.f.*) developed by the dynamo and of a counter-electromotive force (*c.e.m.f.*) generated by the periodical increase in speed which reverses the inductors' polarity and the motor's rotation. He then explained that this inversion happens as soon as the magneto's *e.m.f.* is superior to the dynamo's *e.m.f.* He did not use the scientific paradox concept but rather "reciprocal reactions" as indicated by the article's title. As soon as September 1880 his analysis pushed Gérard-Lescuyer's experiment towards rationality by involving electrical quantities, which would later allow Witz then Janet to provide a thorough explanation of the phenomenon (Fig. 1.2).

1.1.3 Aimé Witz's Geometrical Interpretation of the Paradox

Because of their difference in status Gérard-Lescuyer and Aimé Witz had entirely different approaches: whilst the first focused on practical and technological applications, the second aimed for a purely theoretical approach. Witz was both a Doctor of Science and an engineer from the *Arts and Manufactures* school and taught at the *Faculté libre des sciences* in Lille, where he later became dean emeritus.[9] He was an expert in thermodynamics and electricity. In 1889 and 1890 he published a

[9]For Witz's biography, see for example Charles Lallemand's obituary (1926).

series of articles in which he went back over Gérard-Lescuyer's experiment and offered a heuristic explanation solving the paradox. The first article was a note published in the *C.R.A.S.* on the 6th of July 1889 titled "Polarity reversals in series-dynamo machines" (Witz 1889a). This article explained the principles of geometrical construction explaining the observed phenomenon but the corresponding graphical representation was unfortunately absent due to the format constraints of the *C.R.A.S.* Witz developed his method in a longer and more detailed article published in December 1889 in the *Journal de Physique théorique et appliquée*: "Des inversions de polarité dans les machines série-dynamos" ("Research on series-dynamo machines polarity reversals"), (Witz 1889b). This article provides interesting insight from two different angles. The first one is historical since Witz explained how he came to learn about this experiment:

> This perplexing phenomenon which I had thought unknown had already been observed by Mr. Gérard-Lescuyer, as was pointed out to me by Mr. Hospitalier. With this phenomenon, this eminent physicist had found a paradox, which he gave up on explaining in 1880. Today this fact appears less mysterious, as we will attempt to demonstrate after examining it more thoroughly. (Witz 1889b, 582)

Surprisingly no reference is made to Théodose du Moncel's work either by Witz or Hospitalier[10] despite the latter's apparent familiarity with them. The second angle is theoretical since this article took a second step towards the rational explanation of the phenomenon by using a geometrical construction. Witz's approach was thus based on experiments whereas Gérard-Lescuyer's (1880a) was purely phenomenological. For instance he began his study with the following words:

> Firstly this experiment can be reproduced in a laboratory with any series-dynamo activating a machine with a separate exciter or a magneto-electric machine. (Witz 1889b, 582)

Thus, Witz shows that this experiment was not unique but actually perfectly reproducible, that is to say that its carrying out did not require specific conditions. Following scientific methods Witz measured the voltage and intensities and then analyzed the evolution of these quantities to change the approach from qualitative to quantitative.

1.1.3.1 Principle of Witz's Construction

With this construction, Witz (1889b) provided a geometrical version of du Moncel's analytical explanation (1880). Indeed, he drew the current-voltage characteristic of the dynamo, i.e. the curve representing the variations in voltage at its terminals according to the intensity of the current running through it.[11] Although it appears

[10]See Hospitalier (1881, 64, 98, 196, 220, 224, 237, 255). It should be noted that between the first and third edition of his work, Hospitalier greatly reduced his references to du Moncel.

[11]This is the *e.m.f.* of the dynamo.

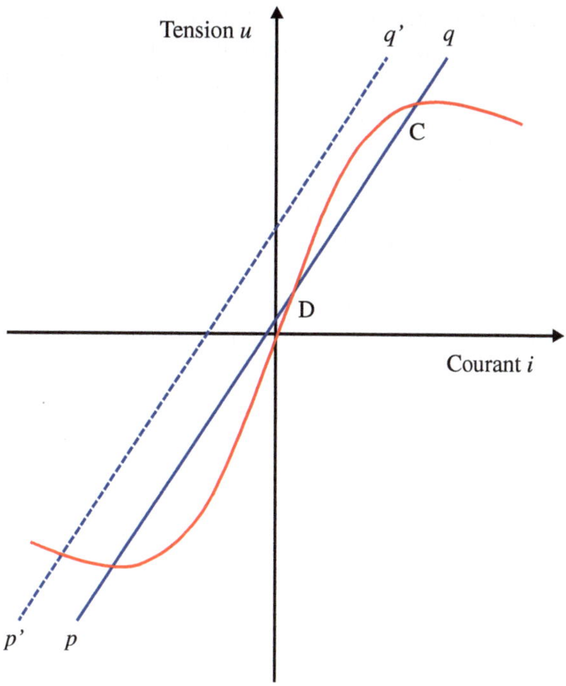

Fig. 1.3 Characteristics of the dynamo (*red*) and the motor (*blue*)

on (Fig. 1.3), which is a simplified version of Witz's original construction, that this curve possesses the characteristics of a cubic plane curve (two extrema and one inflexion point), the next step, i.e. its modeling *via* a mathematical function would require a lengthy process, as it will be demonstrated later. He then drew the characteristic of the magneto,[12] which can be compared to a motor. It is a straight line and its y-intercept depends on the motor rotation speed. This line shifts in parallel with itself according to the values of the motor rotation speed. Thus the two characteristics: the dynamo's (curve) and the motor's (straight line) show a number of intersections. The amount and value of these intersections depend on the straight line's placement in relation to the curve (Fig. 1.3).

> (…) this line intersects the characteristic curve of the generator at points C and D. The abscissa for C indicates the actual intensity in the circuit. This intensity will stay constant as long as the machine's speed is kept the same. But when decreasing the resisting torque of the motor, its speed will instantly increase, the characteristic ordinates go up, and the counter-electromotive force increases. p'q' replaces pq, since points C and D become closer and merge. The secant line has become tangent, then intersects only with the symmetrical branch. The current intensity in the circuit has gone down at the same time, gradually decreasing to zero and suddenly dropping to a negative value. (Witz 1889b, 585)

[12]It is the *c.e.m.f.* generated by the increase in speed. See du Moncel (1880, 352).

His construction shows that the polarity reversal occurs as soon as the motor's *c.e.m.f.* becomes greater than the dynamo's *e.m.f.*, which causes the motor's oscillations. Indeed, the cubic plane curve allows the intensity to take negative values, which would not be possible if it was a straight line. In the circuit the series-dynamo (i.e. a component comparable to a "negative resistance[13]") plays a role similar to a pendulum's friction, but its sign's alternating between positive and negative sustains the oscillations instead of damping them. Therefore, Witz's work in 1889 broke through a second threshold in the understanding of the phenomenon using a graphic representation highlighting the nonlinearity of the dynamo's *e.m.f.* However whilst his construction explains how this polarity reversal observed by Gérard-Lescuyer occurs, it did not provide the cause as he pointed out:

> All in all, by explaining the diagram we can find all the defining features of the phenomenon (...). It is not a paradox anymore. However, let us not delude ourselves: the previous reflections show how things happen, but not why they happen in this way. (Witz 1889b, 586)

Nevertheless, it must be noted that Witz came very close to the explanation provided by Janet (1893) less than four years later:

> Lastly, let us prolong the line tangent to the characteristic curve until it meets the symmetrical branch of this curve: the y-intercept of the intersection gives the value of the electromotive force developed right after the inversion of the rotation and the pole reversals. (Witz 1889a, 1245)

1.1.4 *Paul Janet's* Incomplete Equation Modeling (I)

After graduating from the *École Normale Supérieure* at age 22 and then passing the *agrégation*, Paul Janet, philosopher Paul Janet's son (1823–1899) defended his doctorate dissertation titled "Étude théorique et expérimentale sur l'aimantation transversale des conducteurs magnétiques" ("Theoretical and experimental study on the transverse magnetism of magnetic conductors") in front of the *faculté des Sciences de Paris* in 1890. He was then appointed as *maître de conférences* (lecturer) in Grenoble where he opened the very first course on Industrial Electricity against dean François Raoult's advice. It was met with such success that as soon as the next school year started François Raoult had to request for this course to be made official and for an electrical engineering laboratory to be created. Two years later Paul Janet became the head of the *École Supérieure d'Électricité* he had just founded and kept on teaching there. Therefore, as early as 1893, Gérard-Lescuyer's experiment[14] was

---

[13]The concept of "negative resistance" will be defined in the second paragraph.

[14]Janet did not cite Gérard-Lescuyer's (1880a,b), nor du Moncel's (1880), nor Witz's works (1889a,b). He did not mention Raoul Lemoine's book (1890) either. This book, titled L'Électricité dans l'industrie is reminiscent of his lecture given in 1893, in which the experiment was explained, but attributed to Witz. See Lemoine (1890, 21).

already being quoted by Janet in his *Industrial Electricity* course as an important example then in 1900 in volume I of his *Leçons d'Électrotechnique Générale*:

> If the current produced by a series generator is sent to the armature of a motor with a separate exciter, it will start up and there will be a certain amount of counter-electromotive force developing there. If enough speed is gathered this counter-electromotive force can become greater than the electromotive force of the generator, the current will be reversed as well as the generator's polarity. The motor's armature will then abruptly stop and start in the opposite direction and the same phenomenon occurs with notable periodicity. The frequency depends of course on the motor's excitation. It must be noted that in order to fully explain the phenomenon three electromotive forces must be involved during the variable period: the electromotive force of the generator, the counter-electromotive force of the motor, and the electromotive force of the inductor[(1)].
>
> ---
>
> [(1)](1) We encourage the reader to try and put the problem into equation, which is fairly simple: find either the current's voltage, or the motor speed in relation to time. " (Janet 1900, 222)

He explained that in addition to the *c.e.m.f.* of the motor and the *e.m.f.* of the generator found by du Moncel (1880) and Witz (1889a,b) the *e.m.f.* of the coil inductor must also be taken into account. In this way, as seems to be shown in the footnote[(1)] above, he had already set up the incomplete differential equation characterizing the phenomenon generated by Gérard-Lescuyer's experiment. However the issue concerning modeling the complete equation for this new type of oscillation, which required the modeling of the (*nonlinear*) *current-voltage characteristic*, was still unsolved. At this point in time, a second device was introduced (Fig. 1.4).

Fig. 1.4 Paul Janet in 1923, rue de Staël, from Jacques Boyer/Roger Viollet

1.2 The Singing Arc: *Sustained Oscillations*

1.2.1 William Du Bois Duddell's Revision of Thomson's Formula

At the end of the nineteenth century a forerunner to the incandescent light bulb called *electric arc*[15] was used for lighthouses and street lights. Regardless of its weak glow it had a major drawback: the noise generated by the electrical discharge which inconvenienced the population. In London, physicist William Du Bois Duddell was commissioned in 1899 by the British authorities to solve this problem. He thought up the association of an oscillating circuit made with an inductor L and a capacitor C (F on Fig. 1.5) with the *electrical arc* to stop the noise (see Fig. 1.5). Duddell (1900a,b) created a device that he named *singing arc*.[16] He was able to determine that the frequency of the musical sound[17] emitted by the arc corresponds to the natural frequency of the associated oscillating circuit and it was expressed using Thomson and Kelvin's formula (1853): $T = 2\pi\sqrt{LC}$ ($T = 2\pi\sqrt{FC}$ since $C = F$).

Duddell had actually created an oscillating circuit capable of producing not only sounds (hence its name) but especially electromagnetic waves. This device would therefore be used as an emitter for wireless telegraphy until the triode replaced it.

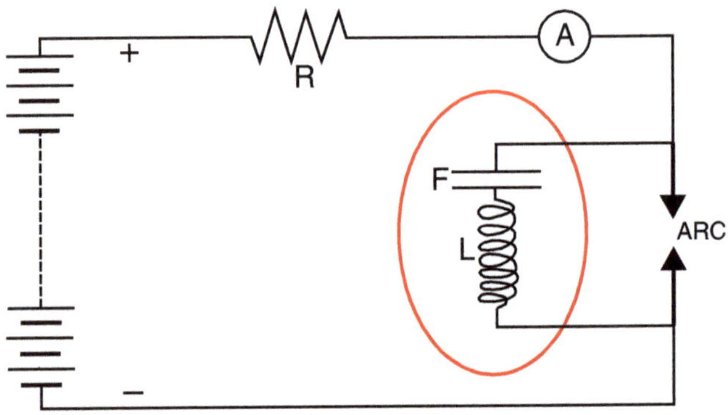

Fig. 1.5 Diagram of the *singing arc*'s circuit, from Duddell (1900a, 248)

[15]The electric arc (artificial, in opposition to lightning's thunderbolts) is associated to the electric discharge produced between the extremities of two electrodes (made of carbon for example), along which comes light emission. It is still used nowadays for theater projectors, thermic plasma, as well as in the metalworking industry for "arc soldering" or metal fusion (electrical arc furnace). See for example Vacquié (1995).

[16]For a *brief history of the arc*, see Hertha Ayrton's book (1902, 19).

[17]If its frequency is within human hearing range.

The *singing arc* or *Duddell's arc* was indeed a "spark gap" device meaning that it produced sparks which generated the propagation of electromagnetic waves shown by Hertz's experiments[18] as pointed out by Poincaré:

> If an electric arc is powered by direct current and if we put a self-inductor and a capacitor in a parallel circuit, the result is comparable to Hertz's exciter. (Poincaré 1907a, 79)

1.2.1.1 Conditions for Starting the Oscillations Sustained by the Singing Arc

After his discovery Duddell kept on studying the *singing arc* aiming to generate electromagnetic waves able to send a signal. Duddell (1900a, 268; 1900b, 310) then showed that in order to establish the oscillation speed, two conditions must be met:

$$\frac{du}{di} < 0 \tag{1.1}$$

$$\frac{du}{di} > r \tag{1.2}$$

meaning that the oscillations occur if: the derivative for the difference in electric potential at the arc's terminals in relation to the current going through it is negative (condition D_1) and if this derivative is greater (in absolute value) than the internal resistance r of the parallel circuit (see Fig. 1.5). This implies that if the *current-voltage characteristic* i.e. the arc's *e.m.f.* is nonlinear and has a "falling" or decreasing part, oscillations may occur (see *infra*). These two conditions have been subsequently analyzed throughout many studies. One of them was a note from Paul Janet (1902b) presented to the *Académie des Sciences*, in which he found Duddell's two conditions through other means concluding with another practical use for the singing arc:

> Duddell's singing arc offers a remarkable way of using a constant electromotive force in order to obtain alternative current. (Janet 1902b, 823)

Duddell's conditions (1.1) and (1.2) were very quickly questioned by several researchers such as Maisel (1903, 1904, 1905) who believed they did not play any role in the phenomenon:

> As a matter of fact this condition does not come into consideration at all in phenomena of this kind, as Maisel has shown by both theoretical and experimental researches. (Ruhmer 1908, 178)

[18]Carried out from 1886 to 1888 by German physicist Heinrich Rudolf Hertz, this experiment, or rather series of experiments proved the existence of electromagnetic waves predicted by James Clerk Maxwell during the previous decade.

The same went for engineer Aimé Williame (1906), who published a very detailed note in the *Annales de la Société Bruxelles* titled "On the singing arc theory" in which he explains:

> *Condition* $\frac{dv}{di} < 0$. – In short, despite both theory and experimentation showing that the consequences ensuing from the condition $dv/di < 0$ cannot be accepted without restrictions, until now nothing has shown that this condition was not necessary from beginning, for $i = 0$. However, our current knowledge, relating to the quotient dv/di, includes consequences that have not been proven to be exact in all cases. (Williame 1906, 188)

These authors also questioned the value of the oscillations' frequency which Duddell modified in 1901.

1.2.1.2 Frequency of the Oscillations Sustained by the Singing Arc

Indeed in response to some comments from the readers of the periodical *The Electrician* (Du Bois Duddell 1901c) recalled that during his first experiments he had overlooked the circuit's resistance (R) and when taking it into account, the frequency of the sound emitted by the arc corresponding to the circuit's specific frequency given by Thomson and Kelvin's formula (1853) had to be modified following another formula which has since then been named after him (see Table 1.1).

The first studies carried out in France then seemed to validate these formulae. Thus for Charles Fabry (1903, 376) the classic formula (Thomson's) seemed to be "indeed verified by Mr. Tissot's experiments (1902)". Whereas Blondel (1905c), in a dissertation "On the singing arc phenomenon", explained by using other experiments that:

> The singing arc's frequency is essentially variable and not properly defined; while in the case of the continuous phenomenon it can be given approximately by Duddell's formula, in the case of the discontinuous phenomenon, it is however quite different from the eigenfrequency of the oscillating circuit. (Blondel 1905c, 102)

Meanwhile, Williame (1906) considered that the quotient dv/di must vary either in a continuous or discontinuous manner and thus he concluded that:

> In both cases it is not proven anymore and it even seems improbable that the current would be sinusoidal in the parallel circuit and that its frequency could be deduced using Thomson's formula. (Williame 1906, 187)

Table 1.1 Duddell's and Thomson's formulae for the singing arc's frequency

Thomson and Kelvin's formula (1853)	Duddell's formula (1901c)
$T = 2\pi\sqrt{LC}$	$T = \frac{2\pi}{\sqrt{\frac{1}{LC} - \frac{R^2}{4L^2}}}$

Two years later in an article entitled "The frequency of the singing arc", which starts with a very thorough bibliographical study of the various research carried out in this field, George Nasmyth (1908, 122), using new experiments demonstrated that the arc's frequency varies in relation to the inductance L and also varies depending on the current flowing through it. Nasmyth (1908, 140) then offered a new formula for the frequency which took into account the derivative for the arc's *e.m.f.* However at this time the researchers had to face another obstacle: "measuring or even defining the resistance of an arc". (Fabry 1903, 376)

1.2.2 *Edlund and Luggin's Work* On the Concept of "Negative Resistance"

Too many variables (diameter, constitution, and width between the carbons of the arc) and the phenomenon called *arc hysteresis* (see *infra*) made it almost impossible to reproduce the experiments exactly. Therefore determining the arc's physical quantities such as its resistance and the voltage at its terminals was a real problem for the scientists at the end of the nineteenth century. However it was observed that the potential difference u on the one hand increases somewhat quickly with the distance l between the carbons and on the other hand it no longer obeys Ohm's law (see Footnote 18, 5). Consequently an empirical relationship, *linear* then *nonlinear*, between the voltage u at the arc terminals, the intensity i running through it, and the distance l between the carbons (see Table 1.2 *infra*[19]) is established. This potential difference expresses the arc's *e.m.f.* To explain the importance of this potential difference between the carbon electrodes, some researchers such as Edlund (1867) then later Duddell (1901a,b, 1904) concluded that there must exist a *counter-electromotive force* in the arc. This led them to consider that the arc's resistance, already seen as a variable, could allow negative values. According to Ayrton (1902, 54) and Child (1909, 233) the concept of "negative resistance" was introduced by Hans Luggin who studied the voltage at the arc terminals using the Wheastone bridge method, which at the time, was used for measuring the

Table 1.2 Relation between the arc's voltage and intensity

Author	Relation	Type
E. Edlund (1867)	$u = (a + bl)\,i$	Linear
S. P. Thompson (1892)	$u = a + \frac{bl}{i}$	Nonlinear
H. Ayrton (1895)	$u = a + bl + \frac{c + dl}{i}$	Nonlinear

[19]In Table 1.2, a, b, c and d represent four constants which depend on the material and diameter of the carbons.

arc's resistance. A slight increase in the electromotive force applied to an ordinary conductor, through which a current flows, produces an increase in voltage at the terminals and results in a difference in electric potential, with a positive sign. Luggin (1888) however showed that on the contrary in the arc can be found an increase in electric current which is followed by a decrease in voltage between the carbons' extremities and consequently the potential difference generated by an increase in the electromotive force has to have a negative sign. From this he concluded:

> (…) der Lichtbogen habe also einen negativen Wiederstand.[20] (Luggin 1888, 568)

In an article published in 1909 in *Physical Review* Child tried to clear up the confusion surrounding this terminology:

> There is no denying that the resistance of the arc is of a negative quantity but of all the uses of the word this is perhaps the least justifiable and to speak of a negative resistance is, to say the least, misleading. The word resistance means primarily something which hinders the movement of some object. An electrical resistance means something hinders the flow of an electric current and the most natural meaning of the expression "negative resistance" would be something which helps the flow of the current. It is needless to say that the resistance of the arc does not help the flow of the current. (Child 1909, 233)

As for Mrs Ayrton (1902, 75–76) she recalled that during his meeting with the *British Association* in Ipswich in 1895, her husband William Edward Ayrton (1847–1908) submitted a study on the "Arc's resistance" in which he got the same results as Luggin despite not knowing about them. Whilst Edward Ayrton did not publish these works[21] they were nonetheless faithfully transposed by Frith and Rodgers (1895) during a presentation in front of the *Physical Society* in May 1896. At the turn of the nineteenth century the electric arc was defined as having three characteristics: "negative resistance", varying electromotive force, and counter-electromotive force. Blondel's research on the subject confirmed the first two and disproves the third entirely.

[20]"(…) hence, the arc possessed a negative resistance."

[21]In 1893 during a stay in Chicago, William Edward Ayrton lost the only copy of an article representing three years of work as seems to be corroborated by the *Proceedings of the International Electrical Congress held in the city of Chicago*, 1893, vol. 1, 258: "The manuscript was partially destroyed by fire through a most unfortunate accident." From his wife, Hertha Ayrton (1854–1923):

> (…) Prof. Ayrton's ill-fated Chicago paper, which, after being read at the Electrical Congress in 1893, was accidentally burnt in the Secretary's office, whilst awaiting publication. (Ayrton 1902, vi).

Hertha Ayrton, who assisted him during his research, took over and published the results in the periodic *The Electrician*, (Ayrton 1895), and in her book *The Electric Arc*, (Ayrton 1902). See also Trotter (1924).

1.2.3 André Blondel's Work and the Non-existence of a c.e.m.f.

In France the electrical installations of the maritime signals and beacons were in disrepair. Seeing this André Blondel, a young engineer assigned to the *Service central des Phares et Balises Balises* (Central Office of Lighthouses and Shore Lights), started researching the electric arc to improve the structures. In order to do this he designed the galvanometric oscillograph.[22] Blondel (1891a,b, 1893a,b,c, 1897) dedicated the first part of his work to this oscillograph which helped "the electric arc theory to take a major step forward" (Bethenod 1938b, 751). In 1883 twenty-year-old Blondel joined the *École Polytechnique* after having graduated from the *École Normale Supérieure* the same year. He then joined the *École des Ponts et Chaussées* in 1885 and graduated first in his class in 1888. Following the acquisition of his licence in Mathematical Sciences in 1885 and in Physics in 1889 he attended Poincaré's lectures, before starting his career as an engineer for the *Service central des Phares et Balises.* At the time, this position fell under the management of the *Direction Générale des Ponts et Chaussées.* This is most likely the reason why he turned to researching the electric arc, "in order to find a specific application of lighting" Blondel (1891a,b, 552) but also and mainly to shed light on the arc's nature. Blondel (1891a,b, 621) then introduced the concept of *oscillation characteristic* of the direct current arc (see *infra*) in order to study its stability. He conducted a series of studies on the direct and alternating current arcs and came to the conclusion that an *electromotive force* did exist in the direct current arc (depicted by one of the relations in Table 1.2), as well as resistance variability:

> (...) the shapes of the periodic curves for the current and the difference in electric potential at the terminals show that the arc's resistance varies in relation with the current (...) (Blondel 1897, 515)

However in 1893 he rejected (see Blondel 1893c, 614) the theory stating that the arc held a counter-electromotive force and in 1897 he managed to disprove it:

> I believe these measures make it absolutely clear that the electric arc markedly behaves like a resistance and that it does not feature what usually defines a counter-electromotive force comparable to the observed potential difference ([2]). It is therefore not caused by and electrolysis phenomenon (Fig. 1.6).
>
> ---
>
> ([2]) This does not mean this resistance is the same as an ordinary resistance, but the resulting effects are equivalent (...) (Blondel 1897, 519)

Blondel then specified in a footnote that in Ayrton's formula (1895), which he nonetheless considered to be the most accurate and adequate, that:

> The term $a + \frac{c}{i}$ would be the one symbolizing the counter-electromotive force. (Blondel 1897, 515)

---

[22]Since Blondel's oscillograph (1893a,b) enabled the study of alternating currents, it was modified and improved by Duddell (1897), then replaced by Braun's cathode-ray oscilloscope (1897).

Fig. 1.6 André Blondel. Stamp drawn and engraved by Jules Piel (14 september 1942)

It therefore appeared that the electric arc's resistance was indeed variable or even negative and that the arc held an electromotive force represented as a nonlinear function determined empirically (see Table 1.2) but that it did not feature a counter-electromotive force.

1.3 The "Arc Hysteresis" Phenomenon: *Hysteresis Cycles or Limit Cycles?*

By observing a simplified diagram of Duddell's layout (1900a, 248) we can notice, as Janet did, (1902b) the superposition of two currents in the arc branch: one is constant and produced by the direct current generator, the other is alternating and produced by the parallel circuit (*LC*).

1.3.1 The Static and Dynamic Characteristics of the Arc

The term characteristic was introduced by Marcel Deprez (1881, 893) (see *supra*) to refer to the curve representing the function relating the voltage at the terminals of a dipole and the intensity of the current traversing it. Blondel (1891a,b, 621) then associated the term with the word oscillation in order to define the stability of the direct current arc. One of the issues with the electric arc is that it can be powered by either direct or alternating current. Moreover, even with D.C., it is traversed by a current which is the result of the superposition of these two types. This led to a differentiation in the arc's characteristics.

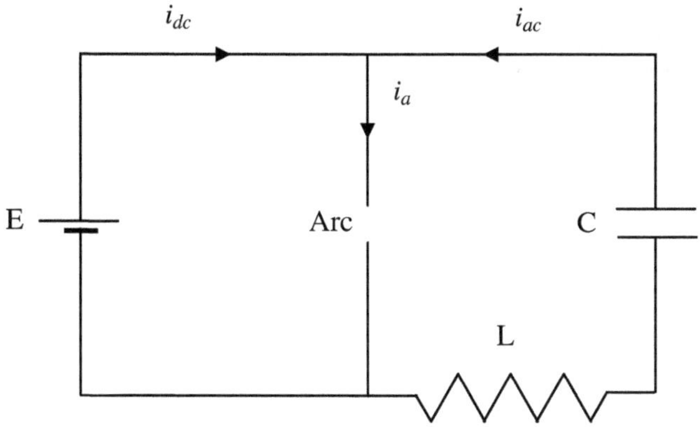

Fig. 1.7 Static characteristic of the *singing arc*

1.3.1.1 Characteristic of the Direct Current Arc

Therefore when the arc is powered by a direct current generator its *e.m.f.*, which can be represented by one of the relations in Table 1.2 is called *oscillation characteristic* or *static characteristic*. As for Thomson's formula (the most commonly used at the time): $u = a + \frac{bl}{i}$ this characteristic is shaped as an equilateral hyperbola (in red on Fig. 1.8).

This representation was already problematic at the time since the use of low-intensity currents invalidated Thomson's formula and made it necessary to "connect" the hyperbola (in red on Fig. 1.8) to another function (dotted line on Fig. 1.8) whose expression had not been yet determined. The characteristic was deemed as possessing a "falling" part, i.e. an increase in current corresponded to a decrease in voltage.

1.3.1.2 Dynamic Characteristic of the Alternating Current Arc

When an arc is powered by an alternating current its *e.m.f.*, which is not represented by any of the relations in Table 1.2, is called *dynamic characteristic*. In this case the potential difference in the arc, i.e. its *e.m.f.*, differs depending on whether the current flowing through it increases or decreases. This phenomenon, which Théodore Simon (1906) called arc *hysteresis*, was discovered by Mrs. Ayrton.

1.3.2 Hertha Ayrton's Works

At the start of the twentieth century a major roadblock in Hertha Ayrton's research was the impossibility of recreating exactly the experiments on the arc.

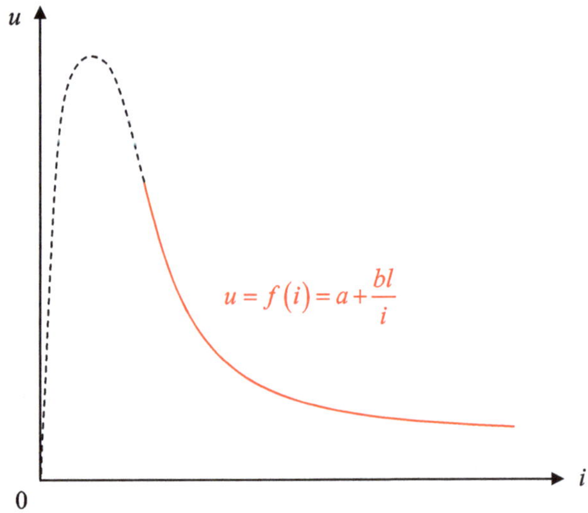

Fig. 1.8 Dynamic characteristic of the arc, from Ayrton (1902, 101)

> When these experiments were first started, at the beginning of 1890, it was not known what were the conditions necessary for the P.D.[23] between the carbons to remain constant when the current and length of arc were both kept constant, and consequently it was found, as had been found by all previous experimenters, that a given current could be sent through an arc of given length by many different potential differences, and that no set of experiments made one day could be repeated the next. (Ayrton 1902, 100)

With a slight increasing then decreasing of the intensity Mrs Ayrton noticed that the potential difference at the carbon terminals of the arc, i.e. its *e.m.f.* was greater when the current increased than when it decreased and that it did not vary in a straight line but rather in a closed curve which is represented on Fig. 1.8 following a dotted line starting at point 1 (Start circled in red on Fig. 1.8) till the finish point. Mrs Ayrton inferred that:

> Hence, from these curves it would be impossible to find any exact relation between P.D. and current for a given length of arc. (Ayrton 1902, 101)

Mrs Ayrton's experiments on the alternating current arc on the one hand brought to light, based on the cyclical aspect of the *dynamic characteristic*, the hysteresis phenomenon, and on the other showed that it is impossible to establish a relation of the same type as the ones in Table 1.2 between the voltage at the arc terminals (its *e.m.f.*) and the intensity traversing it.

[23]Potential Difference.

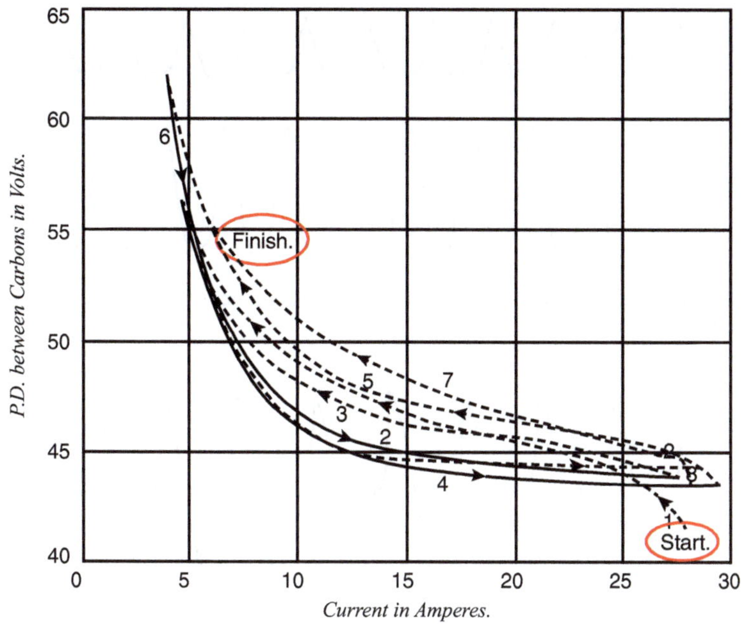

Fig. 1.9 Oscillographic record: first type curve: musical arc, voltage *u* and intensity *i* in the arc, from Blondel (1905b, 78)

1.3.3 André Blondel's Work On the Singing Arc Phenomenon

During the following years Blondel (1905a,b,c) intended to “completely shed light on the phenomenon of the singing arc” (Bethenod 1938b, 752). Therefore almost ten years after his last publication titled “On the electrical arc phenomenon” (Blondel 1897) he submitted the results of his research in a short note to the *C.R.A.S.* on the 13th of June 1905 titled “Sur les phénomènes de l'arc électrique” (“On the singing arc phenomenon”). He then made a presentation on the 7th of July 1905 in front of the *Société française de Physique*. It was better developed but would only be published in February 1906 in the Physique Théorique et Appliquée. In the meantime a detailed and unabridged version of his work was published in the periodical *L'Éclairage Électrique* in July 1905. By modifying the settings of Duddell's singing arc and using a bifilar oscillograph he designed, Blondel (1905a,b,c) found two new opposite types of singing arcs: one continuous, the other discontinuous.

- The first type (Fig. 1.9) “to which corresponds a quite pure sustained sound, and that is strictly speaking, Duddell's musical arc, gives rise to current curves in the arc and capacitor with continuous shapes, almost sinusoidal, without the intensity in the arc decreases to zero or at least remains zero during a considerable amount of time; changes in voltage across the arc are contained within very close limits.” (Blondel 1905b, 79).

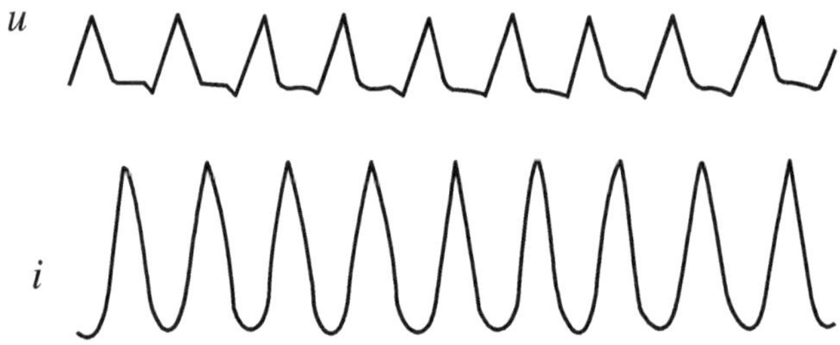

Fig. 1.10 Oscillographic record: second type curve: whistling arc, voltage *u* and intensity *i* in the arc, from Blondel (1905b, 79)

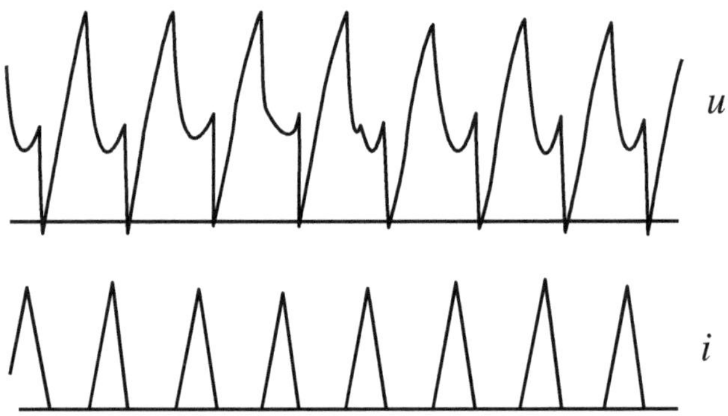

Fig. 1.11 Circuit diagram: *H*, the arc; *C*, the capacitor; *R*, the rheostat; *L* and *l*, self-induction; *ABDF*, the power supply circuit produced by line power; *BCD*, the oscillation circuit, from Blondel (1905b, 77)

- The second type (Fig. 1.10) "to which corresponds a shrill or hissing sound is a discontinuous phenomenon characterized by the fact that the arc current *i* has angular points and appreciable zeros of time during which the current of charge *j* presents ordinarily some flat spots, whilst the voltage between the electrodes *u* undergoes a double oscillation of large amplitude the values of which go often below zero or above the electromotive force of the generator." (Blondel 1905c, 80).

Analyzing these records shows that the first type corresponds to sinusoidal oscillations, as also noted by Blondel, whereas the two time scales appearing on Fig. 1.11 ("double oscillation") are typical concerning oscillations, which Van der Pol (1925, 1926a,b,c,d) called a few years later *relaxation oscillations* (see *infra*). Moreover the way in which Blondel (1905b, 80) went from discontinuous to continuous regime by simply moving the carbons farther apart can be assimilated to 's method (1926a,b,c,d) which enabled him to switch from relaxation oscillations to sinusoidal oscillations by modifying the parameter ε (see *infra*).

Blondel's circuit (1905a,b,c) (see Fig. 1.12) was perfectly identical to Duddell's (1900a, 248), which has been simplified on Fig. 1.7 (see *supra*).

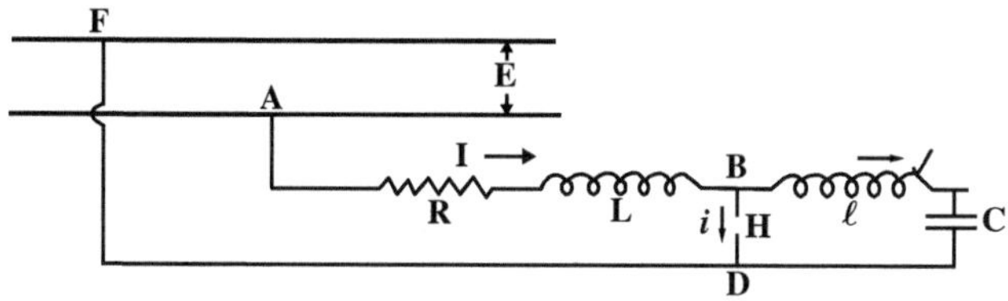

Fig. 1.12 Simplified diagram of Duddell's *singing arc* circuit

Despite the D.C. power supply the current in the arc results from the superimposition of both alternating and direct currents. This is similar to regarding it as being traversed by direct current modulated by a sinusoidal fluctuation. One must therefore consider including these two characteristics.

Blondel hence used on one hand the *static characteristic of the direct current arc*, which he named oscillation characteristic, in order to define the oscillation phenomenon he encountered with these two types of sine waves (BcA curve, symbolized with a red dotted line on Fig. 1.9). On the other hand, the *dynamic characteristic of the alternating current arc* since hysteresis also occurred, each value of the direct current in the arc being modulated by alternating current. An analogy can be made between the conjunction of these two characteristics and Ptolemy's depiction of epicycles.

As a matter of fact the potential difference at the arc terminals, i.e. its *e.m.f.*, obeys a hyperbolic law like Thomson's but the current modulation involves a cyclical evolution between two limit values for each point of this curve.

> These phenomena can be easily explained by the arc's properties between homogenous carbons, in regards to the stability. Let BcA be the theoretical stability curve of an arc (potential difference variation law) when the current is decreased by increasing the power-supply resistance beyond the normal value (corresponding to the power-supply straight line DM_1)). As a consequence of the well-known phenomenon (see Ms. H. Ayrton, *The Electric Arc*) of the continuous-speed delay caused by the heating and cooling of the electrodes, when one modifies the voltage between two limit points I_1' and I_1'', the continuous-speed point M_1 is not exactly a short line but rather a small cycle anbqa (Fig. 1.13). (Blondel 1905a, 1682)

These *hysteresis cycles*[24] were a first experimental depictions of the *limit cycle* concept in the phase plane (i, u), which would allow Poincaré (1908) to provide evidence of the existence of oscillations sustained by the singing arc (see *infra*) a few years later. However they could not be qualified as such since due to the impossibility of exactly reproducing the experiments their claims could not be verified. Using this graphic representation Blondel provided a condition for the oscillations and an explanation for the oscillatory phenomenon, for each noted type of sine wave:

> For the oscillation to be possible the current exchanged between the arc and the capacitor must therefore make up for the difference between $\overline{I_1''I_1'}$ and $\overline{I'I''}$. The experiment for the

[24]It should be noted that Blondel (1905a,b,c) could not use this name, since this terminology was apparently introduced by Simon (1906) the following year.

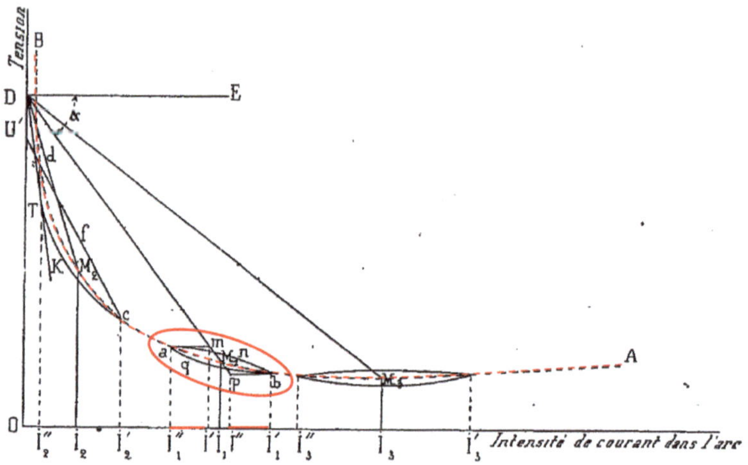

Fig. 1.13 Static and dynamic characteristics of the arc by Blondel (1905a, 1681)

> first type, the musical arc, corresponds exactly to this case. The power-supply current can even still undergo weaker oscillations. When there is little induction in the power-supply circuit, part of the power-supply current serves to compensate the energy losses due to the law of heat loss or other causes in the oscillating circuit, due to the fact that the cycle charging branch *anb* is above the discharging branch *bqa*. The capacitor thus receives more energy than it can return. (Blondel 1905a, 1682)

This description of a system (the arc) in which part of the produced energy was used to compensate for the losses and thus sustained the oscillations, was the early stage of the definition (Blondel 1919d, 120) later provided for a *self-oscillating system*.[25] Lastly he concluded in regards to the oscillation frequency:

> The frequency of the singing arc is essentially variable and ill-defined. With the continuous phenomenon it can be approximately determined using Duddell's formula but that is not the case with the discontinuous phenomenon where the formula has no relation whatsoever with the oscillating circuit's specific frequency. (Blondel 1905b, 54)

This indeed corresponds to what was noted in the case of the *relaxation oscillations* (see *infra*).

1.3.4 Théodore Simon's Work: The Hysteresis Cycle

In an article published in the periodic *Physikalische Zeitschrift*, Simon explains:

> For a conductor to produce stable oscillations in a parallel circuit, its characteristic curve $e = f(i)$ must be falling which means the fall in potential decreases as the current increases

[25] It was then established that this terminology had been introduced by Blondel (1919d). See *infra*.

(...). If the electromotive force varies between two limits, the characteristic[26] takes the shape of a closed loop, in some ways similar to hysteresis loops. (Simon 1906, 435)

Simon corroborated the notion that the oscillation characteristic, i.e. the arc's *e.m.f.* must have a "falling" or decreasing section in order for the stable sustained oscillations to occur. He then carried out the calculations inherent to Blondel's description of hysteresis cycles (1905b, 54) to establish conditions for the oscillation.

1.3.5 Heinrich Barkhausen's Work

The following year (1907) Barkhausen published a fundamental work on the problem of producing oscillations, especially electric oscillations.[27] He (1907, 46) also provided a graphic representation of the *dynamic characteristic of the alternating current arc* in the phase plane (i, e). He thus got a slightly more accurate *hysteresis cycle* than Blondel's (1905a,b,c). The frame at the center of Fig. 1.14 corresponds to the coordinate points (i_0, e_0) where it is met by the *static characteristic* which was not drawn by Barkhausen (and was added in red). He then divided the plane in four sections where he defined a network of equilateral hyperbola $e_1 i_1 = const.$ which enabled him to find the hysteresis cycle and to deduce that if the characteristic is in quadrants II and IV some energy is added to the alternating current whereas the characteristic being in the two other quadrants (I and III) means that energy is taken from the alternating current. This result should be compared to the one Blondel established (1905a, 1682) (see *supra*).

Barkhausen's work (1907) played a crucial role later on in the development of the theory of nonlinear oscillations by the Russian school of thought, especially in the choosing of the terms used to refer to the new type of self-sustaining oscillations, i.e. *selbst schwingungen* (see *infra* Part II).

1.3.6 Ernst Ruhmer's Work

In his work titled "Wireless Telephony" Ruhmer (1908, 148) also displays a diagram denoting the arc's *dynamic characteristic*

> We thus find that on taking oscillographic records of the current and voltage of the arc, we obtain a dynamical characteristic which shows high voltages with increasing currents, and low ones with decreasing currents, forming a hysteresis loop. (Ruhmer 1908, 148)

---

[26]This term refers to the dynamic characteristic. See *infra*.

[27]Mechanical oscillations are only discussed in the last fifteen pages of this hundred-and-twelve-pages long book.

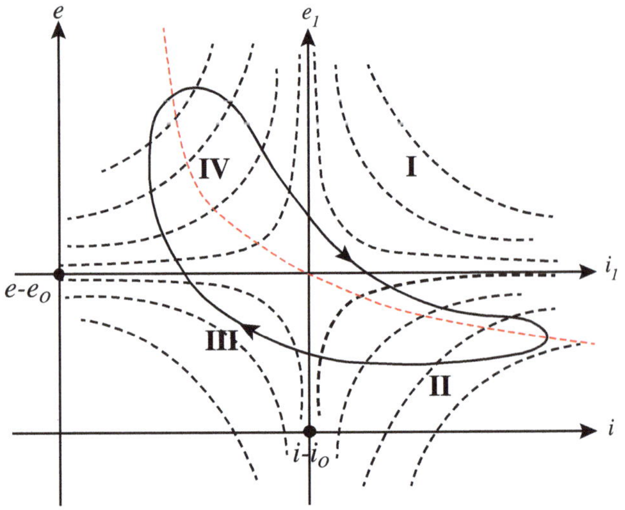

Fig. 1.14 Dynamic characteristic of the arc, from Barkhausen (1907, 46)

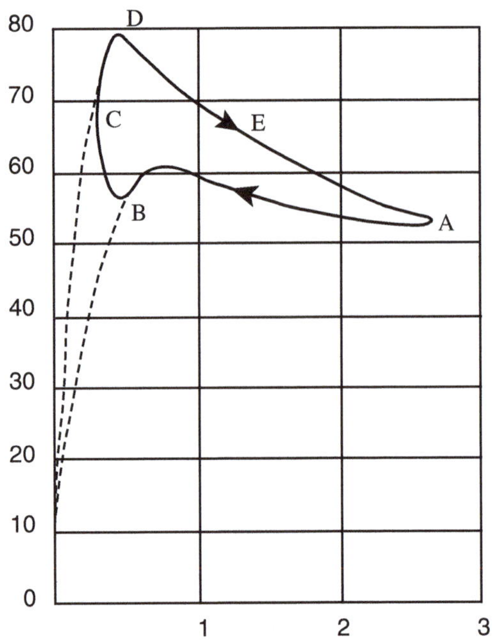

Fig. 1.15 Dynamic characteristic of the arc, from Ruhmer (1908, 148)

He then describes the evolution of the arc's characteristic on the different parts of the hysteresis cycle and explains:

> Whenever the capacity is charged it begins to discharge itself through the arc, thus increasing the latter until the maximum current is reached and the cycle recommences (A, Fig. 1.15). (Ruhmer 1908, 149)

The relation between these hysteresis cycles and the existence of sustained oscillations, i.e. oscillations with limit cycles, therefore seemed definitely established (see *infra*).

The characteristic properties of the arc are hence comparable to those of Gérard Lescuyer's *series-dynamo machine* and the problems to be solved were noticeably similar as a result: isolating the cause of the phenomenon, establishing a model for the arc's *current-voltage characteristic*, and putting these oscillations into equations in order to deduce their amplitude and period. Whilst the existence of an electromotive force in the arc and Ayrton's or Thomson's law of nonlinear variation seemed to provide solutions for these problems in regards to putting the arc oscillations into equations. Henri Poincaré is actually the one who carried it out in 1908 in a little-known study on wireless telegraphy.

1.4 Henri Poincaré's "Forgotten" Lectures: *The Limit Cycles in 1908*

Henri Poincaré already stood out in 1893 by providing a solution to the telegrapher's equation and was soon very involved in the wireless telegraphy development. In 1908 this mathematician and physicist had already published a huge number of articles on the subject (Poincaré 1902a,b, 1907b) as well as a book entitled "La théorie de Maxwell et les oscillations hertziennes. La Télégraphie sans fil"[28] ("Maxwell's theory and Hertzian oscillations. Wireless telegraphy") published in 1904 then translated in English and German. It was according to Blondel (1912, 100) the "first truly scientific presentation" on the subject. As in any field he researched, Poincaré was a unanimously recognized expert, as evidenced by his relations with the two French wireless telegraphy specialists Gustave Ferrié and Camille Tissot,[29] as well as his presence in numerous scientific committees such as the one for the periodic *La Lumière Électrique*. It is therefore not incidental that he was named president of the Development Council of the *École Supérieure des Postes et Télégraphes* (today Sup'Telecom Paris Tech) in 1901. The head of the school Édouard Estaunié even used this in order to restore the school's prestige and

[28]The first edition in 1899 did not include telegraphy, and was simply titled "Maxwell's theory and Hertzian oscillations". Another edition was published in English in 1904 with the title "Maxwell's theory and Wireless Telegraphy", and was published in French in 1907.

[29]We can mention that Poincaré was the examiner for the dissertation of Camille Tissot, defended in 1905, as well as his usual interlocutor on these subjects.

it went far beyond all his expectations.[30] Poincaré's lectures addressed a broad array of subjects:

- Propagation de courants variables sur une ligne munie d'un récepteur (Propagation of varying currents on a line fitted with a receiver),
- Théorie mathématique de l'appareil téléphonique (Mathematical theory of the telephone),
- La T.S.F. et la diffraction des ondes le long de la courbure de la Terre (Wireless telegraphy and the wave diffraction along the curvature of the Earth),
- La T.S.F. et la méthode théorique de Fredholm (Wireless telegraphy and Fredholm's theoretical method),
- La dynamique de l'électron et le principe de relativité (Dynamics of the electron and the principle of relativity).

Poincaré's lectures were held in May and June[31] 1908, and were edited in a series of five editions of the periodic *La Lumière Électrique*, which was at the time seen as a reference in electrotechnics and telegraphy. A very large public was reached, since readers of the periodic were added to the lectures' audience. In short, the published series, in chronological order, addressed the following subjects:

- Saturday 28 November 1908 (257): L'émission d'ondes et l'amortissement (Wave emission and damping),
- Saturday 5 December 1908 (299): Étude du champ dans le voisinage de l'antenne (Study of the field surrounding the antenna),
- Saturday 12 December 1908 (321): Transmission des ondes et la diffraction (Transmission of waves and diffraction),
- Saturday 19 December 1908 (353): La réception des signaux (Signal reception),
- Saturday 26 December 1908 (385): Télégraphie dirigée. Oscillations entretenues (Directed telegraphy. Sustained oscillations).

It must be noted that each lecture was in the headlines of the publication. The editorial covered the contents and the main conclusions.

> In these lectures, the author does not aim to create a complete theory of wireless telegraphy, but he imply intends to explain some mathematical theories likely to facilitate the understanding of these phenomena. (Poincaré 1908, 257)

[30]From Atten et al. (1999, 50): "While the semester Henri Poincaré dedicates, every two years, to especially difficult subjects, brings in a large audience, it is probably not of particular interest to engineers of the Postes et Télégraphes. But this globally renowned mathematician-physicist gives a new prestige to the school, and it is E. Estaunié's aim: "When... I had to reorganize it (the school), it seemed to me that resorting to Poincaré would give me every chance to achieve my goals... he agreed to give a lecture for free on... a question related to electricity of our choosing and never addressed before I must say that the simple announcement of his collaboration brought in numerous outsiders, showing the incredible reputation of the master and the appeal of such a program." Estaunié used his relations in order to invite other renowned scientists such as Pierre Curie.

[31]From Lebon (1912, 67).

Also it is noteworthy that these that these lectures brought as many points of view and theories as the original theoretical elements written by Poincaré. Until now, no traces of them have been found anywhere. The best example of this contribution addresses the question of sustained waves, which was the subject of the last lectures, cited in the editorial:

> He then goes on to the study of sustained oscillations, and establishes four general equations, one of which is differential, in order to determine the stability condition of the regime, as well as the possibility conditions of the problem. From a practical standpoint, the simple inclusion of an arc in the circuit enables sustained oscillations, provided that a specific frequency is not exceeded, as shown in the calculation. (Poincaré 1908, 385)

The same editorial mentioned the issue concerning the arc symmetry and the maintenance of asymmetry that would insure the existence of oscillations on all frequencies. Any reader of the periodic, including specialists in electrotechnics and the then-emerging wireless telegraphy, had therefore access to these writings expressed in perfectly comprehensible, clear and synthetic words.

1.4.1 Setting into Equation the Oscillations Sustained by the Singing Arc

In his last lecture Henri Poincaré looked more specifically into the singing arc's device and the sustained oscillations it produced. The diagram of the circuit shown on Fig. 1.16 is identical to Blondel's (1905b, 77) (see *supra* Fig. 1.11):

This circuit includes a "source of constant electromotive force E, a resistance and an inductor, and in parallel, an arc on one side, and an inductor and a capacitor on the other side." (Poincaré 1908, 390). He then wrote the very first equation for singing arc oscillations, calling x the capacitor charge and i the current in the external circuit. The current intensity in the branch including the capacitor, of capacitance $1/H$, is written as follows:

$$x' = \frac{dx}{dt}$$

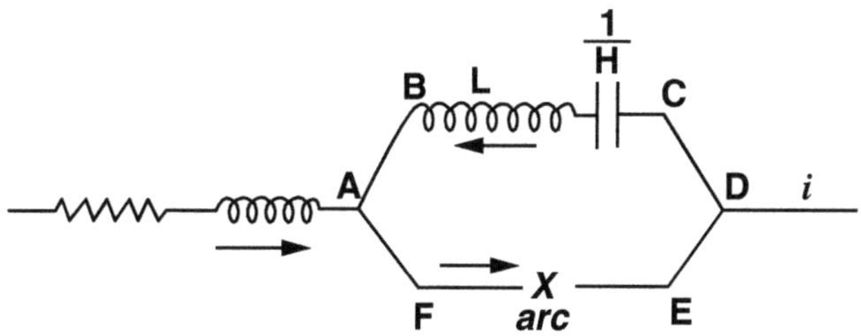

Fig. 1.16 Oscillations sustained by the singing arc, from Poincaré (1908, 390)

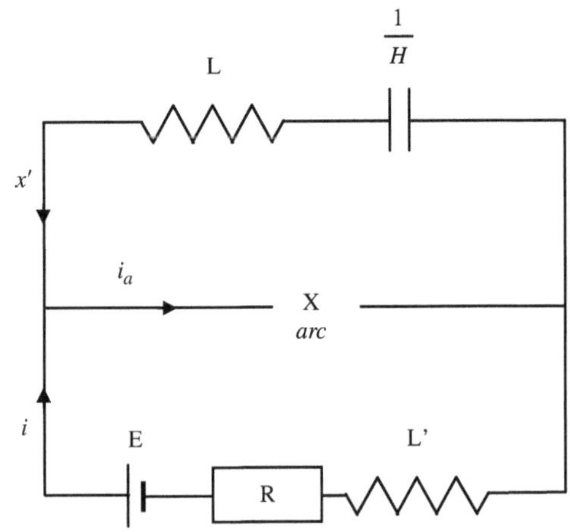

Fig. 1.17 Oscillations sustained by the *singing arc*, simplified version

Let i_a be the current voltage in the arc, by applying Kirchhoff's first law[32] and taking the current direction into account (see Fig. 1.17). Poincaré got the following result: $i = i_a - x'$. Therefore, the current in the arc is $i_a = i + x'$. By expressing, using Kirchhoff's second law,[33] the voltage in the mesh ABCDEF, Poincaré established the second order nonlinear differential equation for the oscillations sustained by the singing arc

$$Lx'' + \rho x' + \phi\left(i + x'\right) + Hx = 0 \tag{1.3}$$

1.4.2 The Singing Arc's Electromotive Force

He specifies that "considering $\rho x'$ is a term referring to the internal resistance of the inductor and other possible causes of damping, including the antenna radiation, $\phi\,(i + x')$ is the term due to the arc." (Poincaré 1908, 390). The latter term represents the electromotive force of the singing arc which should be related to the voltage running through it by an empirically determined relation (see supra Table 1.2). This relation is indeterminable, making it impossible to integrate the equation (1.3), and has been discredited due to the controversies linked with the presence of a counter-electromotive force in the arc and the difficulties inherent to experimentation, as stated by Paul Janet (1919). This did not seem to impede Henri Poincaré's research and he approached the problem as if it had already been resolved. In order to bypass this difficulty, he expresses the tension in the mesh AFED. A simplified version of the circuit is shown below (Fig. 1.17) to visualize the equation he found.

[32]Nodal rule.

[33]Mesh rule.

By neglecting the external inductor L', and equating the tension in the lower and middle branches of the simplified circuit (Fig. 1.17) he wrote: $E - Ri = \phi\,(i + x')$ he then deduced:

$$Ri + \phi\left(i + x'\right) = E \tag{1.4}$$

He then explained that "if the function ϕ is assumed to be known, the equation (1.4) provides a relation between i and x' or between $i + x'$ and x'." (Poincaré 1908, 390). Indeed, it is easy to check that if we choose for this function that Poincaré (1908, 392) uses a few pages later and which is that of S.P. Thompson's[34] (see Table 1.2): $\phi\,(i + x') = \dfrac{a}{i + x'} + b$, we obtain the following equation:

$$Ri^2 + \left(Rx' + b - E\right) i + a + (b - E)\,x' = 0$$

Solving this second order equation for i indeed provides a relation between i and x' but Poincaré's reasoning, probably based on the *implicit function theorem* enabled him to bypass all these calculations. If we assume as he did that the function ϕ is known, the equation (1.4) leads to a function F relating i and x'. We will write out: $i = F\,(x')$ and replace in the equation (1.4) thus obtaining:

$$\phi\left(i + x'\right) = E - RF\left(x'\right) = \theta\left(x'\right)$$

He therefore managed to replace ϕ in the equation (1.3) by $\theta\,(x')$. The difficulty was lifted because the differential equation did not depend on only one variable x anymore. He wrote it as follows:

$$Lx'' + \rho x' + \theta\left(x'\right) + Hx = 0 \tag{1.5}$$

Poincaré thus created the very first incomplete equation modeling[35] of the oscillations occurring in the singing arc. It should be noted that this equation (1.5) corresponds exactly, in accordance to duality, to the one established later by Blondel (1919b) for the triode, and Van der Pol (1920, 1926a,b,c,d). Incidentally, the analogy between the oscillations sustained by the singing arc and the triode was brought to light by Paul Janet (1919).

---

[34]It is interesting to note that Poincaré seemed to know about the latest hypotheses and theories regarding the arc, including Blondel's works (1897), since he did not choose Ayrton's relation (the most recent), but Thompson's, because it does not contain the notion of *c.e.m.f.* See Table 1.2.

[35]The equation (1.5) is nonetheless incomplete, due to the lack of definition of the function $\theta\,(x')$.

1.4.3 Stability of the Sustained Oscillations and Limit Cycles

Poincaré therefore demonstrated more than twenty years before Andronov (1929a) that the stability of the equation (1.5) is related to the existence of a closed curve – a *limit cycle*. To achieve this he used the phase plane, which he introduced in his memoirs "Sur les Courbes définies par une équation différentielle" ("On the Curves defined by a differential equation") (Poincaré 1886a, 168), by writing:

$$x' = \frac{dx}{dt} = y \quad ; \quad dt = \frac{dx}{y} \quad ; \quad x'' = \frac{dy}{dt} = \frac{ydy}{dx}$$

The equation (1.5) becomes:

$$Ly\frac{dy}{dx} + \rho y + \theta\,(y) + Hx = 0 \tag{1.6}$$

Poincaré then provides the following representation:

It should be noted that this closed curve is represented in the phase plane $(x, y) = (x, \dot{x})$, i.e. the current-voltage phase plane[36] (u_a, i_a) of the arc, and that it is only a metaphor of the actual solution, since Poincaré did not use any graphical integration method to find it. The only actual aim of this representation is to specify the direction of the trajectory curve which is a preliminary condition required to demonstrate the proof shown below.

> Curves can be traced in a way that satisfies this differential equation, provided that the function θ is known. The sustained oscillations correspond to the closed curves if there are any. But any closed curve is not suitable; it must meet specific stability conditions which we will study. Firstly, we see that, if $y = 0$, $\frac{dy}{dx}$ is infinite, the curve has vertical tangents. Besides, if x decreases, x' in other words y, is negative, therefore the curve must be traced in the direction of the arrow. (Poincaré 1908, 390).

From the equation (1.6), it can easily be found that:

$$\frac{dy}{dx} = -\frac{\rho y + \theta\,(y) + Hx}{Ly} = -\frac{\rho}{L} - \frac{\theta\,(y) + Hx}{Ly} \tag{1.7}$$

It can be deduced that when y approaches zero the right-hand side of this equation becomes infinite. This closed curve therefore allows vertical tangents represented by dotted lines on Fig. 1.18. As for the direction, the reasoning is based on the fact that the derivative curve of an increasing function is negative. Therefore when x decreases, x' becomes negative. Yet, according to the equation given above: $x' = y$. This implies that y also becomes negative. This is only possible if the curve follows the direction of the arrow. It might seem surprising to look at the direction of

[36]By rotating the i and u axes, we note that this is the same phase plane as that used by Blondel (1905a, 1681) (see page 23). It therefore appears that the hysteresis cycle corresponds exactly to Poincaré's closed curve, i.e. a limit cycle.

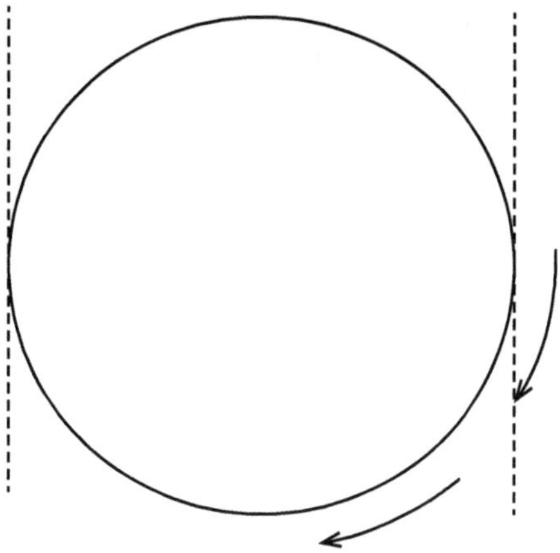

Fig. 1.18 Closed curve, from Poincaré (1908, 390)

the trajectory curve first. The reason is that the search for a condition provided by an inequality whose direction will enable to govern the stability of sustained oscillations. The demonstration shown below uses Green's formula for integration, along a closed curve, which requires knowing the direction of the trajectory curves in order to define their orientation.

Poincaré then described a first version of the stability of sustained oscillations essentially based on the existence of a closed curve.

> *Stability condition.* - Let us then consider another non-closed curve which satisfies the differential equation. It will be a kind of spiral growing indefinitely closer to the closed curve. If the closed curve represents a stable speed, by tracing the spiral in the direction of the arrow, we must be taken back to the closed curve, and this condition alone will enable the curve to represent a stable set of sustained waves, and solve the problem. (Poincaré 1908, 391)

In the *Note on the scientific Works of Henri Poincaré* which he wrote himself in 1886, he defines the concept of *limit cycle*:

> This is how I call closed curves which fulfill the differential equation, for which the other curves defined by the same equation approach it asymptotically without ever reaching them. (Poincaré 1886b, 30)

When comparing this excerpt with the *stability condition* explained in 1908, it appears clearly that the "closed curve", which represents the stable set of sustained waves, is actually a limit cycle as defined by Poincaré himself. The reason he did not write it out explicitly can however be addressed. It can be argued that firstly, this presentation was intended for engineers, and not mathematicians, and secondly, this terminology would have served no purpose.[37] An interesting comparison can be made between Poincaré's "stability condition" and the conclusion of paragraph 8, chapter V, of Andronov's book (1937), where he further explored the results he had obtained in 1928 (see *infra* Part II) and 1929:

[37]The case of the center, which also constitutes a closed-curve solution, appears to have been excluded by Poincaré, insofar as it is a non-conservative system.

> The existence of limit cycles in the description of the provided dynamic system's phase is a necessary and sufficient condition for the eventuality (provided the initial conditions are suitable) of self-oscillations in the system. (Andronov 1937, 293)

Poincaré then seems to avoid the "purely mathematical context",[38] in order to demonstrate that this problem was physically tangible. Beyond the connection between closed curves and sustained oscillations, Poincaré introduced the stability of the closed curve, i.e. of the limit cycle as an inequality.

> *Possibility condition of the problem.* - Let us return to the equation (1.6). We multiply by $x'dt$, and integrate, over one period, the term L and the term x leading to the integration of terms x' and x, disappear, and we find:
>
> $$\rho \int x'^2 dt + \int \theta\left(x'\right) x' dt = 0$$
>
> Yet, the first term is certainly positive, and the function θ must therefore be thus:
>
> $$\int \theta\left(x'\right) x' dt < 0.$$
>
> Is it possible? (Poincaré 1908, 391)

The first integral equation is easy to establish. By following Poincaré's steps, we multiply the equation (1.6) by $x'dt$, taking the fact that $x'dt = dx$ into account, as shown in the equation above. We therefore find that:

$$\int Lydy + \int \rho x'^2 dt + \int \theta\left(x'\right) x' dt + \int Hxdx = 0$$

The first and last terms of this equation, which correspond to conservation of energy ($\frac{1}{2}Li^2 + \frac{1}{2}Hx^2$) cancel each other. The second integral inequation is deduced from the fact that the second quadratic term "is certainly positive" according to Poincaré (1908, 391). The *possibility condition of the problem* was therefore found. The relevance of a comparison between Poincaré's result and Andronov's (1929a) regarding the stability of the limit cycle therefore becomes apparent.

1.4.4 "Poincaré Stability" and "Lyapunov Stability"

Comparing Poincaré's *stability condition* (1908, 391): $\int \theta\left(x'\right) x' dt < 0$ and Andronov's (1929a) requires the modification of the equation (1.3), and writing it as:

$$\begin{cases} \dfrac{dx}{dt} = Ly \\ \dfrac{dy}{dt} = -\rho y - \theta\left(y\right) - Hx \end{cases} \tag{1.8}$$

[38] Diner (1992, 340).

This equation (1.8) can be nondimensionalized by writing one the one hand $x \to \sqrt{\frac{L}{H}}x$ and $t \to \mu t$ with $\mu = \frac{1}{\sqrt{LH}} = \frac{1}{\omega}$, and on the other hand, neglecting the capacitor resistance ρ. We then find:

$$\begin{cases} \dfrac{dx}{dt} = y \\ \dfrac{dy}{dt} = -x - \mu\theta\left(y\right) \end{cases} \tag{1.9}$$

In his note (14 October 1929), Andronov (1929a, 560) considered the following set of differential equations:

$$\begin{cases} \dfrac{dx}{dt} = y + \mu f\left(x, y; \mu\right) \\ \dfrac{dy}{dt} = -x + \mu g\left(x, y; \mu\right) \end{cases} \tag{1.10}$$

where μ is a real parameter, which can be set as sufficiently small. He then specifies:

> When $\mu = 0$, equations (1.10) have a solution $x = R\cos\left(t\right)$, $y = -R\sin\left(t\right)$; the solutions form, in the xy plane, a family of circles. Following Poincaré's methods, it can be seen that for sufficiently small $\mu \neq 0$, the xy plane contains only isolated closed curves, near to circles with radii defined by the equation
>
> $$\int_0^{2\pi} [f\left(R\cos\left(\xi\right), -R\sin\left(\xi\right); 0\right)\cos\left(\xi\right) - g\left(R\cos\left(\xi\right), -R\sin\left(\xi\right); 0\right)\sin\left(\xi\right)]d\xi = 0$$
>
> These closed curves correspond to stable, steady-state motion where the condition (1.11) is fulfilled:
>
> $$\int_0^{2\pi} \left[f'_x\left(R\cos\left(\xi\right), -R\sin\left(\xi\right); 0\right) + g'_y\left(R\cos\left(\xi\right), -R\sin\left(\xi\right); 0\right)\right]d\xi < 0 \tag{1.11}$$
>
> (...) (Andronov 1929a, 561)

This result was actually the first draft of a theorem which was studied in a work titled "On Lyapunov stability" written by Andronov and Witt (1933), later formalized by Pontrjagin (1934). By using Green's formula[39] and using a Cartesian coordinate system again, the stability condition (1.11) is written:

$$\int_\Gamma \left(f\left(x, y; \mu\right)\frac{dy}{dt} - g\left(x, y; \mu\right)\frac{dx}{dt}\right)dt < 0 \tag{1.12}$$

It therefore appears that Andronov's approach is perfectly identical to Poincaré's. A simple comparison between the two differential equations in Poincaré's phase plane can demonstrate this (see Table 1.3).

[39]Green's formula: $\int_\Gamma f\left(x, y\right)dy - g\left(x, y\right)dx = \iint_S \left(f'_x\left(x, y\right) + g'_y\left(x, y\right)\right)dxdy$.

Table 1.3 Poincaré's (1908) and Andronov's (1929a) differential equation systems

Poincaré (1908)	Andronov (1929a)
$\begin{cases} \dfrac{dx}{dt} = y \\ \dfrac{dy}{dt} = -x - \mu\theta(y) \end{cases}$	$\begin{cases} \dfrac{dx}{dt} = y + \mu f(x, y; \mu) \\ \dfrac{dy}{dt} = -x + \mu g(x, y; \mu) \end{cases}$
Voir Eq. (1.9) p. 21	Voir Eq. (1.10) p. 22

When writing: $x' = dx/dt = y, f(x, y; \mu) = 0$ and $g(x, y; \mu) = -\theta(y)$, it can be observed that Poincaré's system (1.9) corresponds exactly to Andronov's (1.10) and, Andronov's condition (1929a) (1.12) is in this case absolutely identical to Poincaré's *possibility condition of the problem* (1908).

$$\int_{\Gamma} \left(f(x, y; \mu) \frac{dy}{dt} - g(x, y; \mu) \frac{dx}{dt} \right) dt < 0 \quad \Leftrightarrow \quad \int_{\Gamma} \theta(x') x' dt < 0$$

It therefore seems that Poincaré (1908) has not only established a connection between sustained oscillations and limit cycles, but has also demonstrated the limit cycle stability using a condition also found twenty years later by Andronov. Aside from the mathematical angle, this conclusion can also be reached by studying the bibliography of Andronov's article (1929a).

Despite all appearances, the actual connection with Poincaré's works is not the one found in his famous essay "Sur les Courbes" ("On curves") (1881–1886), more specifically in the chapter entitled "Limit cycles theory" (Poincaré 1882, 261), but is rather in relation with chapter III titled "Periodic solutions" in the "New Methods on Celestial Mechanics" (Poincaré 1892, 89). In this chapter Poincaré considered a system comparable to (1.10) which possessed a periodic solution for $\mu = 0$, and the following problem arose:

> Under which conditions can we conclude that the equations still have periodic solutions for small values of μ? (Poincaré 1892, 81)

He then demonstrates that:

> If the equations (1) depending on a parameter μ admit, for $\mu = 0$, a periodic solution with no characteristic exponents that are null, they will still admit a periodic solution for small values of μ (Poincaré 1892, 181)

The stability condition (1.11) corresponds to what Andronov and Witt (1930a, 1933) call in their articles the "Lyapunov stability" whilst referring to the chapter titled "Exposants caractéristiques'" ("characteristics exponents") of Poincaré (1892, 162). They define it in 1933 as follows:

> In our case, one of the characteristic exponents is still null,[40] since the equation (1) does not explicitly depend on time. The question of the "Lyapunov stability" is therefore whether the other characteristic exponents have negative real parts. (Andronov and Witt 1933, 373)

[40] Andronov and Witt (1933) refer to Poincaré (1892, 180).

Fig. 1.19 Henri Poincaré

Poincaré's "possibility condition of the problem" (1908) and Andronov's condition of the "Lyapunov stability" (1929a) hence appear to be based on previous works by Poincaré: the negativity of one of the characteristic exponents. Andronov (1929a) seems to have found Poincaré's results (1908) independently with no previous knowledge of them. To prove the existence of a sustained oscillation regime Poincaré then demonstrated that the function ϕ, which represents the electromotive force of the singing arc (see *supra*) is decreasing. He achieved it by hypothesizing that "in the arc, the current $(i + x')$ always flows in the same direction and that the arc does not shut down." (Poincaré 1908, 391). He adds that we can "also assume that the direction of the current changes during an oscillation." (Poincaré 1908, 391). He then addressed a more practical aspect of the realization of the arc, and formulated the following condition for oscillation:

> We can then see that, from the simple presence of an arc in the circuit, the function ϕ becomes decreasing. Therefore, from what has been previously stated, it becomes possible to have sustained oscillations. (Poincaré 1908, 391)

This sentence shows how far he managed to go in the interpretation and comprehension of the phenomenon. The decreasing in the function ϕ is closely linked to the concept of "negative resistance", which plays a crucial part in the sustaining of oscillations. After Henri Poincaré died in 1912, and during the First World War, a new device was developed which played a decisive role for the rest of the conflict in the field of communications: the *three-electrode tube*, or *triode* (Fig. 1.19).

Chapter 2
The Great War and the First Triode Designs: Abraham, Bloch, Blondel, Van der Pol

2.1 The Great War and the Rise of Wireless Telegraphy: *The T.M. Valve and the Multivibrator*

2.1.1 General Ferrié: From Wireless Telegraphy to the Eiffel Tower

Gustave Ferrié[1] joined the École Polytechnique in 1887 at age 19 and chose l'arme du Génie (engineering) afterwards. He became a radio transmission engineer in 1893, specializing in military telegraphy in 1893. In 1897, he was named Head of the École de Télégraphie Militaire that had been created in 1895 at Mont Valérien. From 1899, the young captain showed an interest in wireless telegraphy after witnessing the first experiments carried out by Guglielmo Marconi on short-distance Hertzian links. The same year, The Minister of War, Charles de Freycinet appointed him to the Committee on Wireless Telegraphy research between France and the United Kingdom in order to write a report on the military applications for this communication medium. In October, with commandant Boulanger, he published the first French study[2] on wireless telegraphy. He improved the coherer designed by Édouard Branly, whose lectures he attended, by perfecting an electrolytic detector in 1900. In autumn 1903, Ferrié met Camille Flammarion and told her about one of the major problems with the development of radio transmission at the time – the size of the antenna. Camille Flammarion, being very close to Gustave Eiffel, asked him if Ferrié could use his 300-m high (986 ft) tower for his radio broadcasting tests. The government soon authorized the building of the first experimental military installation, which was based at Champ de Mars, the tower being used to support the antenna (see Fig. 2.1).

[1]See also *Notice sur les travaux scientifiques et techniques de M. Gustave Ferrié*, Ferrié (1921).

[2]See Boulanger and Ferrié (1899). Chapter V was redacted with Blondel's collaboration.

J.-M. Ginoux, *History of Nonlinear Oscillations Theory in France (1880–1940)*, Archimedes 49, DOI 10.1007/978-3-319-55239-2_2

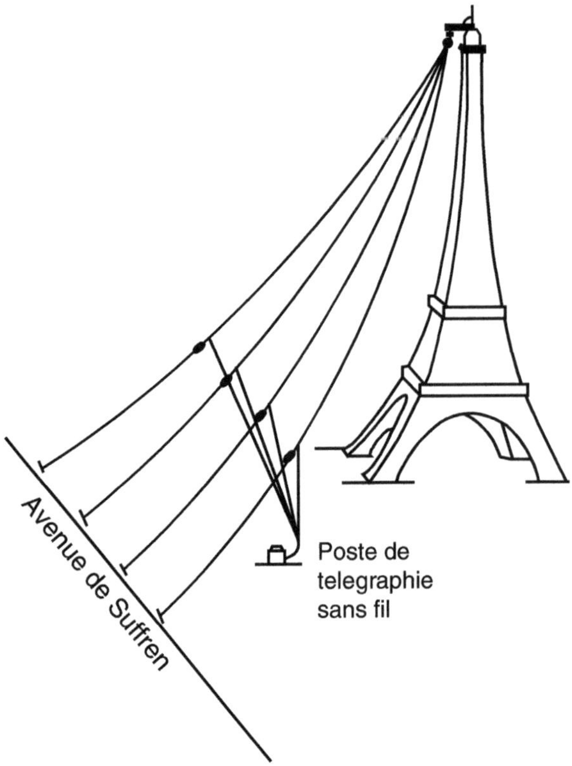

Fig. 2.1 The Eiffel tower's first antenna (1903–1908), from Turpain (1908, 242)

This initiative probably saved the Eiffel tower from being razed, by turning it into the cornerstone of the military wireless transmission network. From 1908 to 1914, *Commandant* Ferrié worked on developing mobile radio communication military units (auto-mobile field stations, planes, and dirigible stations). He created the pendulum comparison method by using wireless telegraphy, making it possible to determine the longitude, within a few meters, of any location as long as it is situated within range of the emitting station. Indeed, in 1911, Ferrié started a series of experiments aiming at accurately determining the difference in longitude between Toulon and Paris, and then between Paris and Washington, using radio signals emitted from the Eiffel tower. Soon after, he installed the time signal emitter used by navigators at the top of the Eiffel tower. During this time, he became a corresponding member for the *Bureau des Longitudes* (1911), the *Comité d'Électricité* (1912), and was appointed as a lecturer on the *Cours de Télégraphie Sans Fil* of the *École Supérieure d'Électricité* (1911). Just before the First World War, Ferrié was promoted to the rank of *colonel* and became the technical director of the *Radiotélégraphie Militaire* department, which would become the *Établissement Central du Matériel de la Radiotélégraphie Militaire* (E.C.M.R.) in 1917. Because it was used to listen into enemy communications, the Eiffel Tower, formerly called the "Dame de Fer" ("Iron Lady"), earned a new nickname, "La grande Oreille" ("The big Ear").

It was in this way, thanks to the interception of a German message,[3] that Joffre was informed of the advance of von Klück's troops, and decided to requisition all the taxis in Paris in order to send soldiers to *La Marne*. At the time, radiotelegraphy equipment was used to receive and emit signals and was based on the singing arc concept described in the previous paragraph. In fact, although the audion[4] had been created in 1907 by Lee de Forest, the triode appeared only in the first days of the war, and then in notably curious circumstances as we will see. A key player was the French engineer Paul Pichon. Pichon had actually deserted the French army in 1900 and migrated to Germany, where he earned a living by teaching French. Among his students were the children of Count von Arco, one of the founders of the Telefunken Company, who then hired him as a technical representative. In March 1913, Abraham visited the United States with him and they met with Lee de Forest, which is how he managed to learn about the latest improvements and applications of audions, which could be used as amplifiers and oscillators from then on.[5] In the summer of 1914, he went to the United States on an assignment from the German company Telefunken to try and gather samples of the most recent valves for wireless telegraphy in order to bring them back and test them in Germany. During his stay, Pichon visited the Western Electric Company, where he was given the latest high-vacuum Audion, and was provided with full information on their use. On the way back from his stay in America, the transatlantic ship stopped in London on the third of August 1914, the day Germany declared war on France. Pichon was then considered as a deserter in France and as an alien in Germany. He decided to go to Calais where he was arrested and brought to the French military authorities which were represented by colonel Ferrié. In October 1914, Ferrié gathered a team of specialists whose mission was to develop a French audion, which should be sturdy, have regular characteristics, and be easy to produce industrially.[6]

2.1.2 The T.M. Valve: Télégraphie Militaire

In October 1915, Ferrié decided to send Abraham to Lyon where he ordered a military emitter to be built that would be capable of replacing the one on the Eiffel tower in case it was rendered unusable. At the same time, a 100 kW emitter was about to depart for Saigon, which should have been conveyed by Captain François Péri of the *Infanterie Coloniale*, chief of the Service Radio of Indochina. Ferrié managed the feat of getting the equipment and men escorting it put at his disposal.

[3]The "emitted free-to-air" message picked up by the Eiffel Tower was the following: "Très bien compris, gagnez rive sud de la Marne. Oberste Heeresleitung (GQG)."

[4]The audion generator was the first triode-type electronic tube. Patent N° 841 386, 15 January 1907.

[5]The audion replaced the singing arc as an emitter, but also a receptor.

[6]See Champeix (1980, 16).

He appointed Péri as chief of the station in Lyon-La-Doua, and the engineer Joseph Bethenod, who had designed this emitter S.F.R.,[7] as head technician. Less than two months later, Péri managed to build barracks and to erect the antenna consisting of 8 towers of 120 m high each, making the radio station of Lyon La Doua operational. Choosing Lyon was not just a strategic move, it was also due to the proximity of the Grammont factory,[8] which produced incandescent valves. The self-taught engineer Jacques Biguet was briefly appointed as director.

Abraham and Péri initially tried to recreate Lee de Forest's audions, but their fragile structure and lack of stability made them unsuitable for military use. After several unsuccessful attempts, Abraham created a fourth structure in December 1914, which was used from February to October 1915 (see Fig. 2.2). A copy of this valve, called the "Abraham lamp" is still in the *Arts et Métiers* museum to

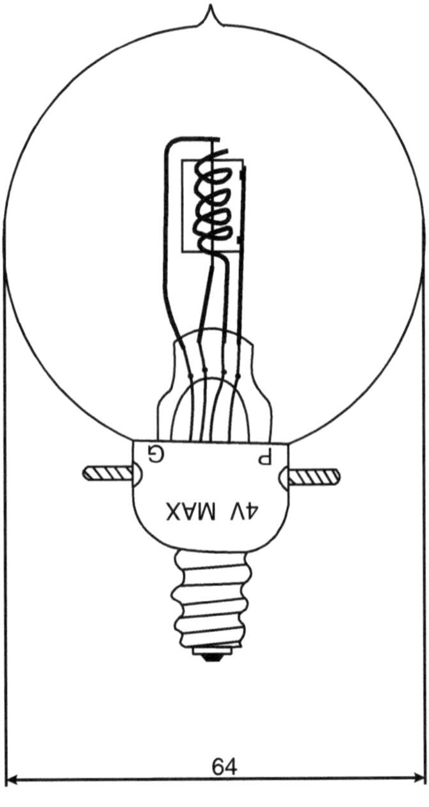

Fig. 2.2 Abraham valve, from Champeix (1980, 15)

[7]Société Française Radioélectrique.

[8]François Grammont, *normalien* like Henri Abraham, was then *Capitaine des Zouaves*. He was demobilized at the beginning of 1915, at the initiative of Ferrié, in order to go back to his post as director of his factory.

this day.[9] It has a cylindrical structure, which appears to have been designed by Abraham. Afterwards he created several new processes for improving the quality of the vacuum inside the lamp, insuring better reliability and stability. However, the relationship between the captain and the physicist soon deteriorated, and a competition arose between them. The atmosphere became so toxic that Ferrié had to call Abraham back to Paris in May 1915. Following his departure, Péri, who possessed extensive skills in the radio-engineering field, resumed his experiments with Biguet to improve the device. He created a valve with a mobile plate and a grid, which made it easier to investigate the characteristics experimentally. This collaboration resulted in the creation the famous T.M. valve (see Figs. 2.3 and 2.4), for which he registered four patents under the names of Mr. Peri and Biguet. The main patent n° 492657 was requested on October 23rd, 1915 and delivered on March 21st, 1919 (Fig. 2.5).

The cylindrical structure of the T.M. valve greatly improved its sturdiness and emission quality. Moreover, the four-pin cap allows quick replacement, as opposed to screw caps (compare Figs. 2.2 and 2.3–2.4). The T.M. valve, also called the "French valve", was refined to such a degree of reliability that it was used by the French and then by the allied armies, and over one million copies were mass-produced over the course of the conflict. It therefore appears that the first triode valve prototype was indeed created by Abraham in December 1914. However, the famous T.M. valve was actually patented by Péri and Biguet in May 1915, after Abraham left. A huge squabble ensued over the invention's paternity. Colonel Ferrié did not forgive Peri for patenting it, as he considered that the credit should have fully gone to Abraham.[10] Hence, when he asked Camille Gutton in March 1918 to write a "Note on the three-electrode valve lamps and their uses[11]", which was published by the *Établissement Central du Matériel de la Radiotélégraphie Militaire* (E.C.M.R.), Péri was not credited as a contributor to the T.M. valve production. This report, n° 412 of E.C.M.R., of more than a hundred and seventy pages and classified as a "military secret", is a remarkable synthesis of the work carried out in France during the First World War, in regard to, on the one hand, the creation of a T.M. lamp, and on the other hand the developing of the multivibrator (see Figs. 2.6, 2.7 and 2.8).

[9]Inventory n° 21204-0000-.

[10]According to Champeix (1980, 20 and following) Abraham refused, in spite of Ferrié's injunctions, to register any patent, and to ask for any compensation.

[11]This E.C.M.R. note, which was destroyed during the rebuilding of the *Service Historique de la Défense* (S.H.D.), was found in a collection: Mr. Jacques Denys's, who agreed to send us a copy. See *infra*.

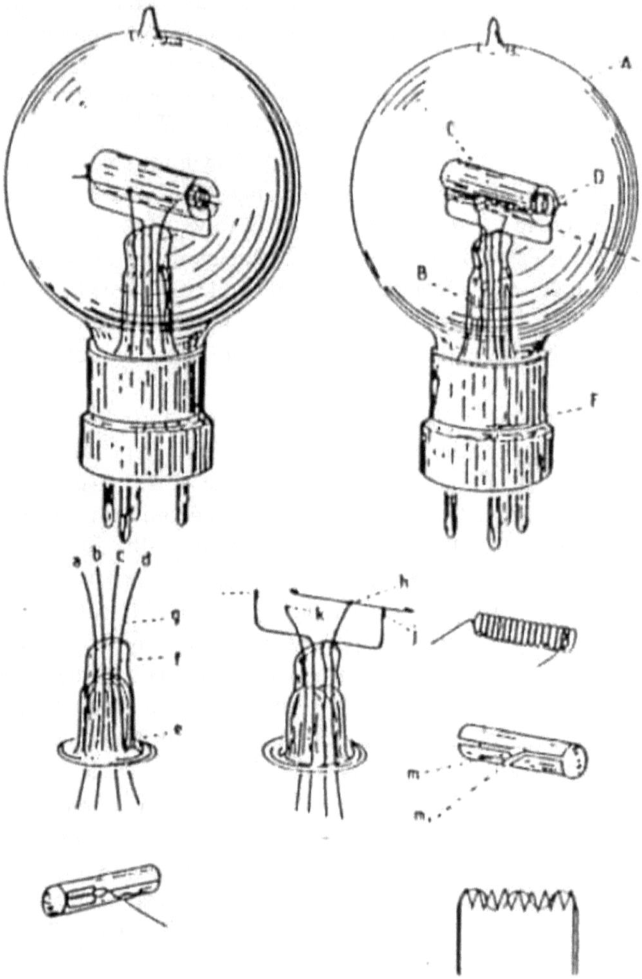

Fig. 2.3 T.M. valve, from Champeix (1980, 18)

On his return to Paris, Abraham joined the Military Telegraphic service's second group, under the supervision of Captain Paul Brenot,[12] along with Maurice de Broglie, Paul Laüt and Lucien Lévy. Abraham then resumed his duties as head of

[12]Paul Brenot (1880–1967) joined the *École Polytechnique* in 1899 and was appointed as attendant to Ferrié (X 1887) from 1904. He played an important part in the development of the *S.F.R.*, created by Joseph Bethenod and Émile Girardeau (X 1902) in 1910, and backed his participation in Military Telegraphy for the construction of high-quality wireless telegraphy materials, both civilian and military.

Fig. 2.4 T.M. valve (Source: Musée des Arts et Métiers)

the Physics Laboratory at the *École Normale Supérieure*, and invented the *astable multivibrator* with Eugène Bloch.

2.1.3 The Multivibrator: From the Thomson-Type Systems to Relaxation Systems

Henri Abraham[13] joined the *École Normale Supérieure* at age 18, and came second in the physics agrégagtion competitive exam in 1889. The following year, after his military service, he went back to the E.N.S., as a "caiman[14]" for the Physics Laboratory, and started a thesis paper titled "Sur une nouvelle détermination du rapport v entre les unités électro-magnétiques et électro-statiques" ("On a new determination of the 'v relation between electromagnetic and electrostatic units") under the supervision of both Jules Violle and Marcel Brillouin, which he defended

[13]Henri Abraham biographies are available, such as: "À la mémoire de Henri Abraham, Eugène Bloch, Georges Bruhat: Créateurs et f Directeurs de ce Laboratoire Morts pour la France", École Normale supérieure, Physics laboratory, École Normale Supérieure, printing house Lahure, Paris, 1948.

[14]Agrégé-préparateur in ENS jargon. Term introduced in 1852.

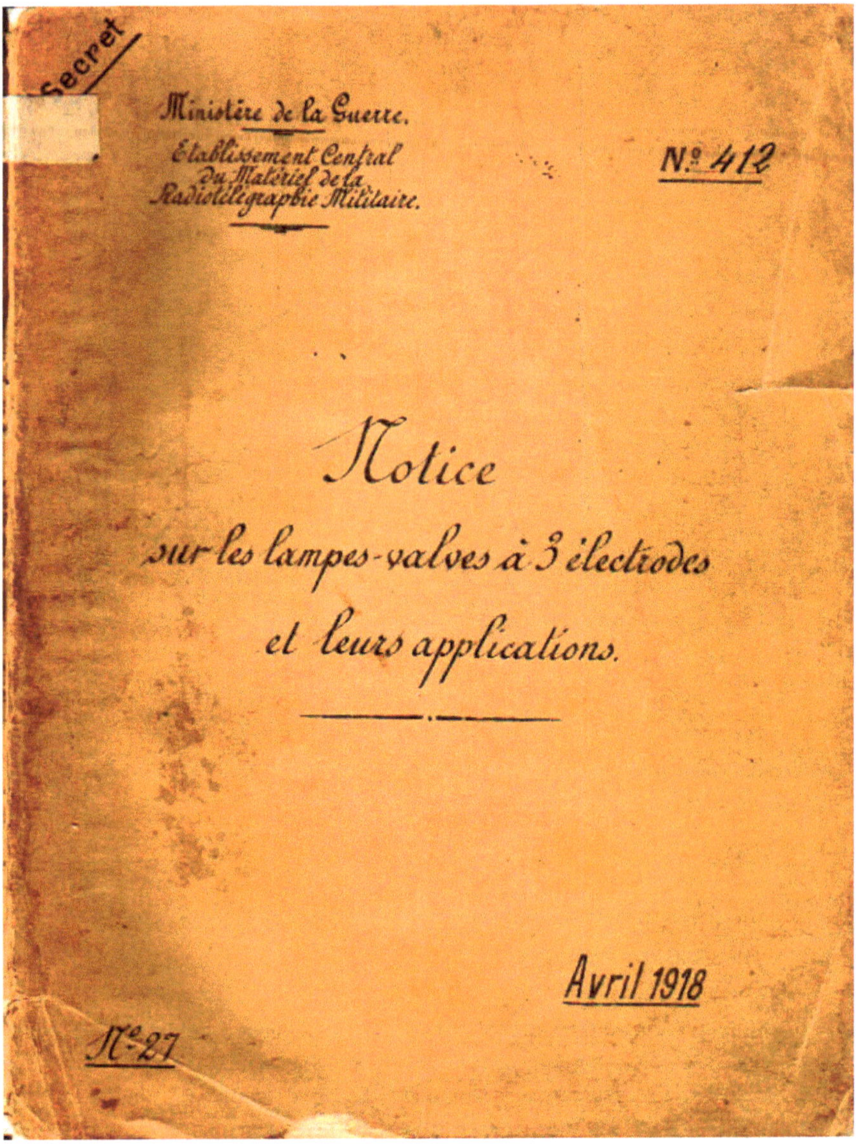

Fig. 2.5 E.C.M.R. report n° 412 (Source Jacques Denys)

barely two years later on the 30th of June 1892. He then started teaching, from 1891 to 1897, at the *lycée* Chaptal, and from 1894 to 1900 at the *lycée* Louis-le-Grand. In 1900, he was appointed associate professor at the E.N.S. then full-professor at the University of Paris. He then became head of the Physics Laboratory. After the war was declared and his six-month long stay at the Lyon-La-Doua station, he was reinstated in Paris in May 1915 and from there carried on with his work on the three-

- 169 -

TABLE DES MATIERES

CHAPITRE I.

Propriétés et fonctionnement de la lampe valve à 3 électrodes. -

CHAPITRE II.

Emploi de la lampe à 3 électrodes comme générateur d'oscillations.

Fig. 2.6 E.C.M.R. report n° 412, table of contents (Source Jacques Denys)

- 170 -

CHAPITRE III.

Emploi de la lampe à 3 électrodes comme détecteur.

Fig. 2.7 E.C.M.R. report n° 412, table of contents (Source Jacques Denys)

-171-

CHAPITRE IV.

Pages

Emploi de la lampe à 3 électrodes comme relai-amplificateur.

CHAPITRE V.

Applications de la lampe à 3 électrodes - Description sommaire de quelques types d'appareils.

Fig. 2.8 E.C.M.R. report n° 412, table of contents (Source Jacques Denys)

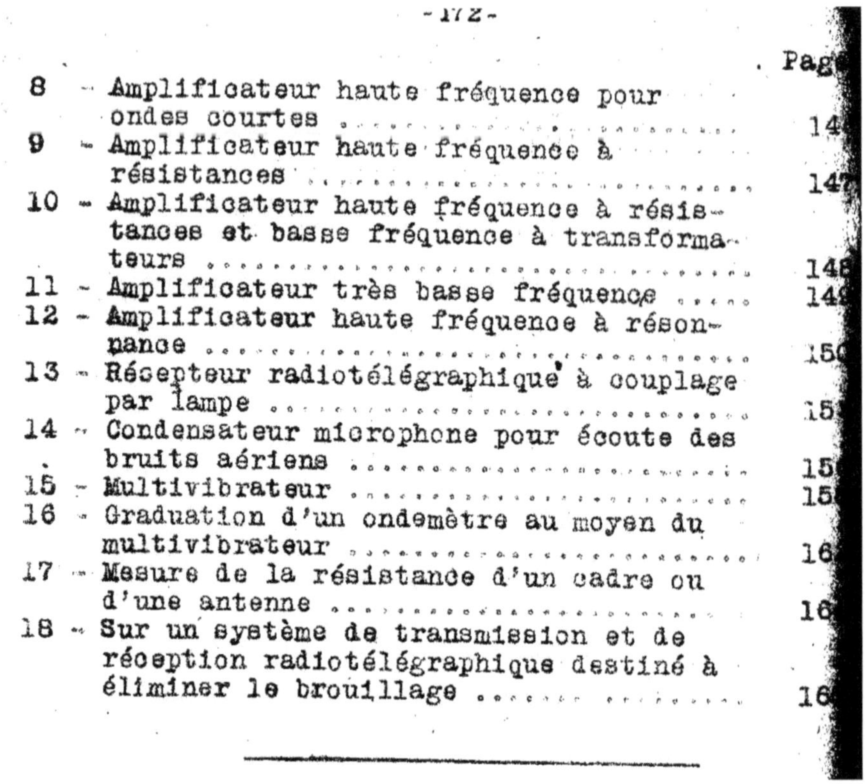
- 172 -

Page

8 - Amplificateur haute fréquence pour ondes courtes 14
9 - Amplificateur haute fréquence à résistances 14
10 - Amplificateur haute fréquence à résistances et basse fréquence à transformateurs 14
11 - Amplificateur très basse fréquence 14
12 - Amplificateur haute fréquence à résonnance 15
13 - Récepteur radiotélégraphique à couplage par lampe 15
14 - Condensateur microphone pour écoute des bruits aériens 15
15 - Multivibrateur 15
16 - Graduation d'un ondemètre au moyen du multivibrateur 16
17 - Mesure de la résistance d'un cadre ou d'une antenne 16
18 - Sur un système de transmission et de réception radiotélégraphique destiné à éliminer le brouillage 16

Fig. 2.9 E.C.M.R. report n° 412, table of contents (Source Jacques Denys)

electrode valve with Jacques Biguet.[15] At the same time, he continued his research with Eugéne Bloch for the Military Telegraphy Service, which became the E.C.M.R. two years later. In November 1917, they invented a device able to measure wireless telegraphy emitter frequencies: the *multivibrator*. They then published classified[16] Notes for the E.C.M.R. with the following titles and publication dates (Fig. 2.9):

- Sur la mesure des longueurs d'ondes de T.S.F. en valeur absolue avec le multivibrateur T.M., novembre 1917, *E.C.M.R.*, n° 2896 (On wireless telegraph wavelengths measurement in absolute value with the T.M. multivibrator),

[15]Abraham and Bloch (1920, 57) reminded that these studies were "done in collaboration with Mr. BIGUET, in 1914–1915 in the incandescent light bulb factory M. A. Grammont in Lyon". Peri's name is mentioned nowhere.

[16]The period during which Abraham and Bloch's notes were "classified" was fifty years long. They were kept until then at the *Service Historique de la Défense* (S.H.D.) in Vincennes, these notes were destroyed during a reconstruction of the premises. The Note n° 412 by Gutton was the only one found as of now. See *supra*.

- Étalonnement en valeur absolue des contrôleurs d'ondes par l'emploi du multivibrateur, novembre 1917, *E.C.M.R.*, n° 2949 (Calibration of radio wave regulators in absolute value by use of the multivibrator),
- Multivibrateur T.M. type A et type B, décembre 1917, *E.C.M.R.*, n° 2900 (T.M. multivibrator type A and type B),
- Ondes entretenues étalons. Mesure des longueurs d'ondes en valeur absolue, 5 juillet 1918, *E.C.M.R.*, n° 4448 (Sustained wave calibration. Wavelength measurements in absolute value),
- Etalonnage d'un diapason en valeur absolue, octobre 1918, *E.C.M.R.*, n° 4148, (Calibration of a tuning fork in absolute value).

After the war, Abraham and Bloch published three articles, which were public versions of the E.C.M.R. notes, and for which the title appears to have been inspired by note n° 4448.

- [1919a] Sur la mesure en valeur absolue des périodes des oscillations électriques de haute fréquence, *C.R.A.S.*, 168 (2 juin 1919), p. 1105–1108 (On high-frequency electric oscillation period measurement in absolute value),
- [1919c] Mesure en valeur absolue des périodes des oscillations électriques de haute fréquence, *J. Phys. Theor. Appl.* 9 (4 juillet 1919), p. 211–222 (High-frequency electric oscillation period measurement in absolute value),
- [1919e] Mesure en valeur absolue des périodes des oscillations électriques de haute fréquence, *Ann. de Phys.* 9 (septembre-octobre 1919), p. 237–302 (High-frequency electric oscillation period measurement in absolute value).

By replacing the classified E.C.M.R. notes, these publications consequently caused a two-year gap as to the date of invention of the multivibrator, which should have been December 1917 rather than 1919, as Abraham and Bloch recalled:

> This method was researched during the years 1916 and 1917, due to the Military Telegraphy's requirements. (Abraham and Bloch 1919a, 1106, e, 244)

> We built various types of amplifiers for the Military Telegraphy (1916). (Abraham and Bloch 1919b, 1198)

> We had to develop this measuring method during the years 1916 and 1917, by researching the causes of specific anomalies in the valve amplifiers of the military telegraphy. (Abraham and Bloch 1919c, 212)

In spite of these references to research being carried out during the war – and even though Abraham and Bloch (1919e, 244) gave a detailed list of all the E.C.M.R. reports – there were no clarifications of the patent Abraham and Bloch had just submitted that same year, 1919, for the invention of the *multivibrator*. Moreover, despite their identical titles, these publications had different contents. Paradoxically, the most complete version, published in the *Annales de Physique*, is also the less quoted. This 65-page article seems to match the E.C.M.R. reports, since the table of contents is comprised of absolutely all the titles of the original reports (see Table 2.1). Paragraph III is dedicated to the description of the multivibrator, a device (see Fig. 2.10) containing two T.M. valves, where each grid is linked to the

Table 2.1 Table of contents of Abraham and Bloch's article (1919e) vs. E.C.M.R. notes (1917–1918)

I. Introduction and principles of the method	
II. Fundamental frequency: calibration of the tuning fork	Calibration of a tuning fork in absolute value, October 1918, *E.C.M.R.*, *n*°4148
III. Production of sustained electric oscillations with many harmonics: multivibrator	T.M. multivibrator type A and type B, December 1917, *E.C.M.R.*, *n*°2900
IV. Practical implementation of the multivibrator	
V. Properties of the multivibrator	
VI. Assembly for the calibration of a wavemeter	Calibration of radio wave regulators in absolute value by use of the multivibrator, November 1917, *E.C.M.R.*, *n*°2949
VII. Action of the wavemeter on the detector-amplifier	
VIII. Mutual reactions of the various circuits. Necessity of using weak couplings	
IX. Role of the heterodyne. Octave method. Harmonics numbering	
X. Operational mode for the calibration of a wavemeter	
XI. Reading the measurements	
XII. Measurement in absolute value of the wavelength of sustained waves received by a wireless telegraph station	On the measurement of wireless telegraph wavelengths in absolute value using the T.M. multivibrator, November 1917, *E.C.M.R.*, *n*°2896 Sustained wave calibration. Wavelength measurement in absolute value. 5 July 1918, *E.C.M.R.*, *n*°4448

other's plate by a capacitor. Such an assembly produces harmonic-rich oscillations. Abraham then explained that this was the reason he called it a *multivibrator*, and described the observed phenomenon:

> The experiment being thus arranged, and the lamps being turned on, we can see that the electric currents flowing through the various circuits are subjected to abrupt periodic variations, and that there can be *no stable steady-current*. The two lamps function alternatively. At one point, the first plate suddenly starts to discharge current, whereas the second one discharges none. A few seconds later, the roles are reversed. And some more seconds later, we go back to the first flow, and so on and so forth, periodically. (Abraham and Bloch 1919e, 255)

The current flow in the second lamp (subscripts 2 on Fig. 2.10) decreases the potential in the G_1 grid, and obstructs the current flow in the first (subscripts 1 on Fig. 2.10), and vice versa. Theoretically, the device can therefore only operate in two distinct states: in which the first lamp is conducting and the second lamp has been cut-off, or the opposite. However, due to the presence of capacitors between the grids and plates of each lamp, these two operating states are unstable, and the device

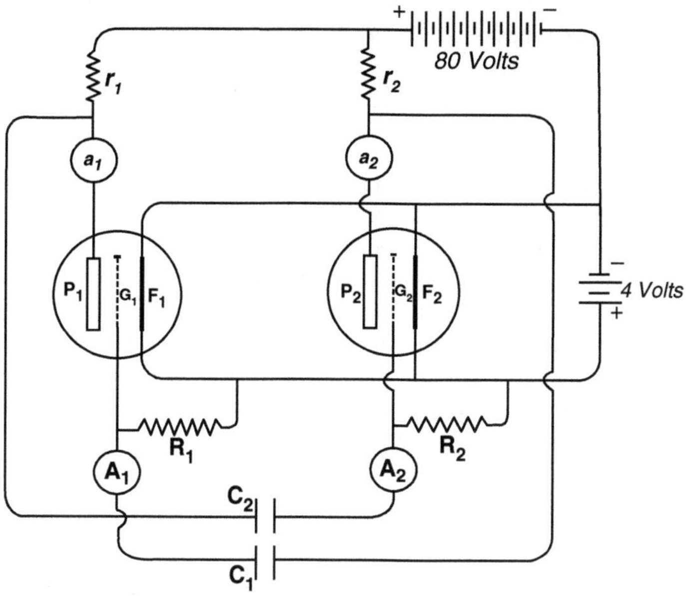

Fig. 2.10 Multivibrator, from Abraham and Bloch (1919e, 254)

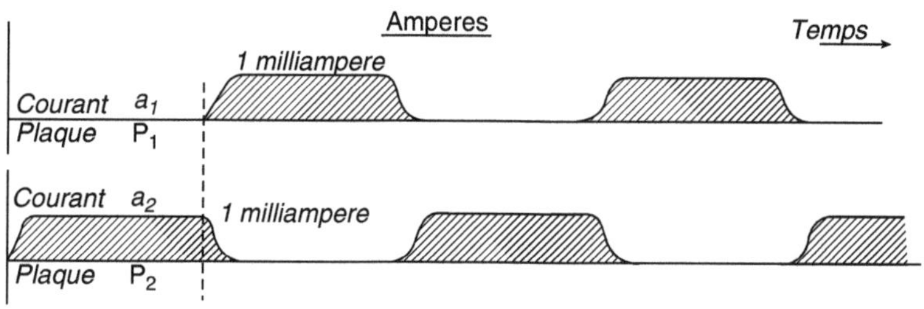

Fig. 2.11 P_1 and P_2 plate current reversals, from Abraham and Bloch (1919e, 256)

oscillates between the two, generating variations in voltage and electric current,[17] as shown on Fig. 2.11.

Meanwhile, researchers calculated the oscillation period of a circuit comprised of a capacitor of capacitance C and an inductor of inductance L by using Thomson's or Duddell's formula (see *supra* Tableau 1.1). However, Abraham demonstrated that the oscillations generated by the multivibrator have a period that does not complied with these formulae, i.e. which were not "Thomson-type". He explained that the

[17]For more details, see the pages 29–32 of the book *Les trois physiciens Henri Abraham, Eugéne Bloch, Georges Bruhat*, éditions Rue d'Ulm 2009.

observed effects on Fig. 2.11 are divided by time intervals corresponding to the charge and discharge times of capacitors C_1 and C_2 through resistors R_1 and R_2.

He deduced the following:

> The system's period is therefore of the order of $C_1R_1 + C_2R_2$. (Abraham and Bloch 1919e, 257)

This reversal time, called the "relaxation time[18]", was then used to refer to the duration of the discharge of the resistor's capacitor. As for the multivibrator, the oscillatory phenomenon is "steered" by the capacitor. The oscillation period is consequently not provided by Thomson's formula anymore, but corresponds to the "relaxation time". Through this invention, Abraham and Bloch therefore brought a new type of oscillating system to light: a relaxation system. It should be noted, however, that neither Abraham nor Bloch used this terminology, which was only introduced a few years later by Van der Pol (1925, 1926a,b,c,d) (see *infra*).

Wireless telegraphy development, spurred by the war effort, went from craft to full industrialization. The triode valves were then marketed on a larger scale. More reliable and stable than the singing arc, the consistency of the various components used in the triode allowed for exact reproduction of experiments, which facilitated research on sustained oscillations. While the singing arc itself became more and more obsolete as time passed, the properties of the oscillatory phenomenon that were discovered using this device did not. In France, Paul Janet and André Blondel set to work on transposing the different results they achieved to the triode, and their work contributed to the nonlinear oscillation theory development. Thanks to Van der Pol's work based on the multivibrator study, a new type of oscillation came into existence: *relaxation oscillations* (Figs. 2.12 and 2.13).

2.2 The Three-Electrode Valve or Triode: *Sustained Oscillations*

2.2.1 *Paul Janet's Work:* Analogy and Incomplete Equation Modeling (II)

In April 1919, Janet published an article of considerable importance on several levels. Firstly, it underscored the technology transfer taking place, in which an electromechanical component (the singing arc) was replaced with what would later be called an electronic tube. This represented a true revolution, since the structure of the singing arc made experiments complex, tricky, and almost impossible to recreate. Secondly, it revealed "technological analogy", based on the *duality principle*

[18]This terminology was introduced by James Clerk Maxwell (1867, 56) as reminded by Colin (1893, 1251) in a note to the *C.R.A.S.* which seemed to be one of the first occurrences of this periodic.

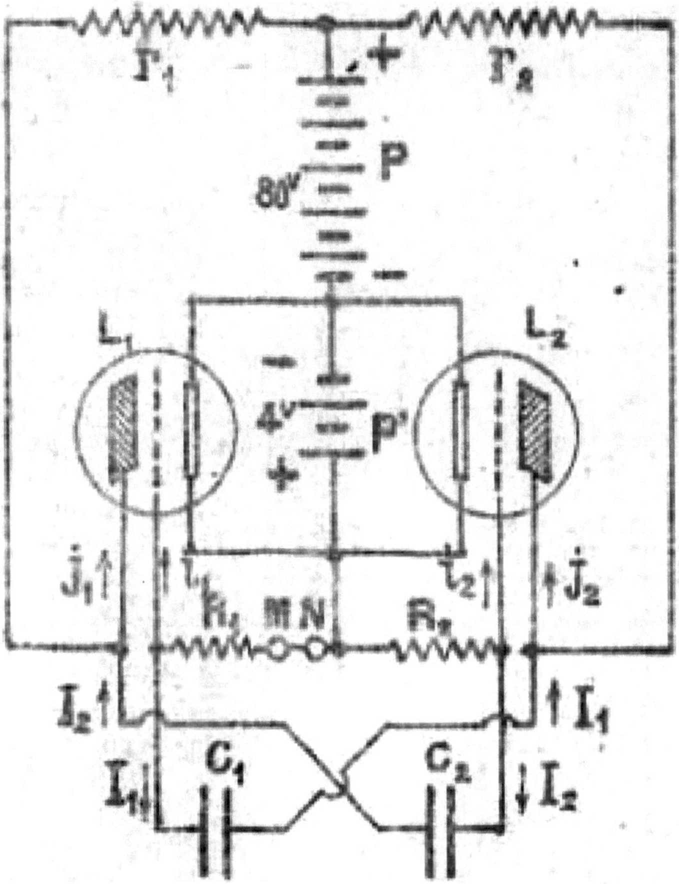

Fig. 2.12 Multivibrator, excerpt from the E.C.M.R. report n° 412 (1918) (Source Jacques Denys)

determined by Sire de Vilar (1901) between sustained oscillations produced by a series dynamo machine like the one used by Gerard-Lescuyer (1880a) and the oscillations of the singing arc or a three-electrode valve.

> I felt it was interesting to note unexpected analogies between this experiment[19] and the sustained oscillations so widely used nowadays in wireless telegraphy, for instance, those produced by Duddell's arc or the three-electrodes lamps used as oscillators. (Janet 1919, 764)

However, in this article, Janet blurs the experiment's paternity by crediting Witz (1889a). He mentioned only this "very old experiment on applied electricity, carried

[19]It is Gerard-Lescuyer's experiment (1880a).

Fig. 2.13 Multivibrator of Abraham and Bloch (Source: Musée des Arts et Métiers)

out in 1880[20]” (Janet 1925, 1193) in his article prefacing studies by the Cartans. Janet justified this “electrotechnical analogy” by basing his reasoning on an older analogy, regarding circuit components.

> Producing and sustaining oscillations in these systems mostly depends on the presence, in the oscillating circuit, of something comparable to a negative resistance. Now, the generating series-dynamo machine acts as a negative resistance, and additionally the separate excitation motor acts as a capacitor. Curiously, these two analogies have been mentioned a long time ago, the first one by M. P. Boucherot[21] and the second one by Mr. Maurice Leblanc.[22] (Janet 1919, 764)

He considered that in order to have analogies in the effects, i.e. in order to see the same type of oscillations in the series-dynamo machine, the triode and the singing arc, there must be an analogy in the causes. In fact, since the series-dynamo machine acts as a negative resistance, responsible for the oscillations, there is indeed an analogy. Consequently, one equation only must correspond to these devices. In this article, Janet appears to be the first to write the incomplete differential equation characterizing the oscillations noted during Gérard-Lescuyer’s experiment:

[20]Gerard-Lescuyer (1880a).

[21]See Boucherot (1904).

[22]See Leblanc (1899).

Writing the equation for the problem, in the case concerning the installation described above, is easy. Let $e = f(i)$ be the series-dynamo electromotive force, R and L be the resistance and self-inductance of the circuit, ω the angular velocity of the separate excitation motor. We, of course, obtain

$$Ri + L\frac{di}{dt} = e - k\omega$$

$$ki = K\frac{d\omega}{dt}$$

hence

$$L\frac{d^2i}{dt^2} + [R - f'(i)]\frac{di}{dt} + \frac{k^2}{K}i = 0 \qquad (2.1)$$

(Janet 1919, 765)

The first equation, which can be rewritten as $e = L\frac{di}{dt} + Ri + k\omega$, expresses what Janet asserted (1900, 222) a few years earlier (see *supra*): in order to completely explain the phenomenon, the following must be taken into account:

(a) the *e.m.f.* of the dynamo: $e = f(i)$,
(b) the *c.e.m.f.* of the motor: $Ri + k\omega$,
(c) the *e.m.f.* of the inductor: $L\frac{di}{dt}$.

By deriving the first equation, and taking the second one into account, he easily established the last one. Then, noticing that the separate excitation machine "acts as a capacitor of capacitance K/k^2" (Janet 1919, 765), he obtained an equation perfectly analogous to the one (Van der Pol 1920, 702) established the following year for the triode. This equation was nevertheless incomplete, as also noted by Janet:

> But the phenomenon is limited by the characteristic's curvature, and regular, non-sinusoidal equations actually occur. They are governed by the equation (2.1), which could only be integrated if we knew the explicit for of the function $f(i)$. (Janet 1919, 765)

Indeed, the question of mathematical representation of the *oscillation characteristic*, i.e. the establishment of the function $f(i)$ draws on the polynomial interpolation of a curve, a concept called observable modeling nowadays. This implies the procurement of a minimal number of points, i.e. a series of facts or measurements, which requires on the one hand the exact reproducibility of the experiment, and on the other hand a measuring device able to provide accurate values. Janet therefore demonstrated, by establishing an analogy between three different devices, that they all fell under the same oscillatory phenomenon, for which he provided the general, albeit incomplete, equation. It is, however, surprising that Janet did not refer Poincaré's work (1908) on the singing arc, even though he cited the older studies conducted by Leblanc (1899) and Boucherot (1904). Nevertheless, in his note, Janet also explained the main obstacle to overcome in order to complete his equation. In November of the same year, 1919, Blondel was the one who solved the problem by establishing, one year before Van der Pol, the equation for the triode and introducing the term "self-sustained oscillations" in order to qualify the phenomenon.

2.2.2 *André Blondel: The Anteriority of the Writing of the Triode Equation*

After he fully solved the question concerning the nature of the electric arc and demonstrated that it does not possess a *c.e.m.f.*, Blondel tackled the oscillatory phenomenon occurring in the singing arc. From 1919, during a transition from the singing arc to the triode, he began his research "by analogy with the already established theory on the singing arc" (Blondel 1919a, 676), transposing most of the results he had obtained.

2.2.2.1 Modeling

As stated by Poincaré (1908), then Janet (1919), the only obstacle preventing the formulation of a complete equation for the oscillations observed in the triode, the singing arc, and Gérard-Lescuyer's experiment, was the modelling of the oscillation characteristic of the nonlinear component, comparable to a negative resistance, present in the three devices. In 1919, Blondel was the first to model the nonlinear characteristic of the triode, using a development "in the form of an uneven terms series" (Blondel 1919b, 946). He therefore established his differential equation a year before Van der Pol (1920). Philippe Le Corbeiller (see *infra* Part II) was the one to point out this apparently unheard-of result.

> Mr. A. Blondel, in 1919, researched in this manner[23] the oscillating triode shown on Fig. 6. (…) and found a third order equation (…)

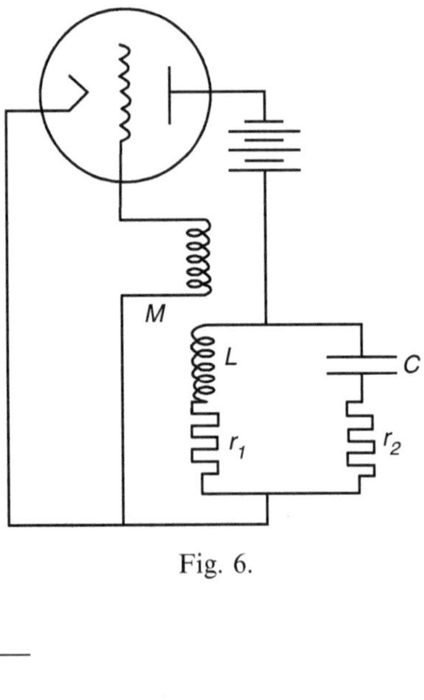

Fig. 6.

[23]See Blondel (1919b).

> In 1920, M. van der Pol also studied this same oscillating triode diagram, but by constructing the oscillating plate circuit with a pure inductance L, a capacitance C and a resistance R in parallel, hence obtaining a second order equation. (Le Corbeiller 1932, 705)

In a note published in the *C.R.A.S.* and presented at the *Académie des Sciences* on the 17th of November 1919, Blondel modeled the oscillation characteristic of the triode. By calling u the voltage in the plate with i being the variation in the plate current of the triode, v the grid potential and k the amplification coefficient of the triode, he explained the relation between u and i in the triode, formed by $i = F(u + kv)$ and "in which F represents a function reflected by a known curve presenting a long inflection around the average value of the static current I (approximately equal to half the value of the saturation current)." (Blondel 1919b, 946). He then hypothesized that we stay in the area where this curve keeps the same form regardless of the value of v and moves only in parallel with itself, by a translation along the u axis when v varies. This led him to model the oscillation characteristic $i = F(u + kv)$ of the triode, and he offered to "develop it as a series of odd terms which must be convergent" (Blondel *Ibid.*, 946). He obtained:

$$i = F(u + kv) = b_1 (u + kv) - b_3 (u + kv)^3 - b_5 (u + kv)^5 - \ldots \quad (2.2)$$

2.2.2.2 Writing the Equation

This note titled "Amplitude du courant oscillant produit par les audions générateurs" ("The oscillating current amplitude produced by generating audions") aimed to calculate an approximation of the amplitude of the oscillations. For this reason, Blondel established the triode's equation and introduced the modeling of its characteristics. Then, by calling i the plate current intensity at the moment t, i_1 and i_2 the intensities in the branches of self-inductance L and capacitance C, with internal resistances $r_1$and r_2 respectively, u the oscillating voltage at the parallel circuit terminals, he obtained the three following equations:

$$i_1 + i_2 = i \quad , \quad r_1 i_1 + L\frac{di_1}{dt} = u \quad , \quad r_2 i_2 + \frac{1}{C}\int i_2 dt = u \quad , \quad h = \frac{kM}{L} - 1;$$

By combining and deriving them, Blondel (1919b, 945) established the triode's differential equation with the form:

$$\frac{d^3u}{dt^3} + \frac{r_2}{L}\frac{d^2u}{dt^2} + \left(\frac{1}{CL} - \frac{r_1 r_2}{L^2}\right)\frac{du}{dt} - \frac{r_1}{CL^2}u - r_2\frac{d^3i}{dt^3} - \frac{1}{C}\frac{d^2i}{dt^2} = 0 \quad (2.3)$$

The presence of internal resistances r_1 and r_2, which Blondel could have neglected, led him to this third order quadratic differential equation. By substituting the expression (2.2) of the intensity i in this equation (2.3), "The final equation for the problem" (Blondel 1919b, 947) took the following form:

$$\begin{aligned}
&\frac{d^3u}{dt^3}+\frac{d^2u}{dt^2}\frac{r_2}{L}+\frac{du}{dt}\left(\frac{1}{CL}-\frac{r_1r_2}{L^2}\right)-\frac{r_1}{CL^2}u\\
&-\left(r_2\frac{d^2u}{dt^2}-\frac{1}{C}\frac{d^2u}{dt^2}\right)\left[b_1\left(h-\tfrac{kMr_1}{L^2}u\right)-3b_3h^3u^2-\ldots\right]\\
&-\left[3r_2\frac{d^3u}{dt^3}\frac{du}{dt}+\frac{1}{C}\left(\frac{du}{dt}\right)^2\right]\left[-\frac{b_1kMr_1}{L^2}-6b_3h^3u+\ldots\right]\\
&-r_2\left(\frac{du}{dt}\right)^3\left[-6b_3h^3-\ldots\ldots\ldots\ldots\ldots\ldots\ldots\ldots\ldots\right]=0
\end{aligned} \tag{2.4}$$

Neglecting the internal resistances r_1 and r_2, i.e. posing in equation (2.4): $r_1 = r_2 = 0$, we have:

$$\frac{d^3u}{dt^3}+\frac{du}{dt}\left(\frac{1}{CL}\right)+\frac{1}{C}\frac{d^2u}{dt^2}\left[b_1h-3b_3h^3u^2-\ldots\right]-\frac{1}{C}\left(\frac{du}{dt}\right)^2\left[-6b_3h^3u+\ldots\right]=0$$

Grouping the terms in b_3h^3, we obtain:

$$\frac{d^3u}{dt^3}+\frac{du}{dt}\left(\frac{1}{CL}\right)+\frac{1}{C}\frac{d^2u}{dt^2}(b_1h)-\frac{3b_3h^3}{C}\left[\frac{d^2u}{dt^2}u^2+\left(\frac{du}{dt}\right)^2(2u)+\ldots\right]=0$$

And noticing that the last term is written as:

$$\left[\frac{d^2u}{dt^2}u^2+\left(\frac{du}{dt}\right)^2(2u)+\ldots\right]=\frac{d}{dt}\left(\frac{du}{dt}u^2\right)+\ldots$$

then integrating once in relation to time, we obtain for this equation (2.4):

$$C\frac{d^2u}{dt^2}-\left(b_1h-3b_3h^3u^2-\ldots\right)\frac{du}{dt}+\frac{u}{L}=0 \tag{2.5}$$

In the same manner, by directly integrating the equation (2.3), we would have obtained:

$$C\frac{d^2u}{dt^2}-\frac{di}{dt}+\frac{1}{L}u=0 \tag{2.6}$$

2.2.2.3 Calculating the Fundamental's Period and Amplitude

By developing the voltage u in the equation (2.4) in a Fourier series, and identifying it term by term, Blondel deduced a first approximation of the period (angular frequency, or pulsation) ω for the oscillations:

$$\omega^2 \approx \frac{1}{CL}\left(1 + \frac{r_2}{\rho}h_0 - r_1 r_2 \frac{C}{L}\right) \text{ with } h_0 = \rho C \frac{r_1 + r_2}{L} \tag{2.7}$$

If we neglect the internal resistances r_1 and r_2 again, this expression is reduced as follows:

$$\omega^2 \approx \frac{1}{CL} \tag{2.8}$$

It should be noted that Blondel found Thomson's formula (1853) as a first approximation (see *supra*). Using the same technique, Blondel obtained a first approximation of the oscillation amplitude A_1:

$$A_1 \approx \frac{2}{h}\sqrt{\frac{-b_1 + \frac{r_1 + r_2}{hL}C}{-3b_3}} \tag{2.9}$$

If we neglect the internal resistances r_1 and r_2 again, this expression is reduced as follows:

$$A_1 \approx \frac{2}{h}\sqrt{\frac{b_1}{3b_3}} \tag{2.10}$$

In June the following year, 1920, Blondel explained these results in a longer and more detailed article. Before that, he looked into the origin of the phenomenon, and offered a classification of various types of oscillations.

2.2.2.4 Classifying Oscillations

In an article published in 1919, in which Blondel suggests a classification comprised of three main categories of oscillations, he is the first to introduce the term self-sustained oscillations. There are no traces of this neologism prior to this publication,[24] which he appears to have formed by associating "self-started sustained oscillations" (Blondel 1919d). In his classification, the first type includes "oscillations sustained by continuous action, or self-sustained oscillations" (Blondel 1919d, 118), with notable examples such as Duddell's singing arc and the triode. The second type, which covers the "divided flow" oscillations, is illustrated by the Tantalus vase, or cup, and the vase culbuteur ("tumbler vase"). However, it is the third type that represents the "long-period oscillations" (Blondel 1919d, 124) which he linked, in reference to Janet, to Gérard-Lescuyer's experiment, rather than the first type:

[24] Nevertheless, as soon as the end of the twentieth century, the *series-dynamo machine* was already called *self-exciting*. See *supra*.

> It seems to me that the electric oscillations produced by the excitation's reversal in a generator, such as the series-excited dynamo, should be classified under this same type (long-period oscillation). Mr. Janet recently mentioned a type of oscillation of this kind once again. (Blondel 1919c, 125)

This is especially surprising since Blondel, with this classification, seemed to have separated the series-dynamo machine from the singing arc, whereas Janet had linked them with an analogy. Although the term "self-sustained oscillations" introduced by Blondel appears to have been initially used in a slightly too constraining way, Andronov's article (1929a) greatly broadens its scope (see *infra* Part II).

2.3 Balthasar Van der Pol's *Equation for the Triode*

In 1916, following his studies in physics and mathematics at the University of Utrecht, Balthasar Van der Pol (1889–1959) went to study under John Ambrose Fleming, an English electrical engineer and physicist, who taught at University College in London. At the time Fleming was the first electrotechnics teacher there, but he was better known as the inventor of the diode, i.e. the first thermionic valve, in 1904. On the 2nd of June 1917, Van der Pol married Pietronetta Posthuma in London (see *infra* Part II) with whom he had a son and two daughters. Then, after spending one year with Fleming, Van der Pol started working with John Joseph Thomson at the Cavendish laboratory of Cambridge. In 1920, he went back to Holland in order to finish his physics doctorate at Utrecht, under Hendrik Lorentz. His dissertation addressed "the influence of ionized gas on the propagation of electromagnetic waves, as applied to wireless telegraphy and ultraviolet radiation measurement". As early as the 1920s, Van der Pol set to work on the production of electromagnetic waves, by using oscillating electric circuits containing a triode instead of a singing arc. However, as recalled by Cartwright (1960, 370) as well as Stumpers (1960, 366), it was not in his famous contribution "On relaxation-Oscillations" (Van der Pol 1926c), but in a previous article, completed[25] on 17 July 1920, and published in November and December of the same year, that Van der Pol (1920) modeled characteristic oscillation of the triode by using a cubic function, and established his differential equation one year after Blondel (1919b).

[25]The footnote in which Van der Pol (1920, 702) referred to an article written by W. E. Eccles published in November 1919 seemed to indicate that he only started the writing after this date.

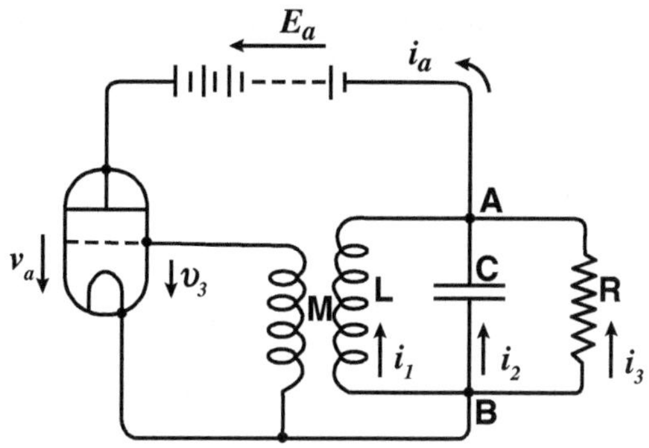

Fig. 2.14 Diagram of the oscillating triode, from Van der Pol (1920, 701)

2.3.1 *Modeling*

It can be observed in Fig. 2.14 that (Van der Pol 1920, 701) assembled the circuit differently to Blondel (1919b). In order to simplify the problem as much as possible, he chose to place the internal resistances of the inductor L and the capacitor C, not in series as it should be, but in parallel, with a resistance R. This choice, justified at the start of the article by Van der Pol, led him to a second-order differential equation:

> When the non-linear terms are retained in the equations the latter, and still more their solutions, soon become very complicated and in order to show clearly and definitely the importance of these terms it seems advisable to treat analytically that system of connections which renders the equations as simple as possible, thus obviating as far as possible, purely analytical complications, and allowing the physical meaning of the formulae to be clearly seen. This is especially the case in locating the resistance of the oscillatory L C flywheel circuit connected to the anode and filament, not in series either which the induction or capacitance but in parallel to both. (Van der Pol 1920, 701)

He called v_a the voltage of the plate corresponding to the variation i_a of the anode current, i.e. the plate current of the triode, v_g the grid potential, and g the amplification coefficient of the triode. He explained the relation between v_a and i_a in the triode, with the form $i_a = \varphi\left(v_a + g v_g\right)$. He then considered that with an unstable stationary state, the plate voltage is reduced to a value $v_{a0} = E_a$, where E_a represents the electromotive force of the generator, the current intensity in the anode therefore being $i_{a0} = \varphi\left(v_{a0}\right)$. He then wrote $v = v_a - v_{a0}$ and $i = i_a - i_{a0} = \varphi\left(v_{a0} - kv\right) - \varphi\left(v_{a0}\right) = \psi\left(kv\right)$, where v and i represent respectively the instantaneous voltage and intensity in the triode's plate. By using a Taylor-McLaurin series expansion limited to the first three terms, he wrote that $\psi\left(kv\right)$ "can be represented by the equation" (Van der Pol 1920, 703):

$$i = \psi\left(kv\right) = -\alpha v + \beta v^2 + \gamma v^3 \tag{2.11}$$

Van der Pol added that, using symmetry considerations for the oscillation characteristic, this expression can be reduced by writing: $\beta = 0$, as noted by Cartwright (1960, 370). Two years later, in order to describe the oscillation hysteresis phenomenon in the triode, Appleton and Van der Pol (1922, 182) had to expand the function $\psi\,(kv)$ to the fifth order, just as Blondel had done (1919b, 946).

2.3.2 Writing the Equation

Van der Pol's aim (1920) in this article was more ambitious than Blondel's (1919b), since he offered not only to calculate an approximation for the free oscillations of the triode, but also for the forced oscillations, as indicated by the title "A theory of the amplitude of free and forced triode vibrations". He expressed the voltage at the terminals of each dipole: $L\frac{di_1}{dt} = Ri_3 = \frac{1}{C}\int i_2 dt = E_a - v_a$ and managed to establish the differential equation for the triode:

$$\frac{di}{dt} + C\frac{d^2v}{dt^2} + \frac{1}{R}\frac{dv}{dt} + \frac{1}{L}v = 0 \tag{2.12}$$

By substituting the expression (2.11) of the intensity i in this equation (2.12), the equation for the triode that Van der Pol obtained is written as follows:

$$C\frac{d^2v}{dt^2} + \left(\frac{1}{R} - \alpha\right)\frac{dv}{dt} + \frac{1}{L}v + \beta\frac{d\left(v^2\right)}{dt} + \gamma\frac{d\left(v^3\right)}{dt} = 0$$

In order to enable a comparison with Blondel's works (1919b), we should put $\beta = 0$ and the resistance R should be disregarded by letting $R \to \infty$:

$$C\frac{d^2v}{dt^2} - \left(\alpha - 3\gamma v^2\right)\frac{dv}{dt} + \frac{1}{L}v = 0 \tag{2.13}$$

2.3.3 Calculating the Period and Amplitude of the Oscillations

In order to calculate the amplitude, Van der Pol offers three methods. The first is "analytical", as he explained (1920, 704), and consists in a singular perturbation expansion used by astronomers. The second resorts to a Fourier series expansion, used by Blondel (1919b), which was generally used by engineers, and led him to the introduction of a first correction for the value of the period (angular frequency, or pulsation ω):

$$\omega^2 \approx \frac{1}{CL} - \varepsilon \quad \text{avec} \quad \varepsilon = \frac{a^2\beta^2}{3C^2}$$

Table 2.2 Grid view of the simplified results by Blondel (1919b) and Van der Pol (1920)

Blondel (1919b)	Van der Pol (1920)
$i = F(u+kv) = b_1(u+kv) - b_3(u+kv)^3$	$i = \psi(kv) = -\alpha v + \gamma v^3$
$C\frac{d^2u}{dt^2} - (b_1 h - 3b_3 h^3 u^2)\frac{du}{dt} + \frac{u}{L} = 0$	$C\frac{d^2v}{dt^2} - (\alpha - 3\gamma v^2)\frac{dv}{dt} + \frac{1}{L}v = 0$
$\omega^2 \approx \frac{1}{CL}$	$\omega^2 \approx \frac{1}{CL}$
$A_1 \approx \frac{2}{h}\sqrt{\frac{b_1}{3b_3}}$	$a = \sqrt{\frac{4}{3}\frac{\alpha}{\gamma}}$

But since the symmetry shows that $\beta = 0$, Van der Pol (1920, 705) therefore obtained:

$$\omega^2 \approx \frac{1}{CL} \tag{2.14}$$

It should be noted that Van der Pol found, as Blondel did, the Thomson formula (1853) as a first approximation (see *supra*).

The third calculation method for the amplitude is geometrical, and apparently based on Witz's construction (1889b), which allowed him, similarly to the two previous ones, to find the following expression

$$a = \sqrt{\frac{4}{3}\frac{\alpha - \frac{1}{R}}{\gamma}} \tag{2.15}$$

By taking all the previously described simplifications into account, it is possible to establish a comparison between the studies accomplished by Blondel (1919b) and Van der Pol (1920), presented in the grid view below (see Table 2.2).

It therefore clearly appears that, by performing simplifications, the equations (2.11), (2.12), (2.13) and (2.14) written by Van der Pol (1920) and (2.2), (2.5), (2.6) and (2.8) by Blondel (1919b) are perfectly identical, the exception being the arbitrarily chosen signs representing the current.

Chapter 3
Van der Pol's Prototype Equation: *Existence and Uniqueness of the Periodic Solution* Cartan, Van der Pol, Liénard

At the beginning of the twentieth century, Du Bois Duddell (1901c) demonstrated experimentally that by taking the circuit resistance R into account, the period of the sound emitted by a singing arc provided by Thomson's formula (1853) should be modified, depending on a relation which he established (see *supra* Tableau 1.1). Twenty years later, Blondel (1919b) and Van der Pol (1920) also demonstrated that in the case of the triode the presence of resistance introduces a correction to the Thomson formula (see *supra*). The analogy between the three devices: the series-dynamo machine, the singing arc, and the triode, proven by Janet (1919), gives a new impulse to the research on the series-dynamo machine, more stable than the singing arc, and probably less expensive than the triode. Among the researches carried out in the 1920s[1] at the *l'École Supérieure d'Électricité*,[2] the study conducted by Trutat and Bouttes (1925) caught Janet's attention. It was an oscillographic analysis of the series-dynamo machine period, in which the results of their experiments showed a complete disagreement with the Thomson formula. Reading this article might have led Janet[3] to submit the problem to Élie and Henri Cartan.

3.1 Janet and Cartan's Work

3.1.1 Janet's Preface

In 1925, Élie Cartan, one of the most influential mathematicians of his time, along with his son Henri Cartan, suggested in a note the idea of calculating the upper and lower limits of the amplitude and period of oscillations sustained by the

[1]See for example Mestraud (1921) and Korowine (1921).

[2]Paul Janet was head of this school from 1895 to 1937.

[3]Interview of Mr. H. Cartan. See C. Gilain.

J.-M. Ginoux, *History of Nonlinear Oscillations Theory in France (1880–1940)*, Archimedes 49, DOI 10.1007/978-3-319-55239-2_3

series-dynamo machine. This unique contribution by the Cartans to the research on oscillations was then prefaced by Paul Janet, who had suggested the study.[4] In his introduction, Janet (1925) first evoked the roots of the Cartans' mathematical work, which can be found in a "very old experiment on applied electricity, carried out in 1880" (Janet 1925, 1193). This was an explicit reference to Gérard-Lescuyer (1880a), whereas Janet (1919, 764) had previously attributed the paternity to Witz (1889a) (see *supra*). He then explained this experiment, and added:

> The succinct explanation of the phenomenon is simple, and was provided by A. Witz in 1889. The complete theory is more difficult. (Janet 1925, 1193)

Janet recalled that by using his construction, Witz had managed to demonstrate the way the polarity reversal happened, which had been sensed by Gérard-Lescuyer, but who had not found its cause. He then re-established the differential equation (2.1) for the oscillations of the series-dynamo machine (see *supra*), and indicated that the electromotive force e of the dynamo is defined by the relation $e = f(i)$ in which the function's explicit form $f(i)$, unknown until then, had a derivative $f'(i)$ from this point forward, with the following properties:

> It does not change when i is changed into $-i$, shifts for $i = 0$ by a maximum value that we will assume to be greater than R (which can still be obtained by making R small enough and giving the series-dynamo a sufficient angular velocity) and vanishes for $i = \pm\infty$. (Janet 1925, 1194)

If this description of the derivative for the dynamo's *e.m.f.* possessing a hyperbolic characteristic, vanishing for $i = \pm\infty$ and also presenting the symmetry property $i \rightarrow -i$ (parity) seems to correspond to the singing arc's, defined by Thomson and Kelvin (1892) (see *supra* Table 1.2), it can absolutely not represent Blondel's or Van der Pol's. This is very surprising, considering that the analogy established by Janet (1919) shows that the equation (2.1) also played a part in the theory of the singing arc and three-electrode lamp. Yet, Blondel's (1919b) and Van der Pol's (1920) work solved the issue concerning the function's explicit form $e = f(i)$ by expanding it as a series in odd powers of the variable, i.e. representing it as a quintic function for Blondel and cubic for Van der Pol (see *supra*). It therefore seems that Janet did not know about their results, or that he chose to ignore them in order to keep more general properties for the function $f(i)$. Second, Janet pointed out the fact that the oscillations sustained by the series-dynamo machine's period was completely at odds with Thomson's formula (see *supra*). He backed his statement with the results of a series of experiments conducted by Mr. Trutat and Mr. Bouttes (1925), which indeed demonstrated that:

> The period always stays almost insensible to the variations in the self-induction, and keeps approximately proportional to the moment of inertia (...) From these experiments, it must be concluded that the calculations lead to results very remote from reality, and, since it is impossible to take all variables into account, it must not be given too much credit. (Trutat and Bouttes 1925, 670)

[4]See interview of Mr. H. Cartan by C. Gilain.

Janet's introduction therefore allowed him to do a "state of the art" of sorts, by recalling on the one hand, the main results, and on the other, the problems to be solved – the calculation of the limits and the amplitude of the period. It should be noted that Janet mentioned neither the work provided by Blondel (1919b) nor by Van der Pol (1920), in which these authors had provided a first approximation of the amplitude's value and the oscillations sustained by a triode period (see *supra*).

3.1.2 *Élie and Henri Cartan's Work:* The Existence of a Periodic Solution

In this "Note on the generating of sustained oscillations," Henri and Élie Cartan presented a demonstration calculating the upper and lower limits of the amplitude and the oscillation period, generalizing Blondel's (1919b) and Van der Pol's results (1920), since the limits they obtained no longer depended on a quintic or cubic function anymore, but on $f(i)$, which can take any form. But they went much further in this study, by demonstrating a crucial result: the existence of a periodic solution for the differential equation (2.1) of the series-dynamo machine's oscillations, and by analogy, for the differential equation of the singing arc's and the triode's oscillations. This result was a necessary prerequisite concerning the existence of bounds for the amplitude and period. The Cartans considered the equation (2.1) in the following form:

$$L\frac{d^2i}{dt^2} + [R - \varphi(i)]\frac{di}{dt} + \frac{1}{C}i = 0 \tag{3.1}$$

in which $\varphi(i) = f'(i)$. They then established that the solution of this equation, i.e. "the curve representing the variation of i is formed by an infinity of sine-waves" (Henri and Élie Cartan 1925, 1197) and that it possesses maxima i_2 and minima i_1, such that "i_2 is a continuous function of i_1" (Henri and Élie Cartan 1925, 1197). They then represent the curve representing i_2 depending on i_1 (see Fig. 3.1):

Using this diagram, which is reminiscent of the one representing a "first return map," the Cartans explained:

> The points H_1, H_2, H_3 where the curve cuts the bisector correspond to periodic solutions (sustained oscillations) the existence of which is therefore demonstrated. They are periodic because starting at a given minimum $-i_1$ the following maximum is equal to i_1, consequently the following minimum is equal to $-i_1$, etc. We can now easily see that any solution stretches toward a periodic solution. (Henri and Élie Cartan 1925, 1199)

This last sentence can be compared to what Henri Poincaré said (see *supra*)

> (…) we must be taken back to the closed curve, and this condition alone will enable the curve to represent a stable set of sustained waves, and solve the problem. (Poincaré 1908, 391)

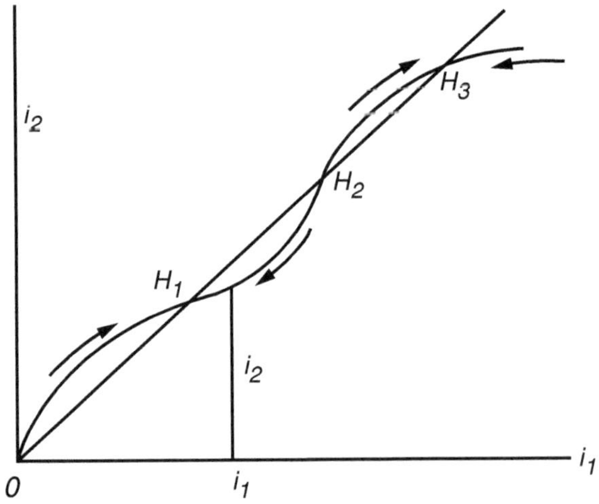

Fig. 3.1 Diagram of i_2 depending on i_1, from Cartan, Élie, and Henri (1925, 1199)

It therefore seems that the Cartans came very close to a connection with Poincaré's work, on the one hand with the diagram, for which Poincaré imagined the construction principle (1882, 251) and on the other hand, with this periodic solution, which is actually Poincaré's limit cycle (1882, 261). After establishing the existence of sustained oscillations, the Cartans determined the limits for the amplitude and period. Their demonstration, based on geometrical and analytical arguments, led them to a lower limit I_0 of the amplitude, in the form of the integral:

$$\int_0^{I_0} [R - \varphi(i)]di = 0 \tag{3.2}$$

and an upper limit: $2\left(\sqrt{2}+1\right)I_0$. For example, by noting that the equation (3.18), which would be established by Blondel in 1926, is absolutely identical to (3.1), we can replace the function $\varphi(i)$ in the equation (3.2) by the one used by Blondel (1926, 901), which represents the derivative of the *e.m.f.* of the singing arc $\varphi(i) = f'(i) = h - 3qi^2$ (see *infra*). Therefore an interval for the amplitude of the sustained oscillations can be found:

$$\sqrt{\frac{h-R}{q}} < A < 2\left(\sqrt{2}+1\right)\sqrt{\frac{h-R}{q}}$$

By comparing with the value of the amplitude established later by Blondel (1926, 902) and by expanding the solution in the form of a Fourier series: $A_1 = 2\sqrt{\frac{h-R}{3q}}$ we can see that it indeed belongs to the set of boundaries given by the Cartan method.[5] Whether it be the Cartans in regard to the work of Blondel (1919b), Van der Pol

[5]Indeed: $1 < 2/2\sqrt{3} < 2\left(\sqrt{2}+1\right) \quad \Leftrightarrow \quad \sqrt{3} < 2 < 2\sqrt{3}\left(\sqrt{2}+1\right)$.

(1920), or Blondel (1926) in regard to the Henri and Élie Cartan's note (1925), it appears here that the mathematicians, as well as the engineers-physicists ignored each other's results.

Using the same method, the Cartans obtained the limits of the period of the sustained oscillations. In the particular case of $\varphi(0) - R = R$ they found, for the supremum:

$$T < \frac{2\pi}{\sqrt{\frac{1}{LC} - \frac{R^2}{4L^2}}} \tag{3.3}$$

This upper bound[6] is none other than Du Bois Duddell's formula (1901c) (see Table 1.1). Hence, where Blondel's process led, in the particular case when the explicit form of the *e.m.f.* is known, to an approximate value of the amplitude, the Cartans' provided, in the more general case, a set bounds for this amplitude. Concerning the time period, the bounds they established, confirmed that Thomson's formula cannot be applied to this type of oscillations. This brings to mind the one established by Duddell a quarter of a century earlier. Nevertheless, the crucial result[7] of this note undoubtedly remains the demonstration of the existence of a periodic solution for sustained oscillation. The uniqueness of this solution, although appearing to have been considered by the Cartans in their conclusion, was established three years later by Alfred Liénard.

3.2 Balthasar Van der Pol's Study *Towards a New Type of Oscillation*

It has already been mentioned that it was not through his famous contribution "On relaxation-oscillations" written in 1926 that Van der Pol established the differential equation for the triode. Nor did he introduce the terminology "relaxation oscillation" in this article either. In the middle of the 1920s, Van der Pol needed, as Blondel did in his (1919c), to clarify the different types of oscillations. Hence, in a short dissertation published on the 19th of November 1925, Van der Pol offered to isolate various causes for electric oscillations, and presented a classification based on three types. The first type includes the forced oscillations (Van der Pol 1925, 791). The second type considers the production of alternating currents using direct currents, carried out with an electric arc (*arc-transmitter*) or a triode (*triode-transmitter*) (*Ibid.*, 791). Concerning the third and final type, Van der Pol explained the problem

[6]The formula (3.3) represents an upper bound of the oscillation period, whereas in the original text, the Cartan, Élie, and Henri (1925, 1204) considered the period of a half-oscillation, because they envisaged the non-symmetrical case.

[7]It seems very surprising that Van der Pol (1926a,b,c,d) did not mention the Henri and Élie Cartan's work (1925), when he had represented this periodic solution using the isocline method.

of frequency demultiplication. Then, Van der Pol (1925, 793) explained that for circuits such as Bloch and Abraham's multivibrator (1919a), the oscillation period cannot be determined by Thomson's formula any more, i.e. it is not proportionate to the square root of the product of induction L and capacitance C. In fact, the approximate period of these oscillations, which do not belong to any of the three types, corresponds to a capacitor's discharge time length, called its "relaxation time" (see *supra*). This is why he called a "relaxation oscillation" what he regarded as a new type of oscillation. The following year, the now famous "On Relaxation-Oscillations" was published, of which four versions now exist: two Dutch, one German, and one English, published in the following chronological order:

- Over Relaxatietrillingen, *Physica*,[8] 6, p. 154–157,
- Over Relaxatie-trillingen, *Tijdschrift Nederlandsch Radiogenoot* 3, p. 25–40,
- Über Relaxationsschwingungen, *Jb. Drahtl. Telegr.* 28, p. 178–184,
- On relaxation-oscillations, *Philosophical Magazine*, serie 7, 2 p. 978–992.

The first, "Over Relaxatietrillingen" is a transcription of a lecture given by Van der Pol (1926a) in front of the *Nederlandsche Natuurkundige Vereeniging*[9] on the 27th of March, 1926. It appears to be the oldest. The second one, "Over Relaxatie-trillingen" corresponds to a conference given by Van der Pol (1926b) on May the 15th, 1926 in front of the *Nederlandsch Radiogenootschap*.[10] The third one, "Über Relaxationsschwingungen" was written by Van der Pol (1926d) as well as the previous one in April 1926, and it was received by the publisher on September the 3rd, 1926. Van der Pol finished writing the fourth one (1926c) on the 6th of May, 1926. "On Relaxation-Oscillations" was only published during the November-December period in the *Philosophical Magazine*. Although these contributions have the same title and almost the same content, there are differences in each one's conclusion regarding the choice of examples illustrating the relaxation oscillation phenomenon (see *infra*).

3.2.1 The Generic Character of Van der Pol's Equation

Within his publications[11] on relaxation oscillations, Van der Pol (1926d) introduced a differential equation that is neither attached to the triode, nor to any other device (series-dynamo machine or singing arc). Using a pedagogical approach, he

[8]The periodic *Physica: Nederlands Tijdschrift voor Natuurkunde* was created in 1921 by Adriaan Fokker, Ekko Oosterhuis and Balthasar Van der Pol. After being restructured in 1934 this periodical was bought by the company *Elsevier* in 1970, and took the form we nowadays know, being entitled *Physica*.

[9]Dutch Physics Society.

[10]Dutch Radio Society.

[11]Following this, the reference Van der Pol (1926d) will be used.

recalled the damped oscillator, then introduced the "negative resistance" concept, considering that if the friction factor sign α is reversed, which happens particularly in the series-dynamo machine, the singing arc and the triode, then the amplitude of oscillations increases indefinitely, which is physically impossible. He then explained that in order to limit the amplitude, we must suppose that the friction factor is a function of the amplitude. He therefore suggested replacing α by the expression $\alpha - 3\gamma x^2$, where γ is a constant, in the equation for the damped oscillator. He obtained the following equation:

$$\ddot{x} - \left(\alpha - 3\gamma x^2\right)\dot{x} + \omega^2 x = 0 \tag{3.4}$$

It should be noted that this equation does have dimensions[12] even though it is not clear here. It corresponds to the equation (3.9), showing the oscillations of the triode, previously established by Van der Pol (1920). The variable x therefore has the dimensions of a potential difference (plate voltage of the triode). In this form, the equation (3.4) possesses three parameters: α the "negative resistance", ω the angular frequency or pulsation, defined by (V_4) and γ a constant. Van der Pol then used a technique introduced by Pierre Curie (1891) without mentioning it.[13]

In the first part of this dissertation, Curie (1891) demonstrated that the differential equation characterizing mechanical oscillations of a damped pendulum, as well as the one for the electric oscillations of a capacitor's discharge in a circuit comprising a resistor and an inductor, can be the same "reduced-form equation", in which the coefficients have the dimension of the inverse of a length of time. This allowed him to establish the very first *electromechanical analogy*[14] between oscillations of a mechanical system and the ones of an electrical system. The changes in variables he carried out gave a generic nature to his "reduced-form equation", and allowed

[12]The dedimensionalized equation only appeared in 1926. See Van der Pol (1926a,b,c,d).

[13]While Curie's name was not mentioned in Van der Pol's article (1926c), the concept of logarithmic decrement introduced by Curie (1891) is mentioned in the opening pages. See Van der Pol (1926c, 978). Blondel, however, seemed to have a good understanding of Curie's work (1891). See Blondel (1893c, 507) and Blondel (1923a, 381).

[14]It must be noted that Janet (1919) (see *supra*) did not refer to Pierre Curie's work (1891) when he established the same *electromechanical analogy* twenty years later. Research in the *Comptes Rendus de l'Académie des Sciences de Paris* showed that the adjective *éléctromécanique* (*electromechanical*) had first been used in 1855 in a note written by Th. du Moncel entitled: *Calendrier électromécanique* (Cf. *C.R.A.S.*, XL, (1855), 1217) whereas in the *Dictionnaire historique de la langue française* the adjective *électrotechnique* (*electrotechnical*) only appeared in 1882. Indeed, this terminology seems to have been introduced during the *Essais Électro-techniques du palais d'Exposition Royal de Munich* which were held from 16 September to 8 October 1882 (Cf. *La Lumière Électrique*, 1re série, (6) 20 (1882), 475–476). We then find the word *électrotechnique* without caesura or hyphen in an article written by Marcel Deprez entitled: *Transmission du travail à grande distance sur une ligne télégraphique ordinaire* (Cf. *La Nature*, N° 490, (21 October 1882) 328).

him to analyze the solution of these differential equations, using the amplitude of the oscillations as a function of the time and a unique parameter. Considering that the differential equation for electric oscillations of the discharge of a capacitor C in a circuit comprising a resistance R and an inductance L:

$$L\frac{d^2q}{dt^2} + R\frac{dq}{dt} + \frac{1}{C}q = 0 \tag{3.5}$$

Curie (1891, 310) wrote:

$$n_t = \frac{b}{2\pi}t \; ; n_a = \frac{a}{b} \text{ with } a = \frac{R}{2L} \text{ and } b^2 = \frac{1}{LC} \tag{3.6}$$

This led him to a classification of various types of oscillations depending on the values of n_a :

Pour $n_a = 0$, pendular oscillatory motion.
For $n_a < 1$, damped oscillatory motion.
For $n_a = 1$, critical aperiodic motion.
For $n_a > 1$, slower and slower aperiodic motion when n_a approaches ∞.
(Curie 1891, 209)

Similarly, in order to dedimensionalize the equation (3.4) characterizing the oscillations of the triode, Van der Pol (1926d, 979) wrote:

$$t' = \omega t \; ; x = \sqrt{\frac{\alpha}{3\gamma}}v \; ; \varepsilon = \frac{\alpha}{\omega} \text{ with } \alpha = \frac{R}{L} \text{ and } \omega^2 = \frac{1}{LC} \tag{3.7}$$

He therefore obtained an dedimensionalized differential equation, which now only depends on one parameter:

$$\ddot{v} - \varepsilon\left(1 - v^2\right)\dot{v} + v = 0 \tag{3.8}$$

With this reduced form, the equation (3.8) is called the "Van der Pol equation". In Table 3.1, which allows on to compare the changes in variables introduced by Curie and the ones carried out by Van der Pol, we can see that not all coefficients

Table 3.1 Comparison of variable changes used by Curie (1891) and Van der Pol (1926d)

	Curie (1891)	Van der Pol (1926d)
Time	$n_t = \frac{b}{2\pi}t = \frac{\omega}{2\pi}t$	$t' = \omega t$
Parameter	$n_a = \frac{a}{b} = \frac{R}{2}\sqrt{\frac{C}{L}}$	$\varepsilon = \frac{\alpha}{\omega} = R\sqrt{\frac{C}{L}}$
Damping coefficient	$a = \frac{R}{2L}$	$\alpha = \frac{R}{L}$
Frequency	$b^2 = \frac{1}{LC}$	$\omega^2 = \frac{1}{LC}$

are equal. This comes from the fact that they can be chosen arbitrarily, as noted by Curie (1891, 271). Hence, the reduced-form approach leads to a generic equation, which is however not unique:

> We can establish an infinity of different reduced-form equations for a same equation. (Curie 1891, 271)

3.2.2 *Graphical Integration and Relaxation Oscillations*

Early on, Van der Pol (1922, 179) realized that the equation (3.8) was not integrable, which he noted here, at the start of his article:

> It has not been found possible to obtain an approximate analytical solution for (3.8) with the supplementing condition ($\varepsilon \gg 1$), but a graphical solution may be found in the following way. (Van der Pol 1926d, 982)

Nevertheless, Van der Pol did not inaugurate the "isocline method" to graphically integrate the differential equation for the *relaxation oscillations*. Indeed, according to Dominique Tournès (2003b, 181), it was Belgian engineer Junius Massau who should be seen as creator of graphical integration. He incidentally reminded that "Massau introduced the term "isocline curve" based on the remark that it is at the place of the points that the integral has a same inclination (Massau, 1878–1887, book VI, 501) (this seems to be the first occurrence of the word "isocline" in this context)" (Tournès 2003a, 465). Le Corbeiller also added that Van der Pol "not hesitating to carry out graphical integrations using the classic but tiresome isocline curves method, managed to bring the whole sustained oscillations field to a vivid light (...)" Le Corbeiller (1931a, 19). Numerous other graphical integration methods existed long before 1926. The one presented by Curie (1891, 272 and following) is noticeable, in the second part of his dissertation, based on the use of an abacus, and allowing the deduction, from empirical measurements, of the value of the damping parameter of the system, as well as its specific period. But there is also the work of William Thomson and Lord Kelvin (1892), Maurice d'Ocagne (1893), Carl Runge (1912), and others. The graphical integration of the differential equation (3.8), however, allowed Van der Pol to deduce the amplitude, and most importantly the relaxation oscillations period.

3.2.2.1 Graphical Determination of the Oscillations' Amplitude

If, contrary to what has been stated until now, the graphical representation of the solution to Van der Pol's equation (3.8) for three values of the parameter $\varepsilon = 0.1$, 1 and 10 (reproduced below: see Figs. 3.2, 3.3, and 3.4) did not lead Van der Pol to establish a symmetry between a periodic solution and Poincaré's limit cycle, it

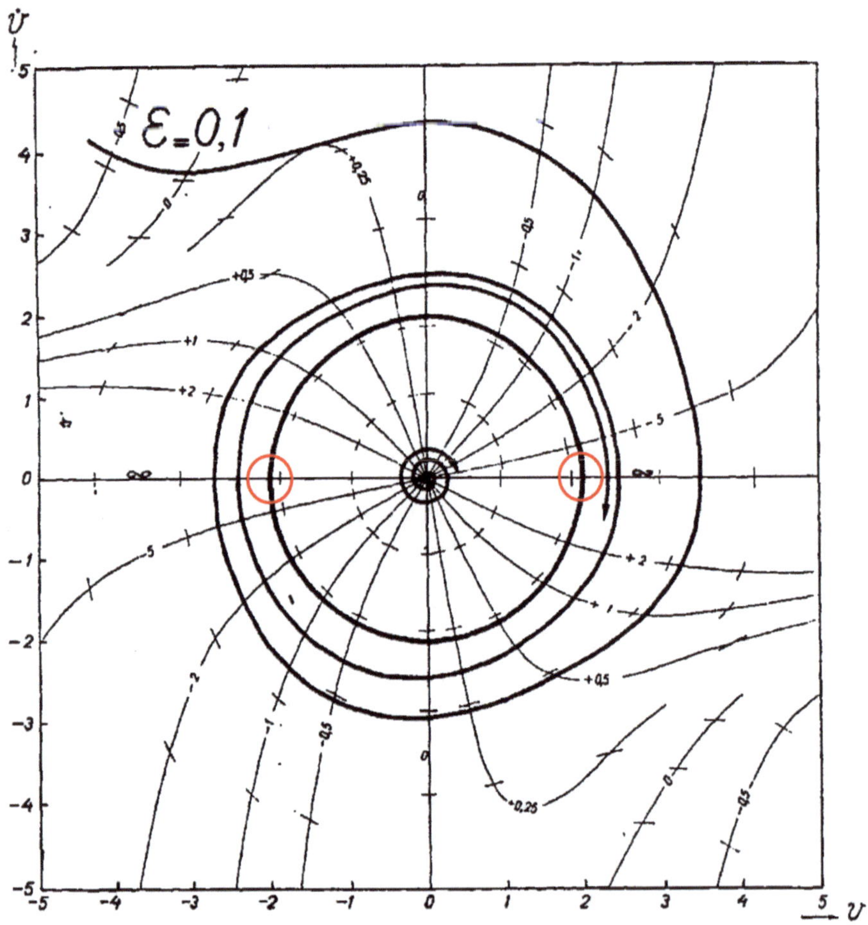

Fig. 3.2 Graphical integration of the equation (3.8), from Van der Pol (1926d, 983)

however enabled him, on the one hand, to obtain the value of the amplitude of the oscillation, which we can read directly on each of these three figures, (inside the red circles, we read $a = 2$) and on the other hand, to provide evidence of the continuous passing from one type of *sinusoidal oscillation* to a new type of *quasi-aperiodic oscillations*, which he later called a *relaxation oscillation* (see *infra*). An equivalence between the various types of oscillations established by Curie and by Van der Pol is shown in Table 3.2.

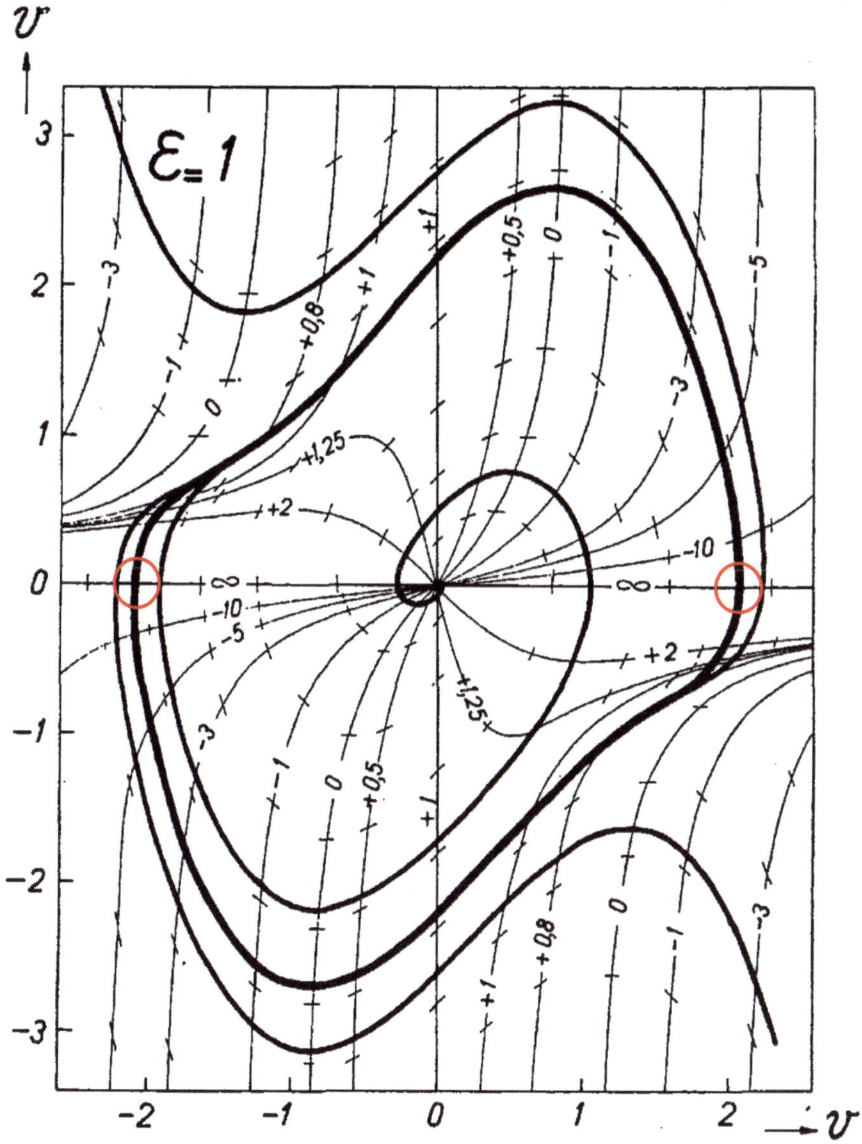

Fig. 3.3 Graphical integration of the equation (3.8), from Van der Pol (1926d, 983)

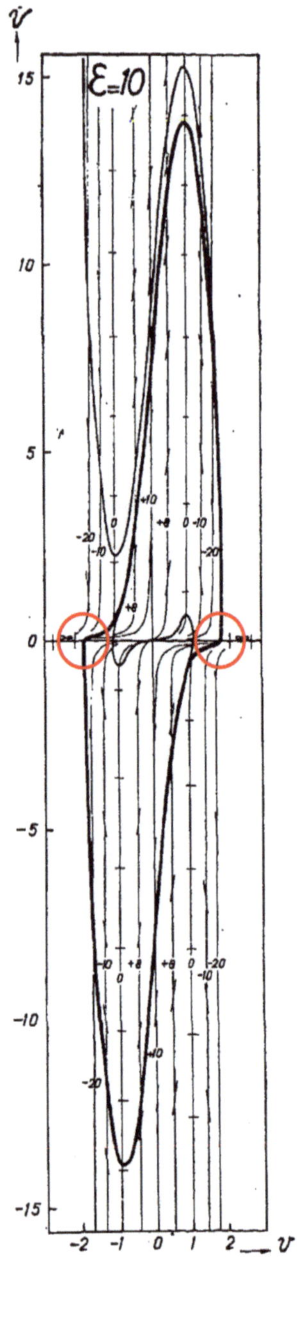

Fig. 3.4 Graphical integration of the equation (3.8), from Van der Pol (1926d, 983)

Table 3.2 Equivalence between different types of oscillations established by Curie (1891) and by Van der Pol (1926d)

Curie (1891)		Van der Pol (1926d)	
$n_a = 0$	Pendular motion	*Sinusoidal* oscillations	$\varepsilon = 0.1$
$n_a = 1$	Critical aperiodic motion	Continuous transition	$\varepsilon = 1$
$n_a > 1$	Aperiodic motion	*Quasi-aperiodic* oscillation	$\varepsilon > 1$

> This case represents a sinusoidal oscillation of gradually increasing amplitude, the value of which is finally steady and equal to 2. (Van der Pol 1926d, 984)

While Fig. 3.2 represents *sinusoidal oscillations* corresponding to the case of $n_a = 0$ in Curie's classification (1891, 209), Fig. 3.3, below, shows a transition toward *quasi-aperiodic oscillations*, which corresponds to a *critical aperiodic motion* for which $n_a = 1$ (see Table 3.2).

We can see (inside the red circles) that the amplitude is still equal to 2. Figure 3.4 represents *quasi-aperiodic oscillations* corresponding, in Curie's classification (1891, 209), to an *aperiodic motion* for which $n_a > 1$ and the amplitude is equal to 2 (see red circles).

The system being self-sustaining, the terminology used by Curie (1891, 209): "slower and slower aperiodic motion" now seems inadequate, since it is supposed to describe a damped motion. This is probably the reason why Van der Pol (1926d, 981) introduced the phrase: "*quasi-aperiodic solution*".

3.2.2.2 Analytical Determination of the Amplitude of Relaxation Oscillations

The integer value $a = 2$ of the amplitude, which can at first seem surprising, comes from the nature of change in variables which led to the reduced-form equation (3.8).

In fact, by comparing the equation for the triode (3.9) that Van der Pol obtained in 1920:

$$C\frac{d^2v}{dt^2} - \left(\alpha - 3\gamma v^2\right)\frac{dv}{dt} + \frac{1}{L}v = 0 \tag{3.9}$$

with the equation (3.8) that he established six years later:

$$\ddot{v} - \varepsilon\left(1 - v^2\right)\dot{v} + v = 0 \tag{3.10}$$

we can easily infer (supposing L = C = 1) that: $\alpha = \varepsilon$ and $3\gamma = \varepsilon$ and $3\gamma = \varepsilon$. By substituting them in the expression (V_5) of the oscillations' amplitude, we obtain (neglecting the resistance R):

$$a = \sqrt{\frac{4\,\alpha}{3\,\gamma}} = \sqrt{\frac{4\varepsilon}{\varepsilon}} = 2$$

It is important to note that this value had already been theoretically calculated almost fifty years earlier by Strutt and Rayleigh (1883, 230). He demonstrated that Lord Rayleigh's equation (3.12) and Van der Pol's (3.8) correspond to each other, following the principle of *electrotechnical duality*. On comparing them:

$$\frac{d^2\theta}{dt^2} + \kappa\frac{d\theta}{dt} + \kappa'\left(\frac{d\theta}{dt}\right)^3 + n^2\theta = 0 \tag{3.11}$$

$$\ddot{v} - \varepsilon\left(1 - \frac{\dot{v}^2}{3}\right)\dot{v} + v = 0 \tag{3.12}$$

$$\ddot{v} - \varepsilon\left(1 - v^2\right)\dot{v} + v = 0 \tag{3.13}$$

we deduce that: $\kappa = -\varepsilon$, $\kappa' = \varepsilon/3$ and $n^2 = 1$. On substituting them in the value of the amplitude (8.7) of these sustained oscillations calculated by Strutt and Rayleigh (1883, 230) we obtain:

$$a = \sqrt{\frac{-4\kappa}{3\kappa' n^2}} = 2\sqrt{\frac{-(-\varepsilon)}{\varepsilon}} = 2$$

It is therefore somewhat surprising that in his publications, Van der Pol (1926a,b,c,d) did not refer[15] to Strutt and Rayleigh's work (1883). It was Le Corbeiller (1931a, 4), then Fessard (1931, 88) who recalled, five years later, that he was at the origin of research into the sustained oscillations. Le Corbeiller (1932, 1936) thoroughly analyzed the symmetry between Rayleigh's equation (3.11) and Van der Pol's equation (3.8). It was only in 1937, during a presentation at the *Palais de la Découverte*, that Van der Pol (1938a, 71) alluded to it.

3.2.2.3 Graphical Determination of the Relaxation Oscillations Period

The graphical integration of the equation (3.8) however, allowed Van der Pol (1926c, 987) to establish that the period of the oscillations in the third case, i.e. for $\varepsilon = 10$ becomes approximately equal to ε, as shown on the diagrams on Fig. 3.5.

On the first curve ($\varepsilon = 0.1$) we count, in the part framed in red, three and a quarter periods for twenty temporal graduations. This provides a period T approximately equal to 2π, as stated by Van der Pol (1926d, 987).

The second curve ($\varepsilon = 1$) makes evident the "relatively perceptible deviation of sinusoidal form." (Van der Pol 1930, 17). It shows the continuous transition between the sinusoidal oscillations ($\varepsilon \ll 1$) and the relaxation oscillations ($\varepsilon \gg 1$).

In the red-framed part of the third curve ($\varepsilon = 10$), we see that the "curve starts to rise exponentially, and after a length of time equal to a period, has already almost

[15] Excepted Van der Pol's publication (1920, 708).

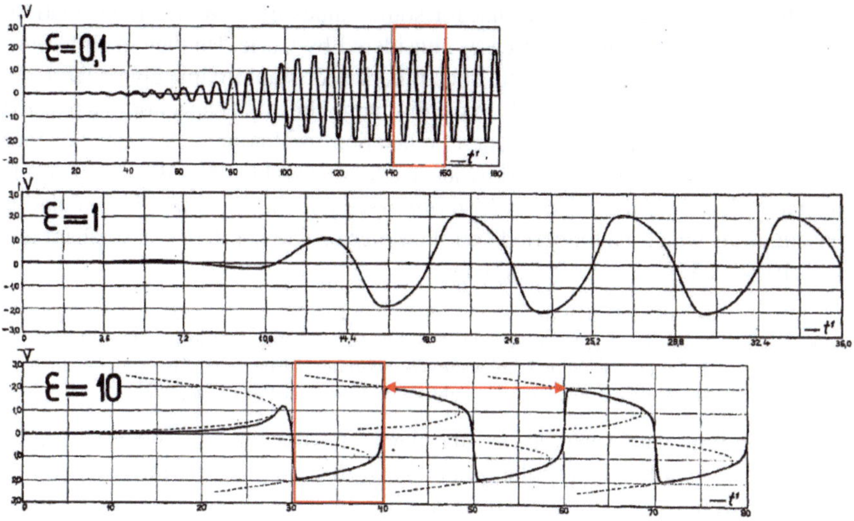

Fig. 3.5 Graphical integration of the equation (3.8), from Van der Pol (1926d, 986)

reached its limit periodic form." Van der Pol (1930, 18). He therefore shows that for $\varepsilon \gg 1$, the value of the period is approximately equal to ε.

3.2.2.4 Analytical Determination of the Relaxation Oscillations Period

The writing of the equation of the triode in reduced form led Van der Pol to introduce the *relaxation oscillation* concept. Indeed, after graphically deducing the oscillations period, he stated that as dedimensionalized variables they are approximately equal to ε:

$$T \doteq \varepsilon \tag{3.14}$$

Van der Pol (1926d, 987) established it analytically. Since ε is an dedimensionalized parameter, it must therefore express the period in a system of temporal units. He applied to it the variable substitution formula (3.7), in other words, $t' = \omega t$ which led to:

$$T = \frac{\varepsilon}{\omega}$$

Then, taking into account the fact that $\varepsilon = \frac{\alpha}{\omega}$, $\alpha = \frac{R}{L}$ and $\omega^2 = \frac{1}{LC}$, he finally obtained:

$$T = \frac{\alpha}{\omega^2} = RC \tag{3.15}$$

In this way, he demonstrated that for some values of $\varepsilon \gg 1$, the oscillations belong to absolutely none of the cases considered by Curie. He then qualified this new case as "quasi-aperiodic", and since this period is approximately equal to the "relaxation time", and corresponds to the discharge time of a capacitor in a resistor, he suggested calling this phenomenon "relaxation oscillations". Van der Pol then recalled that this idea was inspired by an article by Abraham and Bloch (1919a) in which they described the device *multivibrator* (see *supra*), which was the foundation of oscillations which had a period "on the order of $C_1R_1 + C_2R_2$" (Abraham and Bloch 1919a, 257). Van der Pol explains:

> In their original description of the system Abraham & Bloch draw attention to the fact that the time period of the oscillations produced by the multivibrator is approximately equal to the product RC, but, so, as far as I am aware, no theoretical discussion of the way in which the oscillations are *maintained* has been published.[16] (Van der Pol 1926d, 988)

This confirms that it was indeed the work of Abraham and Bloch which led Van der Pol to clarify the oscillation period of the triode as the product *RC* (see Eq. (3.15). Moreover, it seems that their description of the phenomenon, of "sudden" charge, and "slow" discharge, also caused him to draw attention to its duality, by giving these oscillations, which he still called *maintained*,[17] the name relaxation oscillation.[18] Van der Pol's greatest merit is therefore having, on one hand, shed light on the "slow-fast" character of oscillations, which led to a research on *singularly perturbed systems*, under Andronov and Pontrjagin (1937), and on the other, demonstrating that other devices (see *supra*) are governed by the dedimensionalized equation (3.8) and are therefore the basis of relaxation oscillations.[19]

> *Case of ε being very large.*–Here the oscillation curve visibly has a great number of harmonics. In mathematical language, the corresponding Fourier series converges very slowly. It is therefore absolutely unrealistic, in this case, to laboriously calculate the first, second or third terms of the series. One of Mr. Van der Pol's most substantial contributions has consisted in clearly recognizing this fact, giving *a name* to these non-sinusoidal oscillations, and turning them into a tool for research in physics, in the same way as the sinusoidal equations, of which they actually are the counterpart. (Le Corbeiller 1931a, 22)

[16]From 1916 to 1918 Henri Abraham and Eugène Bloch published several reports in the E.C.M.R. (Établissement Central du Matériel de la Radiotélégraphie Militaire) which stayed "classified". See *supra*.

[17]See Van der Pol (1920, 755).

[18]Pierre David, in his analysis of Van der Pol's article (1926d) offered to translate this terminology by "oscillations par décharge" ("discharge oscillation") (David 1927).

[19]During a conference held on the 24th of May, 1928 in Paris, Van der Pol (1928a, 731) provided an exhaustive list of examples for relaxation oscillations, amongst which Janet's experiment on series-dynamo, Abraham and Bloch's multivibrator, but also Volterra's work (1926) on the variation of the number of individuals in coexisting species. Concerning this last point, Israel (1996, 42) considered that the reference is groundless, because Volterra's models do not possess *limit cycles* but a fixed point, *center type*, i.e. for which the time period depends on the starting conditions.

3.2.2.5 Value of the Relaxation Oscillations Period

In the case where ε is very large (see Fig. 3.5, $\varepsilon = 10$, red frame), Van der Pol maintained that the period's value regarding *relaxation oscillation* was $T \doteq \varepsilon$. However, it is shown on the graph (see Fig. 3.5, red arrow) that this period's value is actually twice that: $T \doteq 2\varepsilon$. Such a difference cannot be neglected, something that Van der Pol quickly became aware of. In the second German volume of "über Relaxationsschwingungen", Van der Pol (1927c) offered a modification:

> Die Schwingungszeit. Es folgt aus der in (R[20]) Seite 182, Fig. 4, $\varepsilon = 10$, gegeben graphischen Lösung, daβ der Verlauf von dem Werte $v = 2$ zu dem Werte $v = 1$ (in welcher Zeit eine halbe Periode verläuft) mit guter Annäherung durch unsere Gleichung (1) oder (2)[21] wiedergegeben werden kann, aber ohne das erste Glied. Wie schon in (R), Seite 183, dargelegt wurde, ist der in der Weise berechnete Verlauf der Lösung durch
>
> $$\log_e v^2 - v^2 = \frac{2\omega^2}{\alpha} t + Const.$$
>
> gegeben. Es folgt daraus sofort, daβ der Zeitverlauf $t_2 - t_1$ in dem v von dem Wert $v = 2$ zu $v = 1$ heruntersinkt, durch
>
> $$t_2 - t_1 = \frac{\alpha}{\omega^2}\left(\frac{3}{2} - \log_e 2\right)$$
>
> gegeben ist, so dass die total Periode $T = 2\,(t_2 - t_1)$ annähernd wird:
>
> $$T = 1.61\frac{\alpha}{\omega^2} \doteq \frac{\pi}{2}\frac{\alpha}{\omega^2}$$
>
> Nennen wir deshalb T_{sin} die Schwingungszeit den sinusförmigen Fall ($\varepsilon \ll 1$), und T_{rel} die Schwingungszeit der Relaxationsschwingung ($\varepsilon \gg 1$), so hat man:
>
> $$\left.\begin{aligned} T_{\sin} &= 2\pi\sqrt{L.C} \\ T_{rel} &= \frac{\pi}{2} R.C \end{aligned}\right\}$$
>
> (...)[22] (Van der Pol 1927c, 114–115)

---

[20]The letter (R) refers to the first volume of "Über Relaxationsschwingungen og". See Van der Pol (1926c).

[21]This is the equation (3.8).

[22]The oscillation period. From the graphical integrations presented in (R) page 182, Fig. 4, $\varepsilon = 10$, it results that when we go from a value $v = 2$ to the value $v = 1$ (during a half-period), the solution can be represented with a suitable approximation by the equation (1) or (2) (3.8) without the first member. Since in (R) the form of the solution was given by

$$\log_e v^2 - v^2 = \frac{2\omega^2}{\alpha} t + Const.$$

It immediately follows that the time variation $t_2 - t_1$ during which v goes from the value $v = 2$ to $v = 1$ is given by

In his first publications, Van der Pol (1926b, 37; 1926c, 183; 1926d, 991) established that if the inductance is neglected, i.e. the first member of the equation (3.8) we find ourselves with:

$$-\varepsilon\left(1-v^2\right)\dot{v}+v=0 \quad \Leftrightarrow \quad v=\varepsilon\left(1-v^2\right)\dot{v}$$

This first order differential equation can be integrated without difficulty

$$\int \frac{dt}{\varepsilon}=\int\left(\frac{1}{v}-v\right)dv \quad \Leftrightarrow \quad \log_e v^2-v^2=\frac{2\omega^2}{\alpha}t+Const.$$

Taking into account the variable substitution (3.7) (see *supra*), Van der Pol obtained the following function:

$$\log_e v^2-v^2=\frac{2t}{\varepsilon}+Const. \tag{3.16}$$

for which the representative curve is reproduced below (see Fig. 3.6).

Van der Pol (1926b, 37; 1926c, 183, 1926d, 991) then explained that this curve approximately matches the exact solution to the equation (3.8) as long as the inclination of v is small. Therefore, he superimposed on Fig. 3.5 (see below the

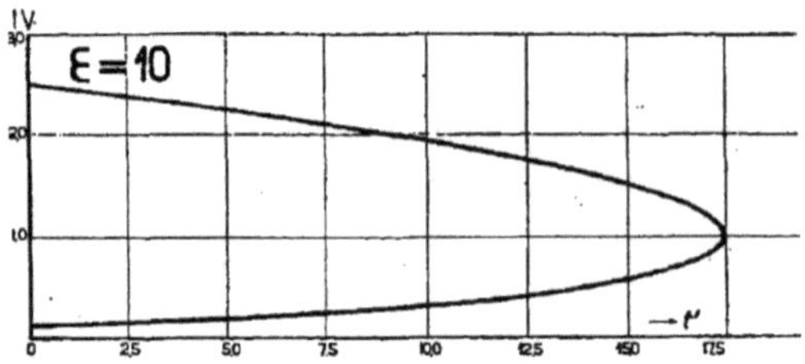

Fig. 3.6 Graphical representation of function (3.16), from Van der Pol (1926c, 183)

$$t_2-t_1=\frac{\alpha}{\omega^2}\left(\frac{3}{2}-\log_e 2\right)$$

so the total period $T=2\left(t_2-t_1\right)$ is about:

$$T=1.61\frac{\alpha}{\omega^2}\doteq\frac{\pi}{2}\frac{\alpha}{\omega^2}$$

This is why we call T_{sin} the oscillation period in the sinusoidal case ($\varepsilon \ll 1$), and T_{rel} the oscillation period of the relaxation oscillations ($\varepsilon \gg 1$), which yields:

$$\left.\begin{aligned} T_{\sin} &= 2\pi\sqrt{L.C} \\ T_{rel} &= \frac{\pi}{2}R.C \end{aligned}\right\}$$

(...) (Van der Pol 1927c, 114–115).

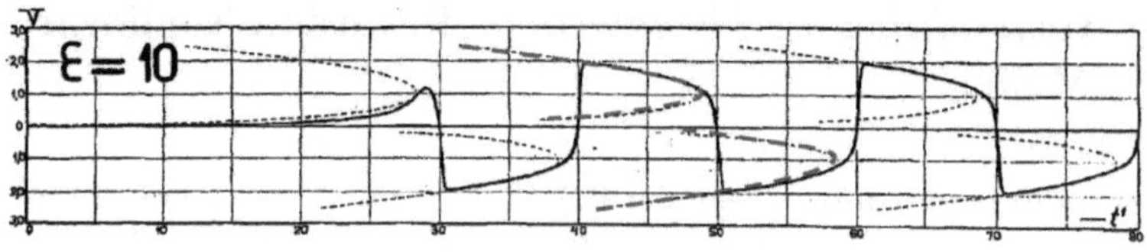

Fig. 3.7 Graphical integration of the equation (3.8), from Van der Pol (1926d, 986)

red dotted line on Fig. 3.7) the graphical representation of function (3.16) with the one for the solution to the differential equation (3.8).

This allowed him to justify the introduction of relaxation oscillation terminology.

> When the inductance is thus disregarded both $\dot{v}$ and $\ddot{v}$ become infinite when $v = \pm 1$, so that the condensers C would acquire a very large charge in an indefinitely small time. But these infinitely sudden changes of current are prevented by the presence of the small residual inductance. (Van der Pol 1926d, 990)

He explained that the charge and discharge of the capacitor are "driving" the phenomenon, which thus becomes "modulated" by the presence of the inductance. The other fundamental aspect of this result is that it allowed Van der Pol to correct the value of the period of relaxation oscillations from the function (3.16). Indeed, by considering that during the time variation $t_2 - t_1$ the variable v (corresponding to the plate voltage of the triode in the dimensional equation) goes from the value $v = 2$ to $v = 1$ he showed, from (3.16)

$$\left[\frac{2t}{\varepsilon}\right]_{t_1}^{t_2} = \left[\log_e v^2 - v^2\right]_2^1 \quad \Leftrightarrow \quad \frac{2\,(t_2 - t_1)}{\varepsilon} = \left(\log_e 1^2 - 1^2\right) - \left(\log_e 2^2 - 2^2\right)$$

that the period $T = 2\,(t_2 - t_1)$ is now written:

$$T \doteq (3 - 2\log_e 2)\,\varepsilon \doteq 1.61\varepsilon \doteq \frac{\pi}{2}\varepsilon \doteq \frac{\pi}{2}RC \tag{3.17}$$

If this result comes close to the graphically deduced value $T \doteq 2\varepsilon$, it was not until the work of Jules Haag (1943, 1944) that a better approximation of the period was given (see *infra*, Part II). Therefore, it is not so much the numerical value, marred by an error close to 20%, as it is the proportionality between the oscillation period and the product RC which allowed Van der Pol to demonstrate that when ε is large, the phenomenon is "driven" by the charge and discharge of the capacitor, as Abraham and Bloch suggested (1919, 257). This persuaded him to use the name relaxation oscillations. While, from a quantitative point of view, the period of this new type of oscillation was not strictly determined, from a semantic point of view, the phenomenon's significance, in other words the precise manner in which it happens, was perfectly defined. Nevertheless, this lack of strictness seems to have greatly influenced the way Van der Pol's work was received in France (see *infra*).

3.2.3 *Generalizing the Phenomenon:* Towards a Nonlinear Model

In his various contributions, (Van der Pol 1926a,b,c,d) Van der Pol demonstrated that the triode and the multivibrator by Abraham and Bloch (1919) were the foundation of relaxation oscillations, and were consequently governed by the generic equation (3.8). Afterwards, his aim was to establish that the phenomenon was actually far more general, and could apparently describe greatly different processes. In fact, out of the three devices studied until then – *series-dynamo machine*, *singing arc* and *triode* – the first two were not subject to the same treatment yet. It was still necessary, on the one hand, to complete their equation, and on the other hand, to categorize their type of oscillation. Nonetheless, Van der Pol was once again closely preceded by Blondel, at contributed to the solving of the phenomenon of the electric arc, and then of *the singing arc*, after the end of the nineteenth century.

3.2.3.1 From the Triode to the Singing Arc

During the 1920s, Blondel's study on the triode progressively turned into what he named a "general theory on self-sustained oscillations" (Blondel 1919d), which aimed at establishing the initial and stability conditions for these oscillations. Blondel, having made good use of his results, went back to the singing arc, for which he completed the differential equation formulated by Poincaré (1908), then Janet (1919). In a note presented on the 29th of March, 1926 in front of the *Académie des Sciences* in Paris, Blondel modeled the arc's current function by writing "the expression of the voltage at the terminals u, depending on the current in the arc i of the form $u = -hi + pi^2 + qi^3 + \ldots$" (Blondel 1926, 900). This allowed him to obtain the following differential equation:

$$L\frac{d^2i}{dt^2} + \left(R - h + 2pi + 3qi^2\right)\frac{di}{dt} + \left(\frac{1}{C} - \frac{1}{s}\right)i = 0 \tag{3.18}$$

It is surprising to notice that Blondel (1926) did not refer here to Poincaré's work (1908) at all, nor to Janet's (1919), when the equations (1.5), (2.1) and (3.18) are identical. Although the form of the arc's oscillation characteristic that he proposed, and the one used by Van der Pol (1920) to explain the *e.m.f.* of the triode (see Eq. 2.11) correspond exactly to each other by duality, he did not mention it either.

3.2.3.2 From the Triode to Gérard-Lescuyer's Experiment

As it was said previously (see *supra*), Van der Pol's contributions (1926a,b,c,d) offer different conclusions regarding the choices of examples illustrating the relaxation oscillation phenomenon. Consequently, in his first three publications on the subject, Van der Pol (1926a,b,c) considered Abraham and Bloch's multivibrator, a fluorescent tube's oscillations, a "Wehnelt" interrupter's oscillations, and a series-dynamo machine's oscillations:

> Een afzonderlijk bekrachtigde motor gevoed door een met constante snelheid loopende serie-dynamo verwisselt automatisch periodisch zijn draaiïngsrichting. Dit verschijnsel is ook een relaxatie-trillingen.[23] (Van der Pol 1926a, 157)

> Ook een afzonderlijk bekrachtigde motor, gevoed door een met constante snelheid draaiende serie-dynamo verwisselt voert relaxatie-trillingen uit, d.w.z. automatisch wisselt de motor periodiek zijn draaiïngsrichting.[24] (Van der Pol 1926b, 39)

> Auch ein von einem mit konstanter Geschwindigkeit drehenden Seriendynamo gespeister fremderregter Motor führt Relaxationsschwingungen aus, d. h. der Motor ändert selbsttätig seine Drehungsrichtung.[25] (Van der Pol 1926c, 184)

We can question the reasons why Van der Pol (1926d) was led to remove, in his last contribution in 1926, this relaxation oscillation example, and to replace it with the heartbeat example. It is difficult to imagine Van der Pol wanting to "take note" of this discovery, which later became the subject of more advanced research with Van der Mark.[26] The following year, in an article published in the *Jahrbuch der drahtlosen Telegraphie und Telephonie*,[27] Van der Pol (1927b,c) explicitly referred to Gérard-Lescuyer (1880a), as well as Janet's (1925) and the Henri and Élie Cartan' work (1925):

> *Relaxations schwingungen die von einem fremd erregten Motor ausgeführt werden, der von einer mit konstanter Geschwindigkeit angetriebenen Seriendynamo gespeist wird.* Dass ein solches system Relaxations schwingungen ausführen kann, wurde in schon kurz besprochen. In einer Verhandlung von M. Janet[(8)] finden wir einer Hinweis auf Gérard Lescuyer (C. R., **91**, 226, 1880) wo diese Erscheinung schon beschrieben steht.[28]

> [(8)] M. Janet, Note sur une ancienne expérience d'électricité appliquée, Annales des Postes, Télégraphes et Téléphones, XIV, N°12, pag. 1193.

> Siehe auch : E. et H. Cartan, Note sur la génération des oscillations entretenues, Annales des Postes, Télégraphes et Téléphones, XIV, N°12, pag. 1196. (Van der Pol 1927c, 116)

He then reproduced, without referring to them, both Janet's demonstration (1919) enabling the development of a differential equation corresponding to the phenomenon observed by Gérard-Lescuyer, and the conclusions Janet had reached.

---

[23] A motor powered by a D.C. series-dynamo also produces relaxation oscillations, that is to say, the rotation of the motor periodically changes direction.

[24] *Idem.*

[25] *Idem.*

[26] See Van der Pol and Van der Mark (1928a,b), Israel (1996), and Israel (2004).

[27] Directory of Wireless Telegraphy, which became Zeitschrift für Hochfrequenztechnik.

[28] Relaxation oscillations produced by a motor powered by a D.C. series-dynamo. The fact that such a system is able to produce relaxation oscillations was already briefly discussed. In an article written by Mr. Janet ([8]), we find a reference to Gérard Lescuyer (CR 91, 226, 1880) where this phenomenon had already been described.

He then hypothesized that the dynamo's oscillation characteristic can be represented by a cubic function, just as in the case of the triode, by using the following function: $e = f(i) = R_0 i - \gamma i^3$. Then, by writing $\gamma = R_0 - R$, he obtained the following differential equation:

$$L\frac{d^2i}{dt^2} - (R_0 - R)\left(1 - i^2\right)\frac{di}{dt} + \frac{1}{C}i = 0 \tag{3.19}$$

Van der Pol thus demonstrated that Gérard-Lescuyer's experiment was the first example illustrating the relaxation oscillation phenomenon, for which he continued the generalization. It is important to notice that the analogy established by Janet (1919) could just as well concern the singing arc rather than the series-dynamo machine. He might have chosen the series-dynamo machine because he was aware of Blondel's work (1926). In fact, by comparing the date when Van der Pol (1927b) finished writing his article: 13th of December, 1926, with the date of Blondel's presentation on his study (1926) in front of the *Académie des Sciences*: 29th of March, 1926, it can be established that Van der Pol could have had access to this document, as well as all the others. That Van der Pol did not make any reference to Blondel is however very surprising, since his studies were deemed to be of great significance when it came to the production of electromagnetic waves using a device comprising first a singing arc, then a triode.[29] Moreover, the similarities between the expansions used first by Blondel and then by Van der Pol to establish the equation for the triode are quite remarkable. However, their process led them to propose a classification of different types of oscillations that is quite dissimilar. Nevertheless, during a series of conferences presented at the *École Supérieure d'Électricité* on the 10th and 11th of March, 1930, Van der Pol recalled:

> When it comes to the electricity field, we have very nice examples of relaxation oscillations, some of which are very old, such as the spark discharge of a disk machine, the oscillation of the electric arc studied by Mr. Blondel, in a famous dissertation ([1]), or Mr. Janet's experiment, as well as others, which are more recent (...)
>
> ([1]) BLONDEL, *Eclair. Elec.*, **44**, 41, 81, 1905. V. aussi *J. de Phys.*, **8**, 153, 1919. (Van der Pol 1930, p. 20)

By recognizing that the singing arc is the basis for relaxation oscillations, Van der Pol made his one and only reference to Blondel's work. His selection of cited articles, even if they might have been suggested by Le Corbeiller (see *infra*), curiously shows a good knowledge of Blondel's work.[30] It should also be noted that Van der Pol (1927c, 116), who initially attributed the merit of having observed relaxation oscillations in the series-dynamo machine to Gérard-Lescuyer, now mentioned "Mr. Janet's experiment". This change might be due to the presence of Mr. Blondel and Mr. Janet, to whom Van der Pol would have thus paid a tribute to.

---

[29] See Bethenod (1938a,b, 754).

[30] It must however be noted that Blondel's second reference is inaccurate: Blondel, *J. Phys.*, **9**, 153, 1919.

Fig. 3.8 Balthasar Van der Pol, from Philips International B.V., Company Archives, Eindhoven

However, Van der Pol confirmed that the phenomenon observed by Gérard-Lescuyer (1880a,b) and Blondel (1905a,b,c) indeed corresponds to what he called relaxation oscillation (Fig. 3.8).

3.3 Liénard's Work and *Skepticism Regarding Van der Pol's Results*

As early as 1927, Van der Pol's work on relaxation oscillations was subjected to analysis in the *Revue Générale d'Électricité* (Van der Pol 1927) by someone who signed the article L. B., and who might be Léon Bouthillon[31] and in *Onde Électrique* by Pierre David (1927). As soon as they were known, Van der Pol's results drew mixed opinions by the French scientific community. The enthusiasm generated by the exhibition of a new type of nonlinear oscillation and the model able to describe them, which led some, such as Philipe Le Corbeiller, initially to pay homage to Van der Pol's work and mention relaxation oscillation theory, quickly dwindled, in the face of the problems that had to be solved in order to characterize its period and

[31]Bouthillon was very involved in radio communication and wireless telegraphy, and published two reference studies in this field (see Petit and Bouthillon 1910, Bouthillon 1919–1921) and several articles in international periodic publications such as the *Proceedings of the Institute of Radio Engineers* (P.I.R.E.). In the middle of the 1920s, Bouthillon (1926, 1927) submitted two notes to the *Académie des Sciences* delivered by André Blondel who was at the time *Président du Comité de Rédaction* for the *Revue Générale de l'Électricité* (R.G.E.). It might have been at this time that Blondel asked him to analyze Van der Pol's work.

amplitude. Moreover, numerous researchers such as Alfred Liénard, Yves Rocard and Jules Haag, seemed to not fully stand by the relaxation oscillations concept which Van der Pol tried to generalize, sometimes in a somewhat excessive manner.[32]

After graduating the *École Polytechnique* at age 20, Alfred Liénard joined the *École des Mines* in Paris, from which he graduated top of his class. He then joined the *Corps des Mines*, and started his career as an engineer in Valencia, Marseille, then Angers. From 1895 to 1908, he was appointed professor at the *École des Mines* in Saint-Étienne. Then, from 1908 to 1929, he was professor of industrial electricity at the *École des Mines* in Paris, and from 1913 to 1922 of Material Resistance and Construction. During the next ten years and the first worldwide war, he became vice-director at the *École des Mines* in Paris, then director in 1929, a position he kept until 1936. During this period, he was vice-president of the *Société française des Électriciens* from 1926 to 1929, then president of the *Société Mathématique de France* in 1933.

3.3.1 Existence and Uniqueness of the Stable Periodic Solution

In May 1928, he was one of the first to take an interest in Van der Pol's work (1926d). In an article entitled "Étude des oscillations entretenues" ("Study of sustained oscillations") published in the *Revue générale d'Électricité*, Liénard (1928) demonstrated, with specific conditions, that the differential equations for the oscillations sustained by a series-dynamo machine, a singing arc or a three-electrode lamp, possess only one periodic solution. To establish this fundamental result, Liénard (1928, 901) considered the following differential equation:

$$\frac{d^2x}{dt^2} + \omega f(x) \frac{dx}{dt} + \omega^2 x = 0 \tag{3.20}$$

where the variable x can represent a length, a current voltage, etc.

The equation (3.20), more general than any other presented before (see Table 3.3), such as Cartan, Élie, and Henri (1925, 1196) (3.1) or Van der Pol's (1926d, 979) (3.4), corresponds to them by analogy or by duality.

Indeed, by writing the equation (3.20): $i = x$, $\omega^2 = \frac{1}{LC}$ and $f(x) = \sqrt{\frac{C}{L}}[R - \varphi(x)]$, we obtain for example Henri and Élie Cartan's equation (1925) (3.1). With simple changes in variables, we can find the others.

To demonstrate the existence and the uniqueness of a periodic solution for the equation (3.20), Liénard (1928, 901 and 904) hypothesized that the function $F(x) = \int_0^x f(x)\,dx$, which represents the oscillation characteristic , must necessarily be an odd function, first negative from 0 to a certain value X_0 de x, then increasing

[32]Regarding other examples of relaxation oscillations proposed by Van der Pol, Israel estimated that "they remained imaginary" (Israel 1996, 42).

Table 3.3 Synoptic of the differential equations

Differential equation	Author	Domain
$Lx'' + \rho x' + \theta(x') + Hx = 0$	Poincaré (1908)	Singing arc
$L\dfrac{d^2i}{dt^2} + [R - f'(i)]\dfrac{di}{dt} + \dfrac{k^2}{K}i = 0$	Janet (1919)	Series-dynamo machine
$C\dfrac{d^2u}{dt^2} - \left(b_1h - 3b_3h^3u^2 - \ldots\right)\dfrac{du}{dt} + \dfrac{u}{L} = 0$	Blondel (1919b)	Triode
$C\dfrac{d^2v}{dt^2} - \left(\alpha - 3\gamma v^2\right)\dfrac{dv}{dt} + \dfrac{1}{L}v = 0$	Van der Pol (1920)	Triode
$L\dfrac{d^2i}{dt^2} + [R - \varphi(i)]\dfrac{di}{dt} + \dfrac{1}{C}i = 0$	Janet (1925) and Cartan, Élie, and Henri (1925)	Series-dynamo machine
$\ddot{x} - \left(\alpha - 3\gamma x^2\right)\dot{x} + \omega^2 x = 0$	Van der Pol (1926d)	Triode
$L\dfrac{d^2i}{dt^2} + \left(R - h + 2pi + 3qi^2\right)\dfrac{di}{dt} + \left(\dfrac{1}{C} - \dfrac{1}{s}\right)i = 0$	Blondel (1926)	Singing arc
$L\dfrac{d^2i}{dt^2} - (R_0 - R)\left(1 - i^2\right)\dfrac{di}{dt} + \dfrac{1}{C}i = 0$	Van der Pol (1927)	Series-dynamo machine

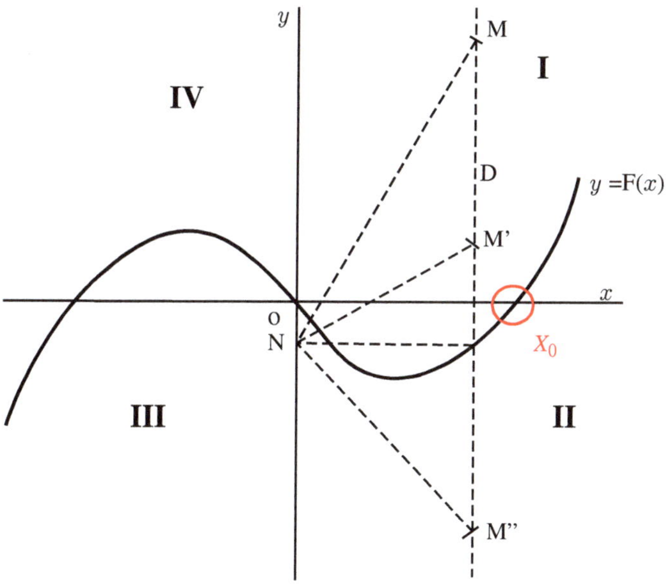

Fig. 3.9 Van der Pol's oscillation characteristic (1926d), from Liénard (1928, 902)

for $x \geq X_0$. This hypothesis is effectively "less restrictive" than the one written by Messrs. Cartan, Élie, and Henri (1925) and Van der Pol (1926d), as shown in Figs. 3.9 and 3.10:

Indeed, in Fig. 3.9, we can see that the oscillation characteristic $y = F(x)$ corresponds to Van der Pol's cubic curve (1920, 703) (see Eq. (V_1)), which is

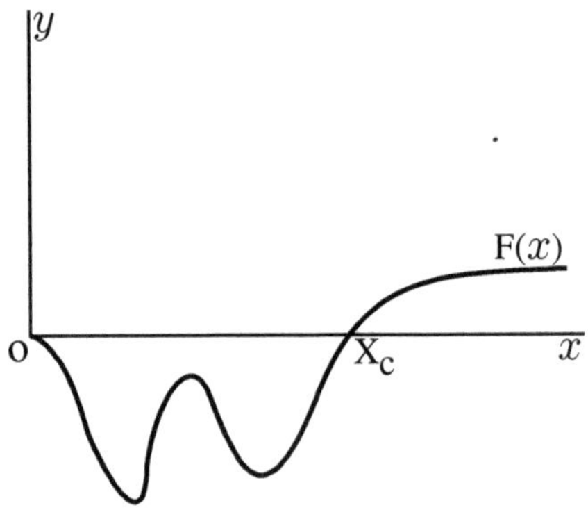

Fig. 3.10 Liénard's oscillation characteristic, from Liénard (1928, 904)

written: $y = F(x) = K\left(x^3/3 - x\right)$ and is negative, from its origin to the point X_0 of the intersection of his representative curve with the x axis (circled in red in Fig. 3.9). First decreasing, this function goes through a unique negative minimum, then increases until the point X_0, where it becomes positive. The *oscillation characteristic* $F(x)$ considered by Liénard (see Fig. 3.10) can, however, show several minima in its negative part before cutting the x axis at the point X_0 as shown in its diagram. He then proceeded to the following variable substitution, allowing him to place himself in what we nowadays call "Liénard phase plane": $\dfrac{dx}{dt} = v$ and $y = \dfrac{v}{\omega} + F(x)$. The equation (3.20) is then written as a first-order differential equation:

$$\frac{dy}{dx} + \frac{x}{y - F(x)} = 0 \tag{3.21}$$

where:

$$xdx + (y - F(x))\,dy = 0 \tag{3.22}$$

Or again, a *symmetrical* form (see Poincaré 1886a, 168):

$$\omega dt = \frac{dx}{\frac{v}{\omega}} = \frac{dx}{y - F(x)} = \frac{-dy}{x} \tag{3.23}$$

Liénard then used a graphical integration method[33] based on the curvature of the trajectory curves which differs from the one used by Van der Pol. Then, to demonstrate the existence and uniqueness of a stable periodic solution, he proceeded step by step.

[33]From Parodi (1942a,b,c, 196), this method was used since 1900.

The first step consisted in showing that the equations system (3.23) possesses a kind of symmetry. Accounting for the hypothesis stating that the function $y = F(x)$ has odd terms, he established that the system invariant by symmetry $(-x, -y, -t)$. He deduced from this:

> To any integral curve arc corresponds a symmetrical arc respecting the origin. (…) We can therefore restrain the discussion to the area of positive x. (Liénard 1928, 902)

He then considered a point P belonging to the y axis, and a point $M(x, y)$ of the integral curve of the equation (3.20 or 3.21), and calculated the distance PM:

$$PM^2 = x^2 + (y - y_P)^2 .$$

By taking the differential of this distance with respect to the time, he obtained, using (3.22):

$$PMd(PM) = [F(x) - y_P]\, dy \tag{3.24}$$

He then placed the point P at the origin, and the equation became:

$$\frac{1}{2} d\left(OM^2\right) = F(x)\, dy \tag{3.25}$$

He then defined the point $C(x_C, y_C)$ as the intersection on an integral curve with the oscillation characteristic. Then, he considered the points where this integral curve meets with the y axis, which he named A_1 and A_2 (see Fig. 3.11) and integrated the equation (3.25) along this integral curve in the direction[34] A_2CA_1. He obtained:

$$\frac{1}{2}\left(OA_1^2 - OA_2^2\right) = \int_{A_2CA_1} F(x)\, dy \tag{3.26}$$

To establish that the equation (3.20 or 3.21) possesses a periodic solution, Liénard had to show that the integral curve is a closed curve, in other words, there exists one, and only one value of x being: $OA_1 = OA_2$.

He then demonstrated that:

- if the point C is on the negative part of the oscillation characteristic, i.e., if C belongs to the interval $0 < x_C < X_0$, then $y = F(x) < 0$ and therefore the integral (3.25) is negative. Consequently, $OA_1 < OA_2$. This means that inside this bounded interval, the curve moves away from the origin (to move closer to the periodic solution).

[34] This integration direction is the direct reverse of the travel direction of the trajectory curve. It is also the reverse of Poincaré's (1908). This is why Liénard's stability condition will have the direction opposite to Poincaré's. See *infra*.

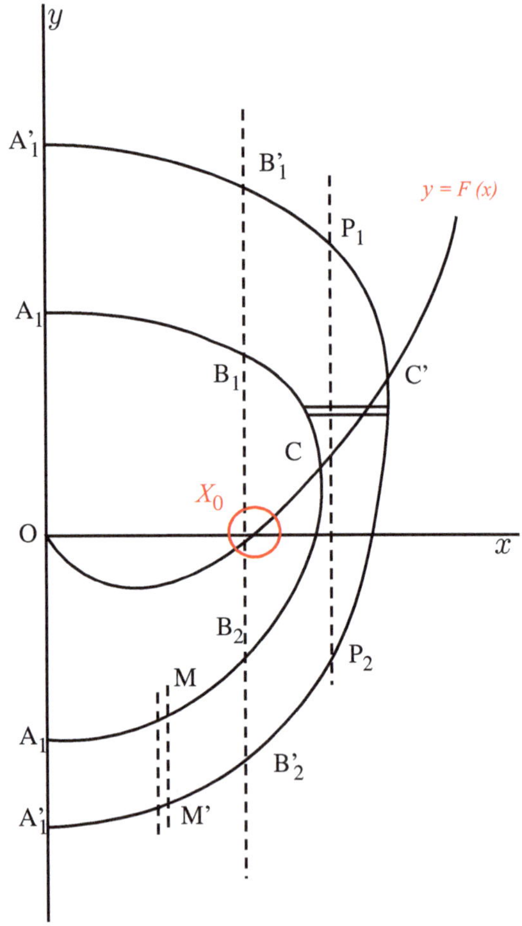

Fig. 3.11 Oscillation characteristic and integral curves, from Liénard (1928, 905)

- if the point C is on the positive part of the oscillation characteristic, i.e., if C belongs to the interval $x_C > X_0$, then the integral (3.26) "increases in negative value to $+\infty$ when x_C varies from X_0 to $\infty(\ldots)$" (Liénard 1928, 905). This implies that the integral (3.26) becomes positive, and that $OA_1 > OA_2$. This means that in this interval, the integral curve moves closer to the origin (more accurately, closer to the periodic solution).

Since the integral, first negative, becomes positive, Liénard therefore concluded that:

> (...) it is obvious that there exists a value of x_C for which the integral vanishes, beside the fact that there is only one, the variation of the integral happening always in the same direction. The equality (3.26) shows that there exists only one value of x_C for which
>
> $$OA_1 = OA_2$$

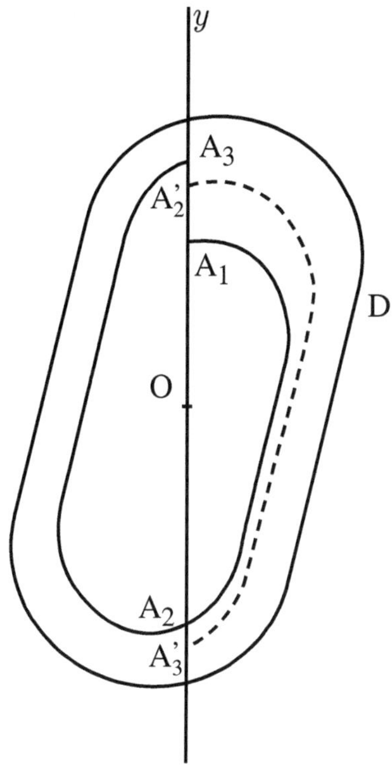

Fig. 3.12 Relative locations of the integral curves and the closed curve D, from Liénard (1928, 905)

> The corresponding integral curve is closed, because, due to it symmetry about the origin, the integral curve will prolong beyond A_2 in an arc identical to the arc A_1CA_2, but turned by 180° around the origin. The integral curve moves back to the starting point. (Liénard 1928, 906)

Liénard then proposed a graphical representation of the closed curve D and its external and external integral curves (see Fig. 3.12).

He explained that for all the integral curves internal (*resp.* external) to the closed integral curve D, the integral (3.26) is negative (*resp.* positive) and OA_2 is greater (*resp.* less) to OA_1. He deduced:

> We thus realize that the integral curve describes a kind of spiral asymptotically approaching the closed curve D. For the integral curves external to the closed curve, it is OA_2 that becomes inferior to OA_1. The curve still approaches the curve D, but from the outside. Due to the fact that these integral curves, internal or external, and traversed in the direction of the increasing times, asymptotically approach toward the curve D, we qualify the corresponding periodic movement as a *stable* movement. (Liénard 1928, 906)

There is an interesting comparison to be made between Liénard's conclusion (1928) and Poincaré (1908) (see *supra*):

> *Stability condition.*–Let us then consider another non-closed curve which satisfies the differential equation. It will be a kind of spiral growing indefinitely closer to the closed curve. If the closed curve represents a stable speed, by tracing the spiral in direction of the

Table 3.4 Poincaré's (1908), Liénard's (1928), and Andronov's (1929a) differential equations

Poincaré (1908)	Liénard (1928)	Andronov (1929a)
$\begin{cases} \frac{dx}{dt} - y \\ \frac{dy}{dt} = -x - \mu\theta\left(y\right) \end{cases}$	$\begin{cases} \frac{dx}{dt} = y - F\left(x\right) \\ \frac{dy}{dt} = -x \end{cases}$	$\begin{cases} \frac{dx}{dt} = y + \mu f\left(x, y; \mu\right) \\ \frac{dy}{dt} = -x + \mu g\left(x, y; \mu\right) \end{cases}$
See Eq. (1.9)	See Eq. (3.23) See Eq. (1.10)	

arrow, we must be taken back to the closed curve, and this condition alone will enable the curve to represent a stable set of sustained waves, and solve the problem. (Poincaré 1908, 391)

If the analogy between the two problems leads to the same result, the similarities between the wording used by Liénard and Poincaré raise the following question:

Was Liénard aware of Poincaré's work?

Finding an answer is difficult. Several hypotheses can however be considered. One the one hand, it is possible that Liénard found this result completely independently (as it seems to be the case for Andronov (1928, 1929a,b). See *infra* Part II). On the other hand, if Liénard knew of Poincaré's studies (1908), why would he not refer to them? Also, Liénard's approach is exactly identical to Poincaré's. This can be demonstrated by simply comparing the two differential equations presented in Poincaré's phase plane (*resp.* Liénard's) with Andronov's differential equation (1929a) (see Table 3.4).

By writing: $\mu = 1, f\left(x, y; \mu\right) = -F\left(x\right)$ and $g\left(x, y; \mu\right) = 0$, we can see that on one hand Liénard's system (3.23) corresponds exactly to Andronov's (1.10), and on the other, that the stability condition (5.7) (see *supra*) of the periodic movement is written:

$$\int_{\Gamma} \left(-F\left(x\right)\frac{dy}{dt}\right) dt < 0 \quad \Leftrightarrow \quad \int_{\Gamma} F\left(x\right) dy > 0 \tag{3.27}$$

It therefore appears that the condition (3.27) used by Liénard (1928) to demonstrate the existence and uniqueness of a stable periodic solution is absolutely identical to the one that was established twenty years earlier by Poincaré (1908). In the second part of his dissertation, (Liénard 1928, 946) confirmed this by recalculating the value of the integral (3.27), placed in polar coordinates. This is also reminiscent of the technique used the next year by Andronov (1929a) to obtain this same condition (5.7) (see *infra* Part II). Nevertheless, if Liénard effectively demonstrated the existence of a stable and unique periodic solution represented by a closed curve which corresponds exactly to Poincaré's definition of a limit cycle (1882, 261) (see *supra*), he did not associate it with this concept, and kept the term "closed integral curve" in this work, as well as all his later studies in this field of research (see *infra* Part II). It must also be noted that Liénard seemed to give little credit to the terminology introduced by Van der Pol, as shown by the title of his article: "Étude des oscillations entretenues" ("Study on maintained oscillations").

3.3.2 Analytical Determination Sustained Oscillations Amplitude

In the second part of his dissertation, Liénard (1928, 946) calculated the amplitude of the sustained oscillations. In order to do this, he followed the very same process that Poincaré had used (1908). He proceeded in the following way. First in equation (3.22) he replaced $f(x)$ by $kf(x)$ and obtained:

$$xdx + ydy - kF(x)\,dy = 0 \tag{3.28}$$

He then noticed that when $k = 0$, this equation is reduced to: $xdx + ydy = 0$, which is reduced to $x^2 + y^2 = constante$. He deduced that:

> All the integral curves are circles, therefore closed curves. But, among this infinity of circles, only one corresponds to the bound of the unique closed curve existing as long as k stays finite. To determine this limit circle, it must simply be noted that the condition $\int Fdy = 0$ characterizing the integral closed curve is not modified by the substitution of F by kF. The limit circle will therefore be the one for which we have $\int Fdy = 0$. (Liénard 1928, 946)

This excerpt can be compared to Poincaré's text (1892):

> If the equations (1) depending on a parameter μ admit, for $\mu = 0$, a periodic solution with no characteristic exponents that are null, they will still admit a periodic solution for small values of μ. (Poincaré 1892, 181)

as well as the terminology Liénard chose, "limit circle", which is confusing in regards to his possible lack of knowledge of Poincaré's work (1881–1886, 1892, 1908).

Liénard (1928, 946) switched to a polar coordinate system: $x = R\sin\varphi$, $y = R\cos\varphi$.

The integral $\int Fdy = 0$ was therefore written, after an integration by parts:

$$\int_0^R f(x)\sqrt{R^2 - x^2}dx = 0$$

By choosing Van der Pol's *cubic oscillation characteristic* (1920, 703): $F(x) = \frac{x^3}{3} - x$, the integral is reduced, taking into account the fact that $f(x) = F'(x) = x^2 - 1$, to:

$$\int_0^R f(x)\sqrt{R^2 - x^2}dx = \frac{\pi R^2}{16}\left(R^2 - 4\right) = 0$$

Which led him to $R = 2$ for the amplitude of sustained oscillations.

Considering that k is infinitely small, and thanks to this switch to a polar system, Liénard clarified the solution of the equation (3.28), which in this case is none other than the equation (3.8) written by Van der Pol (1926d, 979) (see *supra*). He obtained for the expression of the amplitude:

$$R = \frac{2}{\sqrt{1 + Ke^{-k\omega t}}} \tag{3.29}$$

And for the solution of (3.28) or (3.8):

$$x = \frac{2 \sin \omega t}{\sqrt{1 + Ke^{-k\omega t}}} \tag{3.30}$$

He then recalled that on one hand these equations "have been given by Van der Pol without explanation" (Liénard 1928, 948) and on the other, that:

> When t increases indefinitely, R approaches the bound 2, by lesser values if K is positive, by greater values if K is negative. We therefore find the property stating that the closed integral curve, toward which the other integral curves approach in a spiral, has a limit form, for k infinitely small, a circle of radius 2. (Liénard 1928, 948)

He then replaced x in the equation (3.20) by the expression (3.30) and was "surprised to observe that the equation is not satisfied unless we neglect the first order terms in k." (Liénard 1928, 948). He then provided an expression for x which satisfied the equation (3.20) to the second order in k. This highlighted the French community's scepticism in regards to Van der Pol's work (1926d), that indeed originated from the absence of any explanation or demonstration, and especially from the lack of mathematical precision that sometimes occurred, which is even more visible in the calculation of the period.

3.3.3 *Characterization of the Oscillatory Phenomenon Duality*

To characterize the nature of sustained oscillations, Liénard considered the equation (3.28), in which he wrote: $y = kz$, in order to distinguish two cases: k less than or greater than one. He obtained the following equation, which is absolutely identical to (3.20) (up to the factor k), and therefore to the equation (3.8) written by Van der Pol (1926d):

$$[z - F(x)]\, dz + \frac{1}{k^2} x dx = 0 \tag{3.31}$$

When k becomes infinitely large, Liénard recalled that the second term of this equation becomes negligible, and consequently, as shown on Fig. 3.13, that:

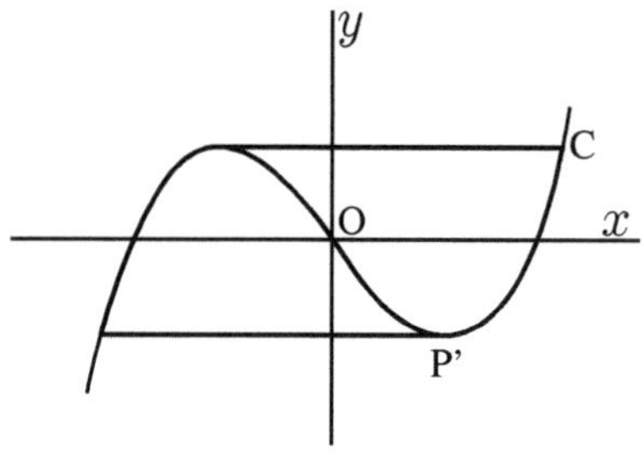

Fig. 3.13 Integral curve for k infinite, from Liénard (1928, 905)

> Any integral curve is comprised of arcs and curves $z = F(x)$ and horizontal parts $z =$ *constante*. If k is very large, the integral curves z will comprise parts very close to the curve $z = F(x)$ and others very close to horizontals. (Liénard 1928, 951)

This excerpt is particularly interesting, since it represents a first version of the definition Andronov (1937) was to give a few years later to "rough systems", for which he elaborated the theoretical frame of the singularly perturbed systems (see *infra* Part II). Figure 3.13 is quite unsettling, since it provides a representation of the solution to the equation (3.31), being none other than Van der Pol's *limit cycle* solution to the equation (3.8) (1926d). In order to clarify the characteristics of these two time scales, in other words, the two types of evolution for the integral curves (one slow, one fast) Liénard rewrote the relation (3.23), accounting for the variables substitutions ($F(x) \to kF(x)$, $y = kz$), in the following form:

$$\omega dt = -\frac{dy}{x} = -k\frac{dz}{x} \tag{3.32}$$

$$\omega dt = \frac{dx}{y - kF(x)} = \frac{dx}{k[z - F(x)]} \tag{3.33}$$

Based on this, he deduced that the integral curves near the curve $z = F(x)$ are traversed at a very small speed on the order of $1/k$, whereas the arcs of the integral curves near horizontals $z =$ *constante* are traversed by a very high speed on the order of k^2.

3.3.4 Analytical Determination of the Sustained Oscillation Period

To calculate the period of the oscillations, Liénard used the considerations he had just established on the "slow-fast" duality of the evolution of the integral curves. Indeed, since k is large, he made the following hypotheses:

- in the horizontal part, the integral curves evolve very rapidly, since their traversing time is proportional to $1/k^2$. This time can therefore be neglected,

- in the part CP', the integral curves are very near the curve We can therefore assimilate the integral curve, solution of (3.31) to the function $z = F(x)$

Taking the fact that $dz = F'(x)\,dx = f(x)\,dx$ into account, (Liénard 1928, 952) thus showed that the duration of a half-revolution $T/2$ can be obtained by integrating the relation (3.32):

$$\omega\frac{T}{2} = k\int_{C}^{P'}\frac{-dz}{x} = k\int_{P'}^{C}\frac{dz}{x} = k\int_{P'}^{C} f(x)\frac{dx}{x}$$

Liénard chose Van der Pol's cubic curve (1920, 703) as the *oscillation characteristic* : $F(x) = \frac{x^3}{3} - x$ with $f(x) = F'(x) = x^2 - 1$. The abscissa[35] of the points P' and C being 1 and 2, he then obtained

$$\omega\frac{T}{2} = k\int_{1}^{2}\frac{\left(x^2-1\right)dx}{x} = \frac{k}{2}\left[x^2 - \log_e x^2\right]_1^2 = \frac{k}{2}\left[3 - 2\log_e 2\right]$$

Then he deduced from this that the period of the oscillations, for k a very large term (the case $\varepsilon \gg 1$ of the relaxation oscillations for Van der Pol (1926d). See *supra*), has a value of[36]:

$$\omega T = k\left[3 - 2\log_e 2\right] = 1.614k \tag{3.34}$$

He then added:

> Although having found that ωt is equal to
>
> $$\omega\frac{T}{2} = \frac{k}{2}\left[x^2 - \log_e x^2\right], \text{(see his formulae (20))}$$
>
> Van der Pol indicates, for an unknown reason, that ωT would be near k ([1]).
>
> ---
>
> ([1]) From the figure 4 of Van der Pol, this would besides be $\omega\frac{T}{2}$ and not ωT which would be near k. The approximation would be much better. (Liénard 1928, 952)

This is precisely what was observed from the analysis of Van der Pol's work (1926d, 986) (see *supra*). In his article, Liénard therefore adopted a somewhat

---

[35]Point C's abscissa corresponds to the maximum value of the amplitude ($x = 2$) and the one for point P to the minimum value of the function $F(x)$, i.e., $x = 1$.

[36]However, this result is an approximation, later refined by Jules (Haag 1943, 1944, 1948). Nevertheless, Liénard's demonstration survived in numerous studies in differential equation theory or dynamical systems theory. See for example Lefschetz (1957, 346), Grassman (1987, 28), Verhulst (1996, 168) or Enns and McGuire (2001, 277). See *infra* Part II.

Fig. 3.14 Alfred Liénard, from Collections E.N.S.M.P.

strict, even critical view on Van der Pol's contribution. The lack of mathematical precision that Van der Pol had shown, as much in the calculation of the amplitude as in the calculation of the period of the relaxation oscillation phenomenon (which is actually close to double what he had graphically deduced) weakened the impact of his results, and decreased his credibility for the French scientific community (Fig. 3.14).

Conclusion of Part I

This first part allowed us to follow the origins of nonlinear oscillations theory, between 1880 and 1928. We saw that three devices – the series-dynamo machine, the singing arc, and the triode – enabled the observation of a new type of oscillatory phenomenon, then called *sustained*, or *maintained oscillation*.

The various researches conducted, particularly in France, then aimed at describing the phenomenon, and isolating its cause. At that time scientists were quickly able to prove, in the series-dynamo machine and the singing arc, the existence of a component presenting a nonlinear behavior, which led to the introduction of the "negative resistance" concept. However, the second step, which was the mathematical modeling of this new type of behavior using a differential equation characterizing these sustained oscillations, proved to be a longer process.

At the beginning of the twentieth century, the rise of wireless telegraphy placed the singing arc on a pedestal, and at the centre of researcher's concerns. After solving the issue of establishing oscillations, it was necessary, in order to allow transmissions using the device, to establish a stable regime for sustained waves. Unexpectedly, this engineering problem was solved by a mathematician: Henri Poincaré. Until now, the historiography has attributed to Andronov, at the end of the 1920s, the link between sustained oscillations and the concept of limit cycles introduced by Poincaré in his second dissertation "Sur les courbes définies par une équation différentielle" ("On curves defined by a differential equation") in the early 1880s. The discovery of forgotten lectures on wireless telegraphy, that Poincaré gave in May and June 1908, twenty years before Andronov's work, led to slight modifications concerning their usual perception. During his presentation, Poincaré indeed demonstrated that the required and sufficient condition for the establishment of operating conditions of sustained waves in the singing arc, was the existence of a stable limit cycle (although he did not use this term).

Following the death of the French scientist in 1912, a third device made its appearance: the triode. Mainly used as a transmitter during the First World War, the triode replaced the singing arc. Easier to use, and more importantly, more reliable, it eased the experimentations and the modeling of nonlinear oscillations. While historiography credits Van der Pol, it appears that André Blondel was the one who – in 1919, a year before the Dutch scientist – first came up with a modeling, leading him *de facto* to the equation for the oscillations sustained by a triode. The same year, Paul Janet showed that the series-dynamo machine, the singing arc, and the triode, were analogous devices governed by the same second-order differential equation, although it was only partially determined. Probably unaware of Poincaré's work, Janet then submitted the problem of the study of the solutions to this equation to the mathematicians Élie and Henri Cartan who, in 1925, demonstrated the existence of the periodic solution, and determined its characteristic.

The following year, Van der Pol published his famous article entitled "On relaxation-oscillations" in which he gave the prototype equation characterizing the oscillations of the triode, and therefore the ones for the singing arc and series-dynamo machine. Even though he represented the solution to this equation, which has the form of a closed curve, by using a graphical integration, he did not identify it as Poincaré's limit cycle. However, Van der Pol's most crucial contribution was his description of the intrinsic nature of the oscillatory phenomenon that he named relaxation oscillation, highlighting the existence of two characteristic time scales. This discovery later plays an essential role in this field of study.

In 1928, the engineer and mathematician Alfred Liénard considered a second-order differential equation, broader than Van der Pol's, and established under specific conditions the existence of a unique periodic solution. He did not, however, associate it with Poincaré's limit cycle concept either. Therefore, nobody until today, and especially in France, seems to have recognized Poincaré's limit cycle in this periodic solution.

Part II
From Relaxation Oscillations to Self-Oscillations

The end of the 1920s was marked by a publication in the *Comptes Rendus de l'Académie des Sciences* de Paris by Russian mathematician Andronov (1929a), in which he established a connection between the periodic solution of a nonlinear differential equation system, and the concept of limit cycle provided by Poincaré (1882). The lack of recognition of Poincaré's work (1908) (see *supra* Part I) led scientists, and then historians of science to consider this result as the starting point of nonlinear oscillation theory. In France, one of the first scientists[1] to take interest was Le Corbeiller:

> [...] the periodic solution[2] will correspond to Poincaré's limit cycle, as noted by Mr. Andronov. (Le Corbeiller 1931b, 211)

Less than five years later, Morched-Zadeh wrote in the first chapter of his doctoral dissertation entitled: "Étude des oscillations de relaxation and des différents modes d'oscillations d'un circuit comprenant une lampe néon" ("A study on relaxation oscillations and oscillation models in a circuit containing a neon tube"):

> From a theoretical point of view, H. Poincaré's cycles play a large part in the self-sustained oscillation theory, as demonstrated by A. Andronov and A. Witt. (Morched-zadeh 1936, 12)

In Soviet Union, Leonid Mandelstam (1879–1944), Andronov's supervisor, recalled that during a presentation entitled "Exposé de recherches récentes sur les oscillations non linéaires" ("Presentation on recent research on nonlinear oscillations") at the radio physics section of the General Assembly of the International Union of Radio Science (URSI) in London between the 12th and 18th of September 1934:

> The links in Poincaré's work as well as by Lyapunov's work, deepened later by Birkhoff, and our physics problems, were shown by one of ours([3]). Here we must distinguish three things. First, that the qualitative theory of differential equations, developed by Poincaré,

[1]Except for Van der Pol (1930, 16) . See *supra* Part I.

[2]On the differential equation characterizing the oscillations of a triode lamp.

[3]Mandel'shtam referred to Andronov (1929a).

> has proved to be very efficient for the qualitative discussion of the physical phenomena occurring in the systems used in radio engineering. But neither the physicist, and even more so, nor the engineer, can settle for a qualitative analysis. Another part of Poincaré's work provided a device able to process our problems quantitatively. Lastly, Lyapunov's work enabled a mathematical discussion of the questions of stability. (Mandel'shtam et al. 1935, 83)

As noted by Mandel'shtam, Andronov (1929a) therefore not only used Poincaré's (1881–1886) papers "On the curves defined by a differential equation" to establish a connection between the periodic solution of a nonlinear oscillator and the limit cycle, he also especially used Poincaré's "New Methods of celestial Mechanics" (1892-93-99), which allowed him to obtain a stable condition for the limit cycle. In the United States at the start of the 1940s, the Russian mathematicians Solomon Lefschetz (1884–1972) and Nicolas Minorsky (1885–1970) started translating part of the work published during 1937 by Andronov and Khaikin (1937), as well as Krylov and Bogolyubov (1937) (see *infra* Part III). In his "Introduction to Non-Linear Mechanics", Nicolas Minorsky wrote:

> Andronov (1929a) was first to suggest that periodic phenomena in non-linear and non-conservative systems can be described mathematically in terms of limit cycles which thus made it possible to establish a connection between these phenomena and the theory that Poincaré developed for entirely different purpose. (Minorsky 1947, 63)

The historians of science were not aware of Poincaré's work (1908), and legitimately followed the scientists' path, just like Amy Dahan-Dalmedico, who stated:

> As shown by S. Diner,[4] Andronov noticed for the first time that in a radio physics oscillator similar to Van der Pol's, which is a *dissipative system* for which the oscillations are sustained by taking energy in non-vibratory external sources, the movement in the phase space is similar to the one in Poincaré's *limit-cycle*. This is the first time that an *attractor*, not reduced to a point but rather *periodic*, is physically evidenced. (Diner 1992, 21–22)

Alexander Pechenkin also corroborated this theory:

> Andronov demonstrated the connection between Poincaré's limit cycles and the above oscillatory phenomena (in tube generators, clocks, etc.). (Pechenkin 2002, 274)

Moreover, he regards Andronov as the founder of the self-oscillations concept.

> The concept of self-oscillations (Russian: *avtokolebaniia*[5]), that was elaborated by Andronov (1928, 1929a,b) and developed by him in collaboration with Mandel'shtam, Papalexi, Vitt and Khaikin, had important implications for the progress of nonlinear oscillation theory. (Pechenkin 2002, 270)

---

[4]In *Chaos et déterminisme* Simon Diner claimed that Andronov "noticed for the first time that in a radiophysics oscillator of the same type as Van der Pol's, which is a non-conservative (dissipative) system [...], the movement in the phase space is similar to the one in Poincaré's *limit-cycle*." (Diner 1992, 340).

[5]A study on the origin of this terminology's introduction by Andronov (1929a) will be carried out in the first paragraph.

In this second part, it will be established that this concept indeed takes root in Andronov's work,[6] and more specifically, in his first article on the subject (Andronov 1928), which was published before his famous note presented to the *C.R.A.S.* in Paris (Andronov 1929a), and for which the historiography only seems to have retained the connection with Poincaré's papers (1881–1886).

The analysis of his work in Chap. 5 will show that this only represents a minor step, whereas the link with Poincaré's works (Poincaré 1892, 1893, 1899) is a crucial step in the process of elaborating a nonlinear oscillation theory, since it represents the "first topological excursion[7]" in a field which was not yet called 'dynamical systems'. Pechenkin considered a reconstruction of this theory using three levels, or three strata:

> Three strata of theoretical reconstruction can be discerned in the self-oscillation concept. The first stratum is engineering: self-oscillations are self-maintained oscillations provided by the control influence on the energy source. As a result, the source operates in accordance with the oscillatory regime determined by the design of the self-oscillatory system. The second stratum is physical: self-oscillations are characterized by their stationary regime, which follows a transition stage of excitation. The important feature of self-oscillations is their stationary amplitude which does not depend on the initial conditions. The third stratum is physic-mathematical: self-oscillations have a stability determined by the system's property and not by initial conditions. (Pechenkin 2002, 271)

He credited Van der Pol for the main contribution at the second level, as well as attributing to him a consequent participation on the third level, he however recalled that he did not analyze the stability of the periodic solution[8] in his equation. He then added, in regard to the problem caused by Abraham and Bloch's *multivibrator*, that:

> It led to the relaxation concept, strongly nonsinusoidal self-oscillations, in which the "slow" and "fast" motions alternated. (Pechenkin 2002, 281)

By following Pechenkin's path, it will therefore be established in Chap. 4 that Van der Pol was the creator of the concept of relaxation oscillations corresponding to the evolution of a system depending on two characteristic time scales: one slow, one fast (see *supra* Part I), which will also play a crucial part in the development of nonlinear oscillation theory.

Chapter 6 will be dedicated to the reception of Andronov's and Van der Pol's work in France and in Europe, which will show that during the 1930s and 1940s, the French scientific community had diverging views concerning Andronov (1929a) and Van der Pol's results (1926d), which at times caused disputes between the concept of self-oscillations or self-sustained oscillations, and the concept of relaxation oscillations. While Andronov's work was simply assimilated, sometimes very discreetly, Van der Pol's work was analyzed, detailed, then bitterly debated.

---

[6] Andronov's thesis dissertation has not yet been located, therefore the information regarding this work will be presented rather than discussed in the first paragraph.

[7] See Chenciner (1985).

[8] It was noted in the first part of this work that until 1930, Van der Pol did not realize it was a limit cycle either. See *infra* Part I.

In January 1933, Van der Pol and Papalexi initiated the very first *Conférence Internationale du Non Linéaire* (*Non Linear International conference*) which was held in Paris. All the details (date, venue, participants) relating to this – until now – little-known event, and the aftermath it should have caused in the process of elaborating nonlinear oscillations theory will be analyzed in Chap. 7.

Some, such as Le Corbeiller (1931a, 1932, 1933a,b, 1934), first stood for the idea of a relaxation oscillation theory, of which Van der Pol would be the author, then recognized the connection established by Andronov as the starting point of a mathematical theory of nonlinear oscillations. The enthusiasm for relaxation oscillation, which was expressed as what could be called a "hunt for the relaxation effect[9]" will be analyzed in Chap. 8. Initiated by Van der Pol, it aimed at providing evidence of this concept in all scientific fields of study (biology, economy, hydrodynamics, acoustics, ...) as shown by Van der Pol's publications (1928b), Bedeau and Mare (1928), Fessard (1931), Hamburger (1931), Hochard (1933), Gause (1935), Eck (1936), Kostitzin (1936), Auger (1938), Bourrières (1939) or Parodi (1942a,b,c), Parodi et al. (1943).

Others such as Liénard (1928, 1931), Haag (1934a,b, 1936, 1937, 1943, 1944, 1952–1955), Rocard (1932, 1935a, 1937a, 1941, 1943) and even Le Corbeiller (1931a, 1932, 1933a,b, 1934, 1936) were in the end reluctant to accept the concept of relaxation oscillations, which they saw rather as a particular case of self-sustained oscillations or self-oscillations, and considered could not, by itself, constitute a theory.

[9]The "hunt for the butterfly effect" by analogy (Witkowski 1995) and "duck hunt" (Benoît et al. 1983).

Chapter 4
Van der Pol's Lectures: *Towards the Concept of Relaxation Oscillations*

The concept of *relaxation oscillations* was introduced in 1925 in an article written in Dutch (see *supra* Part I), then popularized by Balthazar Van der Pol (1926a,b,c,d, 1927, 1928a,b, 1930, 1931, 1934, 1938a,b) in numerous publications.[1] It originated in the oscillatory behaviour of a radio engineering device: Abraham and Bloch's multivibrator (1919a,b,c,e)[2] (see *supra* Part I). Indeed, for some parameter values the period of oscillations in this system, just as with the *series-dynamo machine*, the *singing arc*, the *three-electrodes valve* could not be deduced from Thomson's formula, but rather corresponds to the discharge time of a capacitor in a resistor,[3] called the relaxation time (see *supra* Part I). We therefore switch from a Thomson-type system for which the amplitude and period depend on the initial conditions to relaxation-type systems, for which the amplitude and period are completely independent of the initial conditions. The concept of a relaxation oscillation is strengthened by the ability of the system formalized by Van der Pol (1926a,b,c,d) to describe several apparently very different phenomena, for example heart beats and the oscillations of a triode. Moreover, a duality's existence, for the characteristic time scales ("slow" and "fast") specific to relaxation systems, causes them to fall within another concept, introduced almost simultaneously by Andronov (1928, 1929a,b): the *self-oscillation* concept. In the early 1930s these two concepts played a crucial part in the development of nonlinear oscillation theory.

In his first publications on relaxation oscillations, Van der Pol (1926a,b,c,d) introduced a prototype for a second-order nonlinear differential equation, which he first associated to the *triode*, before its dedimensionalization. He established graphically that the periodic solution of this equation possesses a period that, for

[1]This chapter originally written in 2011 by JM Ginoux has been summarized in a publication entitled: Ginoux, J.M. and Letellier, Ch. Van der Pol and the history of relaxation oscillations: Toward the emergence of a concepts, Chaos 22, 023120 (2012).

[2]It was established in Part I that this device was actually invented in December 1917.

[3]As shown by Abraham and Bloch (1919a, 257) then Van der Pol (1926d, 988).

J.-M. Ginoux, *History of Nonlinear Oscillations Theory in France (1880–1940)*, Archimedes 49, DOI 10.1007/978-3-319-55239-2_4

some values of a parameter, does not conform any more to Thomson's formula, but is expressed as a relaxation time. Van der Pol (1926d, 987) then demonstrated that it also made it possible to describe oscillations occurring in Abraham and Bloch's *multivibrator*.[4] In order to establish a conceptual aspect of this new type of oscillation, Van der Pol attempted, in the conclusion of his articles,[5] to spread it to other phenomena. He therefore wrote in his very first contribution in Dutch on relaxation oscillations:

> Zij komen waarschijnlijk op verschillende gebieden in de natuur voor, zooals b.v. bij de Wehnelt-onderbreker, bij tal van gasontladingen enz.[6] (Van der Pol 1926a, 157)

In his second publication, he recalled that these oscillations occur in the Wehnelt interrupter, and added:

> Ook een afzonderlijk bekrachtigde motor, gevoed door een met constante snelheid draaiende serie-dynamo voert relaxatie-trillingen behoort en wellicht ook de hartslag.[7] (Van der Pol 1926b, 39)

It therefore appears that in his second contribution, Van der Pol had already hypothesized that heartbeats could be represented as relaxation oscillations. In the third version written in German, he used these same examples:

> Des weiteren ist es wahrscheinlich (obwohl es nicht genau geprüft worden ist), daβ die Schwingungen eines "Wehnelt"-Unterbrechers und vielleicht auch der Herzschlag der allgemeinen Klasse von Relaxationsschwingungen angehören. Auch ein von einem mit konstanter Geschwindigkeit drehen Seriedynamo gespeister fremderregter Motor führt Relaxation schwingungen aus, d.h. der Motor ändert selbsttätig seine Drehungsrichtung [...][8] (Van der Pol 1926c, 184)

However, in his fourth and now famous contribution in English, he removed the *series-dynamo machine* example, illustrating his point with the oscillations occurring in the neon-tube case, thus implicitly referring to Righi's work 1902:

> Finally it seems quite likely that, when the total characteristic (including the parts with negative slope) is taken into account, the well-known vibration of a neon-tube connected to a resistance and condenser in shunt may be similarly treated under the heading of relaxation-oscillations. Similarly, (though no detailed investigation has been carried out) it is likely that the oscillations of a "Wehnelt" interrupter belong to the general relaxation-oscillations classification and perhaps also heart-beats. (Van der Pol 1926d, 992)

---

[4] *Ibid.*

[5] It was previously noted that the articles' conclusion were different. See *supra*, Part I.

[6] "These oscillations are probably present but different in various fields in nature, such as the Wehnelt interrupter, in the gas expansion phenomenon, etc." (Van der Pol 1926a, 157).

[7] "A separate excitation motor powered by a series-dynamo machine turning at a constant speed also produces relaxation oscillations as well as, probably, the heartbeats." (Van der Pol 1926b, 39).

[8] "Moreover, it is possible (although it has not been examined in detail) that the oscillations in a Wehnelt circuit-breaker, and maybe even in the heartbeats, belong to the broad classification of Relaxation oscillations. A motor powered by a D.C. series-dynamo also produces Relaxation oscillations, that is to say, that the direction of the motor changes periodically (...)." (Van der Pol 1926c, 184).

At the end of 1926, Van der Pol therefore recalled three times that heartbeats fall under the relaxation oscillation concept. The next year, in two articles in Dutch and German following his first work on relaxation oscillations, Van der Pol (1927b,c) carried out the deduction of the equation for the oscillations of the *series-dynamo machine* (see *supra* Part I) and noted the neon tube example, this time removing the *Wehnelt interrupter*.[9]

4.1 Conferences in France (1928–1937)

From 1928, Van der Pol regularly went to France. He was invited to make a presentation on the results of his research on relaxation oscillations, which were

- on the 24th of May, 1928 in the *Salle de la Société de Géographie*,
- on the 10th and 11th of March 1930 at the *École Supérieure d'Électricité*,
- between the 5th and 12th of July 1932 at the *Congrès International d'Électricité*,
- in October 1937 at the *Congrès du Palais de la Découverte* during the *Réunion Internationale de Physique-Chimie-Biologie*.

He attended the very first *Conférence Internationale de Non linéaire*[10] as an organizer, which was held between the 28th and 30th of January 1933 at the Henri Poincaré Institute. He was also member of the organizing committee for the *Colloque International des Vibrations Non linéaires* which was held between the 18th and 21st of September 1951 on the *island of Porquerolles* (Var, France). During his presentations[11] Van der Pol did his best in order to ensure the rising of the relaxation oscillations concept, which he authored.

4.1.1 Presentation on the 24th of May 1928 at the Société de Géographie

During the year 1928, Van der Pol started his research on heartbeats[12] collaboratively with Jan Van der Mark[13] (1893–1961). This work was originally published in Dutch, then French and English.

---

[9]It is an electrolytic interrupter in which current's flow generates small oxygen bubbles around the electrode, which stops the current's flow, and then disappear, re-establishing it.

[10]See *infra*.

[11]These lectures were inventoried through his complete bibliography (Van der Pol 1960, vol. 2,1331–1339).

[12]For an analysis of Van der Pol and Van der Mark's work on heartbeats, see Israel (1996, 34) and Israel (2004, 14).

[13]There are few biographical elements regarding Van der Mark. See International Telecommunication Union, vol. 28, (1961) 135 (in English), 159 (in French) or 183 (in Spanish).

Fig. 4.1 Société de Géographie, 184 bd St Germain, Paris (source : http://www.socgeo.org/images/facade.jpg)

- De hartslag als relaxatietrillingen en een electrisch model van het hart, *Werk. Genoot. Nat.-, Genees- en Heelk.* **12** (1928), p. 614–618
- Le battement du cœur considéré comme oscillations de relaxation, *Onde Électrique*, **7** (sept. 1928), p. 365–392.
- The Heartbeat considered as a Relaxation oscillation, and an Electrical Model of the Heart, *Philosophical Magazine Supplement*, **6** (1928), p. 763–775.

While these versions all appear similar, if not identical, Van der Pol's publication (1928a) in French is the most interesting on several levels. Firstly, it is the transcription of a lecture Van der Pol held on the 24th of May 1928 in the *Salle de la Société de Géographie* (see Figs. 4.1 and 4.2), 184 boulevard Saint Germain, under the presidency of General G. Ferrié, invited by the *Société des Amis de la T.S.F.*, the *Société Française des Électriciens* and the *Société de Biologie*.

Following this, Van der Pol's short speech (transcript below), which was a preamble to his presentation, shows how he wished to settle his discovery's legitimacy, in the field of *Wireless Telegraphy* as well as Cardiology.

Fig. 4.2 Amphitheater of the Société de Géographie, 184 bd St Germain, Paris (The photography was acquired thanks to the kind permission of President Jean Robert Pitte and with the assistance of Mrs. Sylvie Rivet, Administrative Director)

> I deeply appreciate being honoured to talk in front of you this evening, under the auspices of the *société des Amis de la T.S.F.*, the *Société française des Électriciens* and the *Société de Biologie*. I owe this honour firstly to the kind invitation of Mr. René Mesny, secretary general of the *société des Amis de la T.S.F.*, whose work is well-known and strongly admired in the Netherlands, and whose judgement I personally find invaluable. He is also the one who accepted the responsibility of drawing attention to my work at the famous *Société française des Électriciens*, and I give my thanks to their president Mr. Imbs, for granting me his support during this session. Lastly, to the presence of the Bureau of members of the société de biologie, being there thanks to the interest that the distinguished master of cardiac science, Professor Vaquez, has shown to such a succinct account concerning the first results we obtained on the subject (Van der Pol and Van der Mark 1928a, 365)

At the end of the 1920s, René Mesny, who had been working at the French Naval ministry at the *Établissement Central du Matériel de la Radiotélégraphie militaire* (E.C.M.R.) (see *supra* Part I), and who was then appointed head of the *Laboratoire National de Radioélectricité* (National Laboratory of Radio Electricity), distinguished himself by becoming an emblematic individual in the field of radio transmission in France.

He invented a two-triode-lamp symmetric push-pull oscillator that allowed transmissions using short waves[14] (Mesny and David 1923; Mesny 1924, 1927),

[14]On 28 November 1923 the first transatlantic short wave link was established between Hartford (Connecticut) and Nice.

for which the principle displays all the characteristics of a self-sustained oscillator, as was noted by Édouard Cliquet, who devoted a full chapter to self-oscillating assemblies in the first volume of his work entitled *Émetteurs de petite puissance sur ondes courtes* (*Low voltage emitters on short waves*):

> MESNY's circuit diagram is, from a schematic point of view, a push-pull symmetric oscillator of the Reversed Feed Back type. (Cliquet 1949, t. 1, 147)

Nevertheless, it seems that Van der Pol's tribute to Mesny was limited to being a polite salutation, since in his Selected Scientific Papers (Van der Pol 1960, vol. 1 & 2) before 1928, his work is apparently mentioned only once.[15] The same is true for Louis Henri Vaquez (1860–1936), *professeur agrégé* at the *Faculté de Médecine* in Paris, elected at the *Académie de Médecine*. This cardiology specialist[16] wrote numerous fundamental research projects in this field, the main ones being: *Les troubles du rythme cardiaque* (*Cardiac dysrhythmia*) (Vaquez and Donzelot 1926) and *Les Maladies du cœur* (*Heart pathologies*) (Vaquez 1928). A bibliographic study of these projects, however, did not reveal any references to Van der Pol's work.[17] Moreover, during the various lectures given by Van der Pol in France, he made no other mention of either Mesny or Vaquez.

In the next part of his presentation, Van der Pol gave a definition of what he considered a relaxation oscillator to be, when it actually was a self-oscillator, as later defined by Andronov (1929a) (see *infra*).

> But in the electrical engineering field, and especially in the radiotelegraphy field, we still have at our disposal resistances with negative characteristics. Instead of dissipating energy, these resistances can *provide* energy. Consequently, it is obvious that they are only present in systems comprising an energy source. (Van der Pol and Van der Mark 1928a, 367)

He explained that there are devices in the field of radio engineering in which the presence of a negative resistance results in the providing of energy to the system, and therefore the sustaining of oscillations instead of their damping. He then added:

> Likewise, the electric oscillations produced by an arc transmitter occur thanks to the negative resistance of the arc. (Van der Pol and Van der Mark 1928a, 367)

With this analogy, he demonstrated his perfect knowledge of the operating principle of the arc transmitter, i.e. of the singing arc. This incidentally suggests that he could have read Blondel's work (1905c; 1919d), which he referred to in his second lecture in France (see *infra*). After he recalled that if in a classic circuit RLC, the resistance (positive) is replaced by a negative resistance, the amplitude of the oscillations will increase indefinitely, he therefore offered to express the resistance as a function of the amplitude "as it swaps its sign when the amplitude is greater

[15] In an article published in the *Onde Électrique*, Van der Pol and Posthumus (1925, 536) mentioned Dufour and Mesny's articles (1923).

[16] He is also known for his work on hematology. He indeed discovered and described the blood disorder *polyglogulie vera*.

[17] There are no mentions in Van der Pol's Selected Scientific Papers (1960, vol. 1 & 2) of Vaquez's works.

than a specific constant value determined by the nature of the system" (Van der Pol and Van der Mark 1928a, 368). In the differential equation characterizing the oscillations of a circuit RLC:

$$L\frac{d^2v}{dt^2} + R\frac{dv}{dt} + \frac{1}{C}v = 0$$

he wrote $\alpha = \frac{R}{L}$ and $\omega^2 = \frac{1}{LC}$. Then, he substituted this time (see *supra* Part I) α with the expression $\alpha\left(1 - v^2\right)$ and obtained a variant of (4.2), which is not dedimensionalized either:

$$\ddot{v} - \alpha\left(1 - v^2\right)\dot{v} + \omega^2 v = 0 \tag{4.1}$$

Surprisingly, the approach Van der Pol used to introduce the expression $\alpha\left(1 - v^2\right)$ in the differential equation is undoubtedly a copy of Lord Rayleigh's (1883, 229–230), even if he did not refer to it. Van der Pol then distinguished two cases:

- the case $\alpha^2 \ll \omega^2$ which corresponds to sinusoidal oscillations,
- the cas $\alpha^2 \gg \omega^2$ which corresponds to relaxation oscillations.

The first case represents the *limit operating regime* for which the specific values of the characteristic equation of the differential equation are purely imaginary. In other words, the undamped oscillatory state that Blondel (1919d, 132) had called the "*aeolian regime*[18]". The study of the second case allowed him firstly to demonstrate existence of a new oscillatory state which did not belong in Curie's classification (see *supra* Part I).

> Here, the initial resistance, while negative, is so large that, if it was constant and independent from the amplitude, the *system would be very much aperiodic*. (Van der Pol and Van der Mark 1928a, 369)

By comparing Van der Pol's parameter α^2/ω^2 to Curie's parameter, we indeed find the latter's results: "for $n_a > 1$, slower and slower aperiodic motion when n_a approaches ∞." (Curie 1891, 209). Then, Van der Pol defined the relaxation oscillations far more precisely. He indeed explained that with the condition $\alpha^2 \gg \omega^2$:

> (…) the system first tends to jump from the value zero to a positive value, then gradually decreases, then suddenly jumps to a negative value, and so on and so forth. (Van der Pol and Van der Mark 1928a, 369)

A few pages further on, he indicated, regarding the form of these oscillations:

> (…) it is characterized by discontinuous jumps occurring every time the system becomes unstable. (Van der Pol and Van der Mark 1928a, 372)

[18]"By reminiscence of the aeolian harps, instruments are always ready to vibrate under the god of wind, Aeolus's breath." (Blondel 1919d, 132).

This notion of "jumping", which had already been observed by Witz (1889b, 585) (see *supra* Part I) led Mandel'shtam and Papalexi, a few years later, to suggest the "jump hypothesis[19]", also called a "discontinuity condition[20]" (Mandel'shtam et al. 1935, 97), "Mandel'shtam condition" (Minorsky 1947, 394) or "Mandel'shtam postulate" (Minorsky 1967, 82) (see *infra*). This hypothesis allowed Andronov and Witt (1930b) to explain the relaxation oscillation phenomenon described by Van der Pol in the case of the triode tube, but also for Abraham and Bloch's multivibrator.

After having recalled that in the second case, the period of the oscillations cannot be provided by Thomson's formula anymore but rather by a relation that he established himself (see *supra* Part I), Van der Pol finally clarified the definition and origin of this new concept:

> Consequently, the period of these new oscillations in this specific electric case is determined by the *discharge time of a capacitor*, sometimes called *relaxation* time, and we will consequently call this type of oscillations *relaxation oscillations*. (Van der Pol and Van der Mark 1928a, 369)

It should be noted that on the one hand, this definition is somewhat late, since it took place two years after his first articles on the subject. On the other hand, Van der Pol seems to use, maybe even overuse the analogy, since time – a term he used inaccurately, although it had been perfectly defined by Maxwell (1867, 56) – corresponds to an electrical phenomenon. Yet, in the case of mechanical or biological relaxation oscillations, it does not seem fully justified. We can however object that it was introduced by Maxwell in the context of Thermodynamics, not an electrical one. It therefore seems that the word *relaxation* was used here with a broader meaning, which is more related to a *return to equilibrium*. In order to give a conceptual value to his *relaxation oscillations*, Van der Pol offered and exhaustive list of examples:

> Hence, in the aeolian harp case, as with the case of the wind blowing through the telegraph cables, producing a whistling sound, the period of the sound is determined

This reference to the aeolian harp is of great importance, since it is precisely one of the examples chosen by Lord Rayleigh, who considered that the sustained oscillations could be sorted into two classes: the sustained oscillations, in the contemporary sense of self-sustained oscillations, and the forced oscillations.

> The first class is by far the more extensive, and includes vibrations maintained by wind (organ-pipes, harmonium-reeds, aeolian harps, &c.), by heat (singing flames, Rijke's tubes, &c.), by friction (violin-strings, finger glasses, &c.), as well as the slower vibrations of clock-pendulums and of electromagnetic tuning-forks. (Strutt and Rayleigh 1883, 229)

[19]See Gaponov-Grekhov and Rabinovich (1979, 599).

[20]The term discontinuity seems to be however the translator's interpretation, as he understood it as jump discontinuity.

Van der Pol then resumed his enumeration:

> We can cite numerous other instances of relaxation oscillations, such as: the pneumatic power hammer, the grinding sound of a knife on a plate, a flag flapping in the wind, the buzzing that sometimes comes from a tap, the creaking of a door, the double-acting steam motor with an insufficient wheel, Mr. Janet's experiment on a series-dynamo, Abraham and Bloch's multivibrator ([1]), the bi-grid valve multivibrator ([2]), the periodic sparks obtained by turning a Whimshurst machine, the Wehnelt interrupter, the imminent discharge of a capacitor through a neon tube ([3]), the periodical reappearance of epidemics, and economical crises, the periodical variation of an even number of animal species, living together, with some species being food for others ([4]), flowers' sleep, the periodical appearance of rain after a depression, shivers that come from the cold, menstruation, and finally, heartbeats ([5]).
>
> ([1]) Abraham et Bloch, *Annales de Physique*, **12**, 237, 1919.
> ([2]) Balth. Van der Pol, *Phil. Mag.*, **51**, 991, 1926[21]
> ([3]) Schallreuter, *Ueber Schwingungserscheinungen in Entladungsröhren* (Wieweg), 1923.
> ([4]) Volterra, *Comptes rendus de l'Académie des Lincei*, 1926.
> ([5]) The point of view according to which the oscillations of the heart belong to the relaxation type of oscillation was first expressed two years ago. See Balth. Van der Pol, *Phil. Mag.*, **51**, 992, 1926[22]
>
> In all these examples, the frequency of these periodical phenomena is not determined by the product of elasticity and mass, but rather by some form of relaxation time. (Van der Pol and Van der Mark 1928a, 371)

This list covers almost all of the paragons that Van der Pol (1926a,b,c,d) had introduced in his previous publications: the series-dynamo machine, which he called Mr. Janet's experiment from then on, Abraham and Bloch's multivibrator, the neon tube, Wehnelt's interrupter, and the heartbeats. Van der Pol seems to show an undeniable wish to demonstrate that this new type of oscillation is not limited to these few examples. He also added a series of phenomena to his list in order to show that the relaxation oscillations are much more than a new type of oscillation, and represent a new type of phenomenon omnipresent in nature, in the same way as harmonic oscillations. With this generalization, a daring one since he did not know if it would later be possible or not to demonstrate this type of oscillation in these various phenomena, Van der Pol initiated, in some way, the start of the "hunt for relaxation effect". Some of these examples, such as the periodic reappearance of economic crises, were later recognized as possibly being the seat of relaxation oscillations, and were, contrary to what Israel stated (1996, 42), the subject of numerous studies (Hamburger 1930, 1931, 1934; Le Corbeiller 1933a,b; Rocard 1941; Goodwin 1949, 1951). As for periodical sparks produced by a Wimshurst machine, it seems that Van der Pol referred here to a device which had previously been studied by Janet (1893): the contra-rotating-disk machine (see *infra*). The case of the coexistence of animal species is trickier. As noted by Israel (1996, 42): "this reference has no basis, since Voltera's classic model do not present limit cycles." (see *supra* Part I). Indeed, it was only after Gause (1935), then Kolmogorov (1936)

[21] This reference is not correct. Balth. van der Pol, *Phil. Mag.*, Ser. **7**, vol. 2, issue 11, p. 978–992.
[22] *Idem.*

introduced a limitation in the functional response transcribing the predation, that it became possible experimentally, and then theoretically, to obtain limit cycles in predator-prey types of models. As for the other examples, we can agree with Israel (1996, 42), that they remain fantasies. Van der Pol and Van der Mark (1928a, 372) then addressed the issue of the frequency synchronization and demultiplication (see *infra* Part III), and described the schematic dynamic model of the heart (see Israel 1996, 34, 2004, 14). Before proceeding any further, it must be noted that the English version of this presentation that Van der Pol and Van der Mark (1928b) published in November 1928 is only a translation of this lecture.

4.1.2 Lectures on the 10th and 11th of March at the École Supérieure d'Électricité

These conferences, given barely a few months after Andronov (1929a) published a note in the *C.R.A.S.* where he established a connection between the periodic solution of a self-oscillator of Van der Pol's type and the concept of limit cycle introduced by Poincaré in his papers, stand out on several levels. Firstly, by the mathematical accuracy with which Van der Pol presented his lectures from then on.

4.1.2.1 First Conference

If the starting point of his lecture, which consisted in considering the differential equation for a damped oscillator (RLC), stayed the same, Van der Pol (1930, 246) clarified then that it is a "second-order differential equation", and explained the solution. Likewise, even though he replaced, without any reference, the angular frequency ω "by the classic formula $\omega^2 = 1/CL$" (Van der Pol and Van der Mark 1928a, 366) and wrote: $\alpha = R/L$, he indicated from then on that it was Lord Kelvin[23] who introduced these changes of variables. The condition $\alpha^2 \ll \omega^2$, corresponding to what he had previously called the damped sinusoidal discharge of the capacitor, became the oscillating discharge of the condenser (Van der Pol 1930, 247) and was expressed by the negativity condition of the discriminant of the characteristic equation for the differential equation, i.e. $\alpha^2/4 \ll \omega^2$. He also recalled that if the sign of the resistance (positive) is reversed by a negative resistance the amplitude of the oscillations increases indefinitely. He then explained, using a diagram (see Fig. 4.3), that he did not justify, experimentally or theoretically, that:

> The reason why the amplitude of the oscillations in a system with a negative resistance does not become infinite, is that the *instantaneous resistance R* does not keep the same value for any amplitude, but is on the contrary a *function of the amplitude*. This means that

[23]See William Thomson and Kelvin (1853, 393) alias Lord Kelvin.

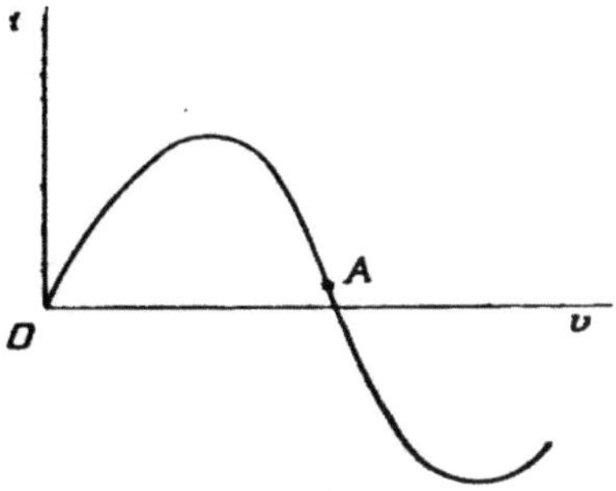

Fig. 4.3 Current voltage characteristic of a triode, from Van der Pol (1930, 249)

> the current-voltage characteristic of the dynatron[24] is not a straight line (Fig. 4.3), and that consequently the instantaneous resistance
>
> $$dv/di$$
>
> varies from one point to the other. (Van der Pol 1930, 250)

He then replaced again α with the expression $\alpha - 3\gamma x^2$, where γ is a constant, and obtained (4.2):

$$\ddot{v} - \left(\alpha - 3\gamma v^2\right)\dot{v} + \omega^2 v = 0 \tag{4.2}$$

He then carried out the same variable substitution as he used in 1926 (see *supra* Part I) and obtained the equation (4.3), which now depends on only one parameter:

$$\ddot{y} - \varepsilon\left(1 - y^2\right)\dot{y} + y = 0 \tag{4.3}$$

He added to this Eq. (4.3) "the next equation which generalizes it" (Van der Pol 1930, 251)

$$\ddot{y} + \varepsilon\psi\left(y\right)\dot{y} + y = 0 \tag{4.4}$$

where $\psi(y)$ is an arbitrary function. He recalled that although this equation (4.4) is non-integrable in the case of $\varepsilon^2 \gg 1$, "it is possible to find and discuss some of its general properties" (Van der Pol 1930, 251) and afterwards referred to Janet's work (1925), Cartan (1925), Liénard (1928) and Andronov (1929a). It clearly appears that the equation (4.4) is none other than Liénard's (see *supra* Part I, Eq. (3.20)), which could hint at the latter's presence during Van der Pol's lecture. Moreover, in the following pages, Van der Pol did his utmost to explain the solution de (4.4) in the integrable case $\varepsilon^2 \ll 1$ up to order one in ε. In his article, Liénard (see *supra* Part I) noted that the solution then proposed by Van der Pol (1926a,b,c,d) satisfied

[24] Audion or triode.

the equation (4.3) "only if the first order terms are neglected" (Liénard 1928, 948). After several lines of calculation Van der Pol (1930, 254) obtained the following expression:

$$y = \frac{2\sin(t+\varphi)}{\sqrt{1 + Ce^{-\varepsilon t} - \frac{\varepsilon}{2}\sin 2(t+\varphi)}} \tag{4.5}$$

By comparing this expression with Liénard's (1928, 948):

$$y = \frac{2\sin\omega t}{\sqrt{1 + Ke^{-k\omega t}}}\left[1 + k\frac{\sin\omega t\cos\omega t}{2\left[1 + Ke^{-k\omega t}\right]}\right] \tag{4.6}$$

we can first see that they are completely different. But by noting that with Van der Pol $\omega = 1$ and with Liénard $\varphi = 0$; $\varepsilon = k$ and $C = K$, then by using a limited Taylor expansion up to order one in ε of the denominator of the equation (4.5) with $\varphi = 0$ we have:

$$\frac{2\sin(t)}{\sqrt{\left(1 + Ce^{-\varepsilon t}\right)\left(1 - \frac{\varepsilon}{2}\frac{\sin 2(t)}{1+Ce^{-\varepsilon t}}\right)}} \approx \frac{2\sin(t)}{\sqrt{\left(1 + Ce^{-\varepsilon t}\right)}}\left[1 - \frac{\varepsilon}{2}\frac{\sin 2(t)}{1 + Ce^{-\varepsilon t}}\right]^{-\frac{1}{2}}$$

$$= \frac{2\sin(t)}{\sqrt{\left(1 + Ce^{-\varepsilon t}\right)}}\left[1 + \frac{\varepsilon}{4}\frac{\sin 2(t)}{1 + Ce^{-\varepsilon t}}\right]$$

By replacing $\sin 2(t)$ by $2\sin(t)\cos(t)$ we finally obtain:

$$\frac{2\sin(t)}{\sqrt{1 + Ce^{-\varepsilon t} - \frac{\varepsilon}{2}\sin 2(t)}} \approx \frac{2\sin(t)}{\sqrt{\left(1 + Ce^{-\varepsilon t}\right)}}\left[1 + \frac{\varepsilon}{2}\frac{\sin(t)\cos(t)}{1 + Ce^{-\varepsilon t}}\right]$$

which yields a perfect identity between the expressions (4.5) and (4.6). This demonstration can seem somewhat "useless", especially since, as Liénard remarked:

> knowing this term does not really add anything essential to Van der Pol's solution. (Liénard 1928, 948)

It seems, on the one hand that it was established as a response to Liénard, which would corroborate the idea that he could have been present at Van der Pol's lecture. On the other hand, he might have incorporated it in order to show more mathematical rigor, which had been lacking in Van der Pol's previous publications, as he himself deplored:

> In 1920, I gave a much shorter, but certainly less rigorous, explanation of the growth of the amplitude from the zero value to the limit value *two* (…) (Van der Pol 1930, 255)

Van der Pol then suggested, in order to justify this amplitude limitation, using a biological analogy. He therefore compared the saturation of the amplitude in a

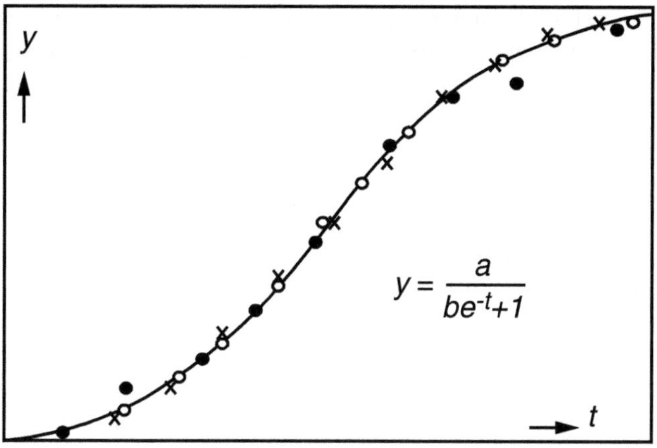

Fig. 4.4 Growth of the square of the amplitude, from (Van der Pol 1930, 255)

triode to the *logistic growth*[25] of a population of *drosophila* (fruit flies) and of an "heliotrope" (sunflower) (see Fig. 4.4).

> It is very remarkable that the equation
>
> $$A^2 = \frac{a}{be^{-t} + 1}$$
>
> which represents the growth of the square A^2 of the amplitude in a slightly more general form, can also be found in biology (²).
>
> (²) V. par ex. LOTKA, Elements of Physical Biology, Baltimore 1925, p. 66. (Van der Pol 1930, 255)

Van der Pol's approach is quite interesting when compared to the one used by Vito Volterra in his article titled: "Variazioni e fluttuazioni del numero d'individui in specie animali conviventi" (Volterra 1926) published the same year as Van der Pol's (1926a,b,c,d). In this article, Volterra (1926, 74) indeed studied the case of an isolated population for which the growth rate depends on the number of individuals. In the exponential growth model of a population with N the number of individuals expressed by Malthus's law of population:

$$\frac{dN}{dt} = \varepsilon N$$

the growth coefficient ε (assumed positive) is constant. But according to this law, the growth would be infinite, and any given population could cover the whole surface of

[25] See Pierre-François Verhulst (1838).

the Earth.[26] Therefore, to limit the growth, Belgian mathematician Pierre-François Verhulst, in 1838, thought of introducing a limitation in the population growth. In this case, the growth coefficient ε is replaced by $\varepsilon - \lambda N$. The Verhulst law (1838) is then written:

$$\frac{dN}{dt} = \varepsilon N - \lambda N^2$$

The second term, limiting the growth, can be interpreted in different ways. For some biologists, the interaction of a part N of the population with another part N corresponds to cannibalism, for others, it represents the struggle for existence, which would be analogous to "a kind of "friction[27]" inside the population" (Israel 1996, 90). The logistic growth law, which continued to be regarded as valid until the 1930s, was applied by Verhulst to symbolize the evolution of demographic growth in France and Belgium, then by Pearl and Reed (1920) for the population of the United States, with great statistical agreement.[28] Volterra (1926, 74) integrated Verhulst's equation, and obtained:

$$N = \frac{C\varepsilon e^{\varepsilon t}}{1 + C\lambda e^{\varepsilon t}}$$

Multiplying the numerator and denominator of this equation by $e^{-\varepsilon t}$ yields:

$$N = \frac{C\varepsilon}{e^{-\varepsilon t} + C\lambda}$$

Factoring at the denominator $C\lambda$ then simplifying by C it yields:

$$N = \frac{\varepsilon/\lambda}{1 + e^{-\varepsilon t}/C\lambda}$$

By writing $a = \varepsilon/\lambda$ and $b = 1/C\lambda$, we find the equation that Van der Pol (1930, 256) wrote on Fig. 4.4. In this diagram, he superimposed the amplitude saturation of the triode by using small white circles, whereas the black circles represent growth of the *drosophila* and the black crosses, growth of the sunflower. Even if we admit that the analogy can represent these three completely different phenomena, their association can seem surprising, as Van der Pol himself admitted:

> The deep similarity between these phenomena, which differ physically, but are mathematically analogous, cannot be denied. (Van der Pol 1930, 256)

[26]Metaphor used in Vladimir A. Kostitzin and Jean Painlevé's film, titled: "Mathematical images of the Struggle for Life", 1937, *Médiathèque du Palais de la découverte*, Paris. This invaluable document was found by Prof. Giorgio Israel. See *infra*.

[27]The Newton-type friction force is proportional to the square of the speed.

[28]See Israel (1996, 22 and 90).

This analogy is rather surprising on several levels. Firstly, it took almost four years for Van der Pol to publish an experimental representation of the triode's amplitude evolution during the time that could have been used a priori as a starting point for modeling of the triode's characteristic with a cubic function. Secondly, he admitted a posteriori that this evolution behaves perceptibly like the solution to Verhulst's equation, in which the saturation is represented not by a cubic but by a quadratic expression. It is however important to note that in the 1920s a new mode of research emerged, which was not yet called modelling. This is evidenced by numerous papers in a wide variety of fields, such as the work of Pearl and Reed (1920) in demography, Lotka (1925) in chemical kinetics, or Volterra (1926) in population biology. It is important to note that Volterra was invited by Émile Borel (1871–1956) to present his results during the winter of 1928–1929, in a series of lectures given at the Henri Poincaré Institute. In his first lecture on the 10th of March, 1930 Van der Pol thus appeared to be trying to give his work some mathematical legitimacy and a new-found rigour, but also to benefit from the keen interest for this new emerging research method: the modeling method. He finished his lecture by recalling that the equation he had introduced could also be the subject of forced oscillations, giving him the opportunity to use the expression *nonlinear*.

4.1.2.2 Second Lecture

During the second conference, on the 11th of March 1930, Van der Pol (1930, 293) covered most of the process he followed in his previous publications on relaxation oscillations (Van der Pol 1926a,b,c,d), although this time he added the subject of Figs. 3.2, 3.3 and 3.4 (see *supra* Part I):

> We can see a closed integral curve on each of these three figures. This is an example of what Henri POINCARÉ called a *limit cycle* ([1]), since the integral curves approach it asymptotically.
>
> ---
>
> ([1]) See for instance: *A. Andronow*, Les cycles limites de Poincaré et la théorie des oscillations auto-entretenues, C. R. **189**, 559 (1929). (Van der Pol 1930, 294)

As previously noted, Van der Pol reacted quickly. Less than six months after Andronov's publication (1929a), Van der Pol took note of his result, which he immediately integrated into his own results in order to improve them. However, we can see that while he referred to Poincaré, he only did it indirectly. During his lecture, as in the previous one, Van der Pol made cautious progress into the territory of mathematical rigor, as shown by this sentence about the period of the oscillations:

> I ask for permission to keep, in order to refer to this time constant, the phrase "relaxation time", which I usually use, and to call this type of oscillation which widely differs from the sinusoidal form, and for which the period is a relaxation time, "*relaxation oscillation*". (Van der Pol 1930, 299)

It seems that Van der Pol was somewhat cautious, even defensive, in front of this audience, since he asked to be allowed to use the terminology he had introduced in 1926 (Van der Pol 1926d, 987) and justified in 1928 (Van der Pol and Van der Mark 1928a, 369), saying that it was simply for personal convenience and by habit. He nevertheless kept his aim, which was to establish the *relaxation oscillation* form as a new concept able to describe apparently widely different phenomena governed by a same generic equation: the equation (4.3).

> As we come across discharge phenomena in various fields of physical and natural sciences, it is clear that the corresponding relaxation oscillations, which have a period depending on a relaxation time for which we have provided the prototype in the previous considerations, must also be very widespread. (Van der Pol 1930, 299)

He then completed the definition of what he considered to be relaxation oscillations, which were in fact self-oscillations as defined by Andronov (1928) (see *infra*).

> A relaxation oscillation generally occurs any time a mechanism, containing a continuous-energy source, allows an essentially aperiodic phenomenon to repeat itself for an indefinite number of times. (Van der Pol 1930, 300)

Van der Pol then resumed his enumeration of relaxation oscillations examples (see *supra*) by clarifying some and adding others.

> In the field of electricity, we have very nice examples of relaxation oscillations, some of which are very old, such as the spark discharge of a Wimshurt disk machine, the oscillation of the electric arc studied by Mr. BLONDEL, in a famous dissertation ([1]) or Mr. M. JANET's experiment, as well as others, more recent, such as ABRAHAM and BLOCH's multivibrator ([2]) which consists in an amplifier with a two-stages capacitor and resistance coupling, for which output terminals are connected to input terminals.
>
> ---
>
> ([1]) BLONDEL, *Eclair. Elec.*, **44**, 41, 81, 1905. V. aussi *J. de Phys.*, **8**, 153, 1919.
> ([2]) ABRAHAM et BLOCH , *Ann. de Phys.*, **12**, 237, 1919. (Van der Pol 1930, 20)

The example of the electric arc has already been discussed (see *supra* Part I). It should however be noted that this was Van der Pol's first reference to this device. The French expression "machine à plateau" ("rotating-disks machine") referred to both the Wimshurt machine, to which (Van der Pol and Van der Mark 1928a, 371) already referred, but also to a rotating circuit-breaker used by Janet (1893) (see Fig. 4.5) in order to study medium-period electric oscillations. This rotating-disks machine, similar to Wimshurt's, and which was later used by Abraham and Bloch (1919e, 250) (see Fig. 4.6) with the name *microalternator* to calibrate a tuner, might be the foundation of relaxation oscillations.

The rotating-disks machine (Fig. 4.5) and the microalternator (Fig. 4.6) seem to be similar, if not in the operating principle, at least in the way they periodically produce sparks. This implies that they might be the foundation of relaxation oscillations, as hinted at by Le Corbeiller (1933b, 329). Janet's experiment and the multivibrator, as well as the oscillations of a neon tube, then became the subject of a detailed analysis by Van der Pol, in order to evidence the characteristic from of relaxation oscillations, by using oscillograms. He thus presented the following curves:

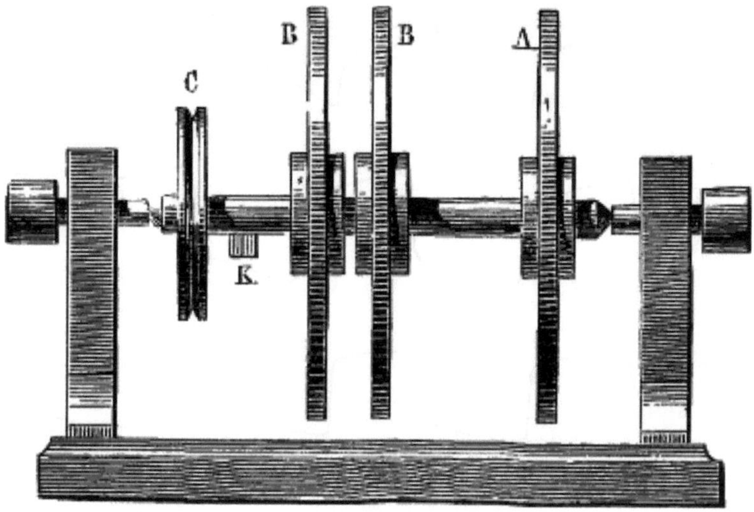

Fig. 4.5 Rotating-disks machine, from Janet (1893, 383)

Fig. 4.6 Microalternateur, from (Abraham and Bloch 1919e, 250)

On Fig. 4.7 Van der Pol drew the dynamo's oscillogram with a periodically-reversing polarity (1), the theoretical curve representing the relaxation oscillations (2), and the oscillogram of the current in a multivibrator (3). He then made an important breakthrough, by stating:

> The great likeness of these three curves clearly shows that the equation (V9) is sufficient to represent these two phenomena, so different at first glance (...) (Van der Pol 1930, 303)

Indeed, by preceding what we nowadays call "topological equivalence" by almost fifty years, Van der Pol considered that the analogy presented by the time series of these devices demonstrates that they are governed by the same equation.

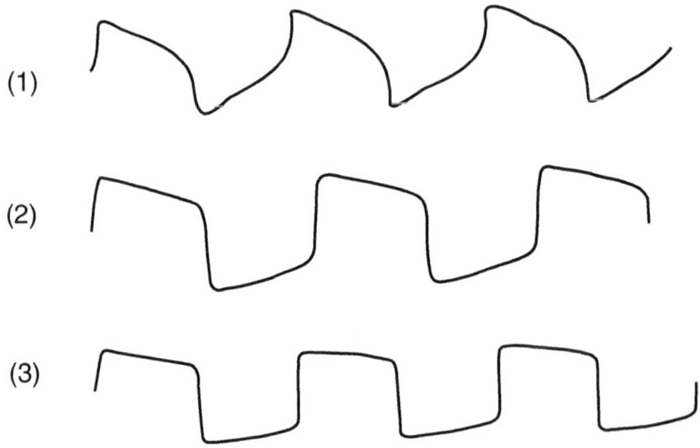

Fig. 4.7 Oscillograms characterizing the relaxation oscillations, from (Van der Pol 1930, 303)

This approach is the same as the one that Janet proposed (1919) for the series-dynamo machine, the singing arc, and the triode lamp (see *supra* Part I). It is, however, surprising that the latter did not make an oscillographic reading as Van der Pol did, when Blondel's oscillograph was available and already efficient. The distinctive form of the relaxation oscillations (see Fig. 4.7 (2)), which Van der Pol (1926d, 986) put forward in his first publications (see *supra* Part I), was also subject to a new denomination. François Bedeau described it in a note published in the *C.R.A.S.* in 1928, its title "Stabilisation des oscillations de relaxation[29]" ("Stabilization of relaxation oscillations") contains one of the first occurrences of the expression 'relaxation oscillations' in this period.

> With such an assembly, we obtain, in the plate circuit of the last triode, a current I of period T equal to the one of the tuning fork, and represented depending on the time by a "Greek pattern" curve. (Bedeau and Mare 1928, 209)

Comparing Van der Pol's (1930, 303) oscillogram (2) which characterizes relaxation oscillations, and the famous "Greek key", a decorative pattern used in architecture and jewelery, the form being reminiscent of Ancient Greece Art, shows their obvious resemblance (Voir Fig. 4.8).

In Signal Processing, this oscillation form is called a "square wave". Among the numerous examples of relaxation oscillations, Van der Pol (1930, 304) again offers the periodical variation of the number of individuals in two different species, where one eats the other, borrowed from Lotka (1925) and Volterra[30] (1928). Following

[29]The original title: "Stabilisation des oscillations de relation" had a printing error.

[30]It must be noted that this time Van der Pol referred to the English translation of Volterra's original article (1926).

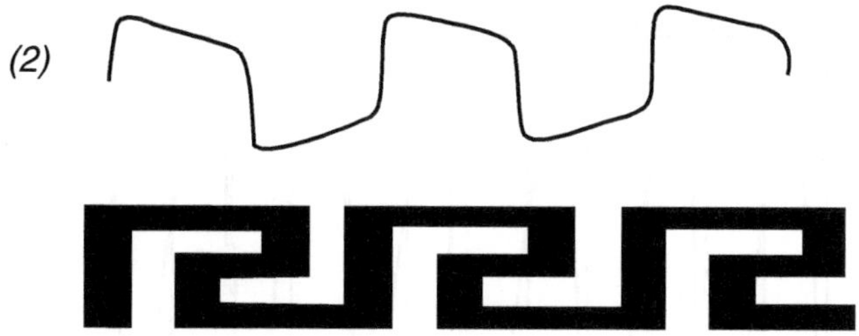

Fig. 4.8 Form of the relaxation oscillations, from Van der Pol (1930, 303)

electrical engineering[31] with the series-dynamo machine and the multivibrator, and biology with the predator-prey model, Van der Pol (1930, 305) described a hydraulic device, which might be the most representative illustration of the relaxation oscillations phenomenon.

> Let us place under a tap a small device comprised of a tank with two compartments, able to tilt around a horizontal axis, and maintained on each side by a stop block. The tank resting first on the left block, the right compartment is progressively filled. When the water reaches a certain level, the tank tips, the compartment on the right empties itself, and the compartment on the left finds itself under the tap, and is filled in turn. When it is filled, the tank tips in the opposite way, and the cycle begins again. The oscillatory period of the phenomenon depends on a filling time, which in turn is the product of two factors, the volume of the tank (playing the role of a capacitance C), and the obstruction brought by the tap (playing the role of a resistance R). (Van der Pol 1930, 307)

It is worth noting that Van der Pol was not the first to present hydraulic analogy for an oscillatory phenomenon. Over ten years earlier, Blondel wrote:

> This kind of divided-flow oscillation is not specific to electrical phenomena. It occurs many times concerning hydraulics or pneumatics. We can mention, for example, some types of water metering devices, in which the liquid used for measuring pours into a small canvas sheet assembled on a system of levers causing the tilting and draining when a certain water weight is in the sheet. (Blondel 1919d, 124)

Lastly, to demonstrate that using the sleep of flowers as an illustration is not delusional, Van der Pol explained:

> Not long ago, detailed research on the subject were published by Ms. Kleinhoonte in a dissertation ([1]) from which fig. 19 was taken. This figure clearly shows the vertical motion of the leaves during a twenty-four hours period, when they are subjected to natural light. The author showed that the plants placed in the dark, or under a perfectly constant light, therefore not periodical, kept moving with a period of roughly twenty-four hours, for which the phase, depending on the time of the day, could have any possible value. The influence of

[31] The terminology electronic was not yet created.

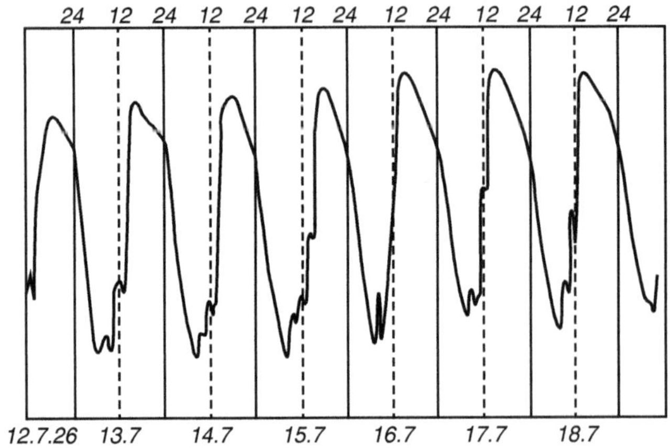

Fig. 19. — Le sommeil des plantes est encore une oscillation de relaxation. (La courbe ci-dessus montre le mouvement diurne d'une feuille de *Canavalia*).

temperature, conductivity or light being therefore entirely ruled out, it becomes obvious that these motions can be seen as relaxation oscillations, with a period determined by a duration of the same nature as a diffusion time, since a mass action is out of the question in this case.

(1) Mlle A. KLEINHOONTE. – *De door het licht geregelde autonome bewegingen der* Canavalia *bladeren*[32] – Delft, 1928. (Van der Pol 1930, 312)

Van der Pol therefore based his statements on the thesis work of Miss Antonia Kleinhoonte, nowadays seen by Chandrashekaran and Subbaraj (1996) or Daan (2010) as fundamental, since it corroborates the discoveries of the botanist Jagadish Chandra Bose (1858–1937) on the diurnal motions of the plants's leaves and circadian rhythms (Bose 1919). We can also mention Labrique's research (1968).

When the Great Depression started taking its toll in Europe, Van der Pol recalled in his conclusion that his works on *Économie*, for which he had the support of Dr. Ludwig Hamburger de La Haye:

> I think that the simple conception of economic cycles as relaxation oscillations can provide a rational and sufficient basis to explain these phenomena. (quoted in Van der Pol 1930, 312)

4.2 The Third International Congress for Applied Mechanics

During summer 1930, Balthazar Van der Pol participated, along with Alfred Liénard and Philippe Le Corbeiller in the Third Congress For Applied Mechanics, which took place in Stockholm between the 24th and 29th August. For Van der Pol

[32]Canavalia's autonomous motion of leaves controlled by light. Van der Pol's reference is once again inaccurate: *De door het licht geregelde autonome bewegingen der* Canavalia *bladeren*, Arch. Neer. Sci. Exca. Natur. Ser. III, B, t. V, p. 1–110, (1929).

(1931), this was the opportunity to present his work on relaxation oscillations to a new audience and in a new context. The way he established the differential equation (4.3), characterizing the relaxation oscillations, however, shows that he perfectly integrated Liénard's results (1928), to the point of appropriation.

> The general properties of relaxation oscillations were first investigated by the writer of the differential equation
>
> $$y'' + f(y) \cdot y' + \omega^2 y = 0$$
>
> This equation is of fundamental importance for the general study of maintained oscillations, either electrical, mechanical or biological. (Van der Pol 1931, 178)

Indeed, it was using this equation,[33] more general than Van der Pol's, that Liénard (1928, 901) he demonstrated the existence and uniqueness of a stable periodic solution, in his study on sustained oscillations (see *supra* Part I). After recalling that the function $f(y)$ is written: $f(y) = -\alpha + \beta y + \gamma y^3$, in which the second term can be assumed equal to zero for symmetry purpose, Van der Pol (1931, 179) introduces a variable substitution which allowed him to find the equation (4.3). He then explains:

> It is found that (4.3) provides an interesting instance of general theorems, first derived by POINCARÉ, concerning limit cycles.[1]
>
> ---
>
> [1]ANDRONOW, Les cycles limites de POINCARÉ et la théorie des oscillations auto-entretenues, C.R. 189, 559, 1929. (Van der Pol 1931, 179)

Here, Van der Pol once again takes note of Andronov's result, which he nonetheless considered as an "interesting example". He then adds:

> This periodic solution, giving the prototype of relaxation oscillations, is characterized by the fact, that the time period is not any more given by 2π, but now is proportional to ε, a relaxation time. (Van der Pol 1931, 179)

thus[34] reasserting that the equation (4.3) constitutes the prototype of the relaxation oscillations. As for the examples of systems which are the seat of relaxation oscillations, Van der Pol naturally presented the ones which belong to the field of Mechanics, and deliberately left the heartbeats out. In the last sentences of his presentation's proceedings, he offered a rather surprising illustration for relaxation oscillations:

> Finally a slow-motion demonstration of a relaxation oscillation produced through Coulomb friction is obtained in the following way. When a weight is placed on a horizontal table, a piece of rubber being attached to this weight, and when further the other end of the piece of rubber is moved in a horizontal direction with a constant speed, the weight will move discontinuously with jumps, owing to the non linear behavior of the solid friction terms in the equation, the period being again determined by a relaxation time or, strictly speaking, by a degeneration thereof. (Van der Pol 1931, 179)

---

[33]This equation is slightly different from Liénard's (1928, 901): $x'' + \omega f(x) x' + \omega^2 x = 0$.

[34]See Van der Pol (1930, 299) . See *infra*.

Indeed, this problem became, a short time later, one of the most emblematic examples of relaxation oscillations (see *infra* Andronov and Khaikin (1937, 159), Minorsky (1947, 420), Abarbanel et al. (1993, 36). It must also be noted that in the *Proceedings* of the congress, Liénard's presentation (1931) and Van der Pol's (1931) are consecutive when it comes to the pagination.

The analysis of lectures Van der Pol gave in France at the end of the 1920s has hence allowed to demonstrate how he created the concept of relaxation oscillations. To reach this goal, Van der Pol used several steps.

Firstly, he demonstrated the existence of self-sustained systems, i.e. comprising a component analogous to a negative resistance sustaining the oscillations instead of damping them. Then, he showed that the existing models based on the representation using a second-order linear differential equation did not allow the description of such systems any more. He therefore modified this equation in order to account for the feed-back that enabled the oscillations to be self-sustaining, established the differential equations nowadays named after him, and then demonstrated (graphically) that the period of the oscillations could not be provided by Thomson's formula any longer, but by an expression corresponding to the relaxation time, in other words, to the discharge time of a capacitor through a resistance. This allowed him to introduce a new terminology in order to describe the oscillatory phenomenon occurring in these self-sustained systems: *relaxation oscillations*. Proceeding by analogy, he then explained that the oscillations are present in a wide variety of scientific fields of research, such as electricity, biology, economics …and are governed by the same equation, for which he had established the prototype.

Van der Pol's approach, which consisted in providing an objective and stable, abstract, and general mental representation of a new type of oscillations, to which he added a specific terminology, therefore resulted in the emergence of the concept of relaxation oscillations. It is nevertheless important to clarify that self-sustained systems are at the heart of this concept. Van der Pol managed to describe perfectly the relaxation oscillations occurring in these systems,[35] he was however not able to establish their stability. Andronov (see *infra*) was the one to provide, twenty years after Poincaré (see *supra* Part I), the stability condition for self-sustained oscillations, and consequently also for relaxation oscillations. He then defined the concept of (dissipative) self-sustained systems, which covers a broader class of systems, and includes the concept of relaxation oscillations. He then demonstrated that Van der Pol's relaxation oscillator is a self-oscillator, but that other systems,[36] which are not governed by Van der Pol's equation prototype, also admit a periodic solution: Poincaré's limit cycle.

---

[35]Indeed, when the parameter has a small value Van der Pol's system is actually self-sustained but not the seat of relaxation oscillations anymore, as was clarified by Van der Pol (1926d, 983).

[36]See *infra* Kolmogorov (1936) and Andronov and Khaikin (1937).

Chapter 5
Andronov's Notes: *Toward the Concept of Self-Oscillations*

After completing his secondary education, in 1920 Aleksandr Aleksandrovich Andronov joined the electrical Engineering Department of the Moscow Higher Technical Institute, which offered specialist courses in radiotechnics. The following year, he attended lectures on physico-mathematics at the Moscow State University. In 1924 he was appointed assistant at the State Educational Institute in Moscow, where he taught Mechanics and Theoretical Physics. In 1925 he obtained a degree in theoretical physics at the Moscow state University, and in 1926 began working on a doctoral dissertation under the supervision of Leonid Isaakovich Mandel'shtam (1879–1944). This charismatic man was at the origin of what he himself called "nonlinear thinking" (Mandel'shtam 1947–1955, vol. 3, 52) and considered that everything in the universe is made of vibrations. His philosophy, which has its source in the work of Henri Poincaré, had a decisive influence on the direction Andronov's research took. Contrary to what Diner (1992, 339) and Dahan Dalmedico (1996, 21) stated, Andronov's famous note (1929a) published in the *C.R.A.S.* of Paris, which was incidentally preceded by a presentation by Andronov (1928) at the Sixth Conference of Soviet Physicists in Moscow between the 5th and 16th of August 1928, only contained the title and summary of his graduating dissertation, as Andronov explained in an autobiographical notice:

> Эта работа, краткое изложение которой было опубликовано порусски в 1928 г. и в другой редакции по-французски в « Докладах Парижской Академии » в 1929 г., определила область моей дальнейшей научной деятельности - теория колебаний и смежные вопросы математики и теоретической физики.[1] (Andronov 1943, 1)

[1]This work, a summary of which was published in 1928 and then in another version in the French *Comptes Rendus* of the Academy of Paris in 1929, has defined the field of my future research – the theory of oscillations and related issues mathematics and theoretical physics. (Andronov 1943, 1)

J.-M. Ginoux, *History of Nonlinear Oscillations Theory in France (1880–1940)*,
Archimedes 49, DOI 10.1007/978-3-319-55239-2_5

His thesis's defence seems to have taken place in 1929, as corroborated by Andronov:

> В 1929 г. окончил аспирантуру. Тема моей заключительной диссертации - « Предельные циклы Пуанкаре и теория Колсбаний ».[2] (Andronov 1943, 1)

However, the full text of his dissertation, as well as the reference that should be linked with it, appears to be lost. There is been no hint of it in either in his *Theory of Oscillations* (Andronov and Khaikin 1937) or in his collected works (Andronov 1956).[3] However, we found traces of an earlier publication to this famous note at the *C.R.A.S*.

5.1 The Lecture of Soviet Physicists of 1928: *From Limit Cycles to Self-Oscillations*

In the summer of 1928, the young Andronov presented a summary of the progress of his work and research at the VIth congress of Soviet Physicists in Moscow. This congress held between 5th and 16th August gathering prestigious Russian scientists, among which were Mandel'shtam, Krylov, and Bogolyubov. In this presentation, entitled: "Poincaré's limit cycles and oscillation theory", Andronov laid fundamental groundwork for what would later become nonlinear oscillation theory. He started out by giving the definition of a system able to sustain its own oscillations[4]:

> Существует ряд устройств, могущих генерировать незатухающие колебания за счет непериодических источников знергии.[5] (Andronov 1928, 23)

He then added that a mathematical theory had not yet been provided for these systems:

> Однако до сих пор нет достаточно строгой и общей теории таких автоколебаний. Между тем имеется адекватная математическая конструкция, созданная вне всякой связи с теорией колебаний, позволяющая установить общую точку зрения на все подобные процессы для случая одной степени свободы Эта конструкция-теория « предельных циклов » Пуанкаре.[6] (Andronov 1928, 23)

---

[2]In 1929, I was graduated from the Higher Studies School. The theme of my final dissertation was the "Poincaré's limit cycles and the oscillation theory". (Andronov 1943, 1)

[3]According to Boyko (1983, 30) this document has not been kept.

[4]Andronov (1928, 24) called such systems self-oscillators.

[5]There are a number of devices capable of generating sustained oscillations from non-renewable energy sources. (Andronov 1928, 23)

[6]However, there is not yet sufficiently rigorous general theory for such oscillations. Meanwhile, there is a pattern or adequate mathematical model, created with no connection to the theory of oscillations, which enables a common view of all these processes to the case of a degree of freedom. This concept is that of the "theory of limit cycles" of Poincaré.

(Andronov 1928, 23)

So he linked the study of these devices with Poincaré's work (1882, 261). Andronov (1928, 23) then provided the nonlinear differential equation that represented the evolution of systems he defined:

$$\frac{d^2x}{dt^2} = F\left(x, \frac{dx}{dt}\right) \tag{5.1}$$

He then moved to the phase plane, writing: $\frac{dx}{dt} = y\,;\ \frac{d^2x}{dt^2} = y\frac{dy}{dx}$ and found the following equation:

$$\frac{dy}{dx} = \frac{F(x, y)}{y} \tag{5.2}$$

If the variable substitution is identical to the one used by Poincaré (1908, 390) (see *supra* Part I), it nonetheless seems improbable that Andronov had access to this work. If that had been the case, he would certainly have quoted it in order to reinforce his approach, which is only justified by a reference to Poincaré. Andronov then explained that:

> При некоторых ограничениях, наложенных на $F\left(x, \frac{dx}{dt}\right)$, эти замкнчтые решения уравнения (5.2) могут быть только двух типов: или это так называемые « центральные » решения, или это « предельные циклы ».[7] (Andronov 1928, 23)

He then clarified that the "center-type" solution corresponds to the ideal case of a conservative mechanical system, and added:

> Если же уравнение (5.2) имеет предельный цикл, т. е. если все кривые, находящнеся в некоторой области, стремятся при $t \to \pm\infty$ к предельному циклу, то это значит, что соответствующие решения уравнения (5.1) стремятся к периодическому решению, амплитуда и период которого не зависят от начальных условий.[8] (Andronov 1928, 23)

He therefore evidenced the fundamental difference between the *center* type of solution of a conservative system such as the frictionless pendulum and the *limit cycle* type of a dissipative system such as the oscillations of a triode. These characteristic were later qualified by Mandel'shtam et al. (1935, 82) as being *extrinsic* and *intrinsic*.

---

[7]Under certain conditions imposed on $F\left(x, \frac{dx}{dt}\right)$, closed curves, solutions of the equation (5.2), can only be to two types: either a solution called "center" or a "limit cycle". (Andronov 1928, 23)

[8]If the equation (5.2) has a limit cycle, i.e., if all the curves which are within a certain area tend to limit cycle as $t \to \pm\infty$, this means that the corresponding solutions of equation (5.1) tend to the periodic solution whose amplitude and period are not dependent on the initial conditions. (Andronov 1928, 24)

Andronov's conclusion, which is reminiscent of Poincaré's own conclusion (1908, 391), provided him with an opportunity to introduce a neologism that enabled him to define this type of system:

> Стационарные движения, которые устанавливаются в приборах, способных к автоколебаниям, всегда соответствуют некоторым продольным циклам.[9] (Andronov 1928, 24)

5.2 Note to the *C.R.A.S.* of 1929: *From Self-Oscillations to Self-Sustained Oscillations*

On Monday the 14th of October 1929, the French mathematician Jacques Hadamard presented to the *Académie des Sciences* de Paris a note given to him by Aleksandr Andronov. The fact that Hadamard initiated this presentation is to be expected, since on one hand he was in charge of the mathematical analysis section of the *Académie des Sciences* and on the other he had been an associate of the Russian Academy of Sciences since 1922 and foreign member of the Soviet Academy of Sciences since 1929. This work, which he certainly appreciated, is surprisingly entitled: "Poincaré's limit cycles and self-sustained oscillation theory" (Andronov 1929a). The neologism *self-oscillations* introduced by Andronov (1928) in the Russian version of his article was therefore changed to *self-sustained oscillations*. The analysis of Andronov's original note with corrections by experts, does not however include any change regarding the title. The origin of the use of this terminology – introduced by Blondel (1919c) (see *supra* Part I) – appearing in the title and first sentence of the note can not therefore be explained for now. Regarding the formation of the neologism self-oscillations, Pechenkin offered an interesting hypothesis (2002, 272), suggesting that Andronov could have been influenced by German research. The last sentence of Andronov's note (1929a) seems to supports this idea:

> It is certain that a series of characteristic phenomena coming with these oscillations ([1]) must be found in self-oscillatory, mechanical, or chemical systems.
>
> ---
>
> ([1]) For instance, the phenomenon called *Mitnehmen* by the German, see H. BARKHAUSEN, *Elektronen-Röhren*, 3, 32, (Leipzig, 1929). (Andronov 1929a, p. 561)

Moreover, hints can be found in Heinrich Barkhausen's work (1907, 59, 1929, 46), using terms such as "selbst Schwingungen", i.e., "self-oscillations":

> Für die Stabilität gegen Schwingungen ist entscheidend, ob die den Schwingungen während einer ganzen Periode zugeführte Energie größer ist als die durch die Dämpfung entzogene oder nicht. Im ersteren Falle ist die Lage labil, im anderen stabil. Bei Labilität entstehen von

[9]Stable motions that are present in devices capable of self-oscillations must always correspond to limit cycles. (Andronov 1928, 24)

selbst Schwingungen, die so Lange anwachsen, bis die zugeführte und entzogene Energie gleich groß werden.[10] (Barkhausen 1907, 59)

Andronov (1929a) then considered oscillating systems that are not conservative:

Let us cite, for the case of equations with partial derivatives, the already-old issue of string being excited by a bow, as well as the problem of Cepheids, as dealt with by Eddington ([1]), and for the ordinary differential equations, in mechanics the Froude pendulum ([2]), in physics the triode oscillator ([3]), in chemistry the periodical reactions ([4]). Similar problems are found in the field of biology ([5]).

([1]) EDDINGTON, *The internal constitution of stars*, p. 2oo (Cambridge, 1926).
([2]) Lord RAYLEIGH , *The theory of sound*, London 1, 1894, p. 212.
([3]) Voir par exemple VAN DER POL, *Phil. Mag.*, 7e série, 2, 1926, p. 978.
([4]) Voir par exemple KREMANN, *Die periodischen Erscheinungen in der Chemie*, p. 124 (Stuttgart, 1913).
([5]) LOTKA, *Elements of physical biology*, p. 88 (Baltimore, 1925). Voir aussi les récentes recherches de M. Volterra. (Andronov 1929a, 560)

Choosing these examples seems to indicate that Andronov sought to show, on all scales found in the Universe (from the most remote stars to the smallest micro-organisms), the existence of "nonlinear processes[11]". Nevertheless, while the triode associated with Van der Pol's name is prominently featured, Andronov did not mention the concept of relaxation oscillations at all. Moreover, the reference to Lotka and Volterra's work is once again groundless. However, if we compare this statement with the one found in Andronov's article (1928, 24), it appears that the concept of a self-oscillator comprises both conservative and dissipative systems. Andronov then offered to represent these self-oscillating systems with a second-order differential equation (5.1) which he transformed into a system of two first-order differential equations, thus moving to Poincaré's phase plane:

$$\frac{dx}{dt} = P(x, y) \quad ; \quad \frac{dy}{dt} = Q(x, y) \tag{5.3}$$

He then explained that the periodic solution of this system (5.3) is expressed in terms of Poincaré's limit cycles:

We can easily demonstrate that, to these periodic movements satisfying these conditions there correspond *isolated* closed curves in the *xy* plane that are approached in spirals, from the exterior and the interior (for *t* increasing), by the neighbouring solutions. This results in self-oscillations which emerge in systems characterized by the equation of type (5.3) corresponding mathematically to Poincaré's stable limit cycles ([3]). It is therefore clear that the period and amplitude of the stationary oscillations do not depend on the initial conditions.

([3]) POINCARÉ, Œuvres, 1, p. 53, Paris (1928).(Andronov 1929a, 560)

[10]To study the stability of the oscillations is to determine if the energy supplied during a period is greater than the energy lost by damping or not. In the first case, the situation is unstable, the other is stable. This stability can result in self-oscillations which develop over a long period until the energy supplied and the energy lost are equal. (Barkhausen 1907, 59)

[11]See *supra*.

Firstly, it must be noted that Andronov does not demonstrate the existence of *limit cycles* in the systems he defined using (5.3), but rather established a "connection" by using an *analogy*, which played a crucial role in the whole story. He then recalls, as he previously did (see *supra*), his systems's fundamental property, i.e. the independence of the period and amplitude with regard to the initial conditions.

Lastly, he refers to Poincaré's collected works and specifically to Chapter I, titled "Théorie des cycles limites" ("Limit cycle theory") rather than to the famous dissertations "Sur les courbes définies par une équation différentielle" ("On the curves defined by a differential equation") (Poincaré 1882, 261), probably because the first volume of Poincaré's collected work had been published the year before. It is, however, possible that Andronov had access to Poincaré's famous essays (1881–1886), of which he had the translation published in Russian in 1947.

He then added:

> The general theory ([4]) of the integral curves for the equations of type (5.3) allows in numerous cases the study of these equations qualitatively, and to come to conclusions regarding the existence, number, and stability of limit cycles.
>
> ---
>
> ([4]) POINCARÉ, *loc. cit.* — BENDIXSON, *Acta Mathematica*, 24, 1900, p. 1. (Andronov 1929a, 560)

Andronov quoted Ivar Bendixson's work (1861–1935) allowing the demonstration of the existence of limit cycles in this type of system and nowadays called Poincaré-Bendixson theorem, but did not however refer to the work of the Cartans (1925), nor to Liénard's (1928), as this would have limited the impact of his results, reducing them to Liénard-Van der Pol-type systems. This is all the more surprising as his approach was very similar to Liénard's (1928) (see *infra*). He then offered to study, as an example, a particular case of system (5.3), borrowed from Poincaré (1892, 89):

$$\frac{dx}{dt} = y + \mu f\,(x, y : \mu) \quad ; \quad \frac{dy}{dt} = -x + \mu g\,(x, y : \mu) \tag{5.4}$$

where he finds a real parameter that can be chosen at a sufficiently small scale. By following the process developed by Poincaré (1892) in Chapter III of Volume I of the *Méthodes Nouvelles de la Mécanique Céleste* entitled "periodic solutions", Andronov (1929a, 560) demonstrates on one hand the existence of a periodic solution for small values of the parameter, and on the other, the stability of the periodic solution, i.e. of the limit cycle for this type of system (5.4). However, to reach these results, Andronov used an approach entirely similar to Liénard's. The analysis presented in the next paragraph therefore aims at highlighting the exceptional likeness between Poincaré's, Liénard's and Andronov's demonstrations, which led to the same result, i.e. the stability of the limit cycle.

5.3 On Limit Cycle Stability: *From Poincaré, to Liénard, to Andronov*

In the second part of his article, Liénard (1928, 946) offers to study, in the case where k is very small, the following equation:

$$xdx + ydy - kF(x)\,dy = 0 \tag{5.5}$$

By moving to Liénard's phase plane (1928, 901), we can then transform the equation (3.20) into a system of two first order differential equations corresponding to Andronov's (1929a, 560) self-sustained system (5.4). (See Table 5.1)

Liénard then explains:

> If we write $k = 0$ in this equation,[12] it is reduced to
>
> $$xdx + ydy = 0$$
>
> or $x^2 + y^2 = constante$. All the integral curves are circles, and consequently, closed curves. (Liénard 1928, 946)

It is interesting to then compare this sentence to Andronov's:

> When $\mu = 0$, the equations (5.4) have a solution $x = R\cos(t)$, $y = -R\sin(t)$. The integral curves form a family of circles in the xy plane. (Andronov 1929a, 560)

In this case $k = 0$ (resp. $\mu = 0$), Liénard's equation (5.5) and Andronov's equation system (5.4) (presented in the Table 5.1), are reduced to the equation for a harmonic oscillator, for which the solution is a circle. While Liénard exhibits this solution in a cartesian form, Andronov made it explicit in a parametric form. This allowed him to introduce a change of variables: the transformation of Cartesian coordinates into polar coordinates. However this change in variables possesses a distinctive particularity. It is common to write: $x = R\cos(t)$, $y = R\sin(t)$, but Andronov's y-coordinate is negative in such case. This stems from his accounting for the travel direction of the trajectory curve, developing in the limit cycle, counterclockwise (see Table 5.2), as had been demonstrated by Poincaré (1908, 390) and this amounts to replacing t by $-t$ in a transformation into polar coordinates (see *supra* Part I). Liénard then added:

Table 5.1 Liénard's (1928) and Andronov's (1929a) differential equations systems

Liénard (1928)	Andronov (1929a)
$\begin{cases} \dfrac{dx}{dt} = y - kF(x) \\ \dfrac{dy}{dt} = -x \end{cases}$	$\begin{cases} \dfrac{dx}{dt} = y + \mu f(x, y; \mu) \\ \dfrac{dy}{dt} = -x + \mu g(x, y; \mu) \end{cases}$

[12] It's the equation (5.6).

Table 5.2 Transformation into polar coordinates

Transformation into polar coordinates		
Classical	Liénard (1928)	Andronov (1929a)
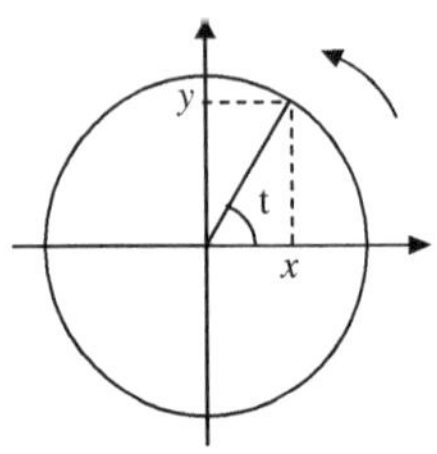	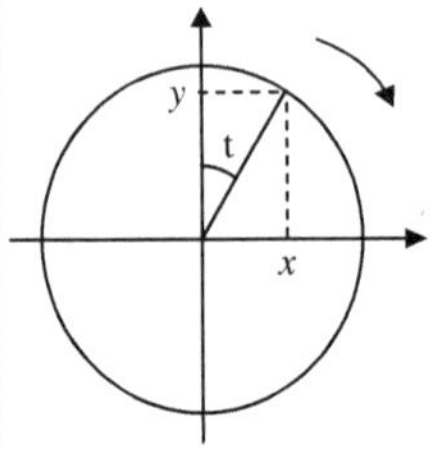	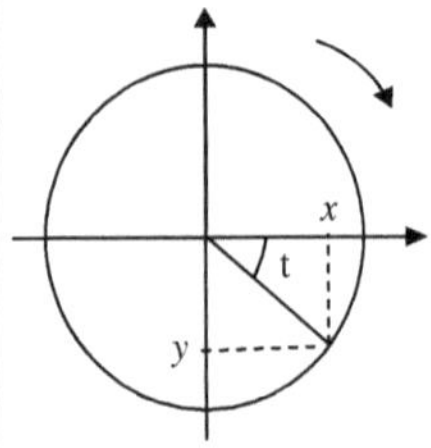
$\begin{cases} x = R\,cos\,(t) \\ y = R\,sin\,(t) \end{cases}$	$\begin{cases} x = R\,sin\,(t) \\ y = R\,cos\,(t) \end{cases}$	$\begin{cases} x = R\,cos\,(t) \\ y = -R\,sin\,(t) \end{cases}$

> But, among this infinite number of circles, only one corresponds to the limit of the unique closed curve existing as long as k stays finite. To determine this circle, it must simply be remarked that the condition characterizing the integral closed curve is not modified by the substitution of F with kF. The limit circle will therefore be the one for which we have $\int Fdy = 0$. (Liénard 1928, 946)

As for Andronov, he wrote:

> By using Poincaré's methods, we see that for the case where $\mu \neq 0$ is sufficiently small, Only isolated closed curves stay in the xy plane, close to circumferences that have a radius determined by the equation
>
> $$\int_0^{2\pi} [f(R\cos(\xi), -R\sin(\xi); 0)\cos(\xi) - g(R\cos(\xi), -R\sin(\xi); 0)\sin(\xi)]d\xi = 0$$
>
> (...) (Andronov 1929a, 560)

We see that Liénard uses the term "limit circle", and not "limit cycle", since he explains that at the limit $k \to 0^+$, the closed curve takes the form of a circle. It is also notable that Andronov's approach almost exactly follows Liénard's. The only difference resides in the function F, and in the change of variables. Indeed, contrary to Andronov, Liénard chose to write: $x = R\sin(t)$, $y = R\cos(t)$. However, as it can be easily verified by using Table 5.2, these two changes of variables are perfectly equivalent. By writing: $k = \mu$, $f(x, y; \mu) = -F(x)$ and $g(x, y; \mu) = 0$, we see that Liénard's system corresponds exactly to Andronov's (5.4), and that the conditions allowing to find the radius of Liénard's "limit circle", in other words, of Andronov's "closed curve", are absolutely identical. In the following pages, Liénard (1928, 906–907) determined the radius R of this circle, for the case of the function F corresponding to the *oscillation characteristic* of Van der Pol's triode (1920, 703), and therefore found the value of the amplitude ($a = 2$) calculated by Van der Pol[13] (1926d, 980).

The aim of Liénard's article (1928) was to demonstrate the existence and uniqueness of the periodic solution to the Liénard-Van der Pol type of equation, under the condition that the function representing the oscillation characteristic verifies specific conditions. In order to do this Liénard (1928, 906) then established a stability condition for the "closed curve", i.e. the limit cycle (see *supra* Part I), which is expressed as an inequality:

> For all the integral curves external to the closed curve D, $\int Fdy$ is negative, and OA_2 is superior to OA_1. The integral curve is closer to the closed curve at the point 5.2 than at 5.1 (...). We therefore see that the integral curve draws a kind of spiral asymptotically approaching the closed curve D. For the integral curves external to the closed curve OA_2 is the one becoming inferior to OA_1. The curve still approaches the curve D, but from the outside. Because of the fact that all the integral curves, internal or external, and traversed in the direction of the time increases asymptotically approach the curve D, we say that the corresponding periodic movement is a *stable* movement. (Liénard 1928, 906)

We can deduce the stability condition of the limit cycle for Liénard (see *supra* Part I):

$$\int_{\Gamma} F(x)\, dy > 0 \tag{5.6}$$

Andronov also provided a stability condition for the "closed curves", in the more general case of the self-sustained systems, in the form:

> These closed curves correspond to stable steady-state motions, when the condition (A_5)
>
> $$\int_0^{2\pi} \left[f'_x \left(R\cos(\xi), -R\sin(\xi); 0\right) + g'_y \left(R\cos(\xi), -R\sin(\xi); 0\right) \right] d\xi < 0$$
>
> is met (...) (Andronov 1929a, 561)

By using Green's formula[14] and moving back to a Cartesian coordinate system, Andronov's stability condition (A_5) is written as

$$\int_{\Gamma} \left(f(x, y; \mu) \frac{dy}{dt} - g(x, y; \mu) \frac{dx}{dt} \right) dt < 0 \tag{5.7}$$

By writing: $k = \mu, f(x, y; \mu) = -F(x)$ and $g(x, y; \mu) = 0$, we observe that Andronov's stability condition (5.7) is perfectly identical to Liénard's (5.6):

[13] Van der Pol also deduced it graphically. See *supra* Part I.

[14] Green's formula: $\int_{\Gamma} f(x, y)\, dy - g(x, y)\, dx = \iint_{S} \left(f'_x(x, y) + g'_y(x, y) \right) dxdy$.

Table 5.3 Synoptic of the systems and stability conditions

Poincaré (1908)	Liénard (1928)	Andronov (1929a)
$\begin{cases} \dfrac{dx}{dt} = y \\ \dfrac{dy}{dt} = -x - \mu\theta(y) \end{cases}$	$\begin{cases} \dfrac{dx}{dt} = y - kF(x) \\ \dfrac{dy}{dt} = -x \end{cases}$	$\begin{cases} \dfrac{dx}{dt} = y + \mu f(x, y; \mu) \\ \dfrac{dy}{dt} = -x + \mu g(x, y; \mu) \end{cases}$
$\int_\Gamma \theta(x')\, x' dt < 0$	$\int_\Gamma F(x)\, dy > 0$	$\int_\Gamma \left(f(x, y; \mu) \dfrac{dy}{dt} - g(x, y; \mu) \dfrac{dx}{dt} \right) dt < 0$

$$\int_\Gamma \left(-F(x) \frac{dy}{dt} \right) dt < 0 \quad \Leftrightarrow \quad \int_\Gamma F(x)\, dy > 0$$

It has been established in Part I, that Andronov's stability condition (5.7) is identical to Poincaré's (1908), and it therefore appears that the three stability conditions: Poincaré's (1908), Liénard's (1928) and Andronov's (1929a) are completely identical. Those results are summarized in Table 5.3.

Andronov's approach is broader than Liénard's and Poincaré's, since it is not limited to the study of a Liénard-Van der Pol type of equation, and contains a whole category of self-sustained systems. In his demonstration of the stability of the limit cycle, however, Andronov used each steps of Liénard's and Poincaré's arguments. While he cited Poincaré, it seems he did not refer to Liénard before his book was published in 1937. It was only a year later that Andronov (1933) published a demonstration of the limit cycle based on the use of "characteristic exponents" introduced by Poincaré (1892, 176) in his *Méthodes Nouvelles de la Mécanique Céleste*.

We hence find several results which were already presented in the first part of this work, that is to say, that Andronov was most likely unaware of Poincaré's article (1908), and that he must have built his demonstration based on Poincaré's work (1892). As for Liénard's, however, (1928), the two approaches are so alike that he must have known of his work.

Lastly, the analysis of this note allows us to confirm that the fundamental connection that Andronov established with Poincaré's work is in regards to chapter III: "Solutions périodiques" and chapter IV: "Exposants caractéristiques" of Poincaré's *Méthodes Nouvelles* (1892, 89 and 176), rather than to his memoirs *Sur les courbes* (1881–1886), and more specifically to the chapter entitled "Théorie des cycles limites" (Poincaré 1882, 261).

In fact, the link with the *perturbation theory* and *Analysis situs* is the true start of a *theory of nonlinear oscillations*, whereas the connection between periodic solutions and limit cycles is only a minor step in the process of developing these theories, as also shown by Andronov's conclusion:

> The theory of self-oscillations, for which non-rigorous approaches have almost always been used until now, has therefore been provided with a solid mathematical basis, at least for the

simplest case. The electrical self-oscillations are the most suited to experimental research. It is certain that a series of characteristic phenomena accompanying these oscillations ([1]) must be found in chemical or mechanical self-oscillating systems.

([1]) For example the phenomenon that Germans call *Mitnehmen*, see H. BARKHAUSEN, *Elektronen-Röhren*, 3, p. 32, (Leipzig, 1929). (Andronov 1929a, 561)

This mention of the *mitnehmen* phenomenon, i.e. synchronization, prefigures the prospect of the results of Andronov's second note (1930a) which he presented less than three years later to the *C.R.A.S.* with Aleksandr' Witt,[15] and in which they addressed the stability of periodic or quasi-periodic solutions for a self-oscillating system with two degrees of freedom.[16] This shows the remarkable advantage that Russian scientists had over the developments that had been carried out until then in Europe. While some tried to provide evidence of the existence of limit cycles in various types of autonomous systems, the School that Andronov founded in Gorki[17] with Artemij Majer (1905–1951), Mikhail E. Leontovich (1903–1981) and Nikolai N. Bautin (1908–1993) and in parallel, the one founded in Kiev by Nikolai M. Krylov (1879–1955) and his student Nikolai N. Bogolyubov (1909–1992), already researched quasi-periodic solutions for non-autonomous systems, as well as problems of entrainment and synchronization. It will be established in Part III of this work, that while Van der Pol cannot be viewed as the author of a relaxation oscillation theory, he is however, as noted by Krylov and Bogolyubov, at the origin of entrainment and synchronization theories (Fig. 5.1).

Through numerous researches, Mr. Balth. van der Pol has elaborated the theory of entrainment, as well as the theory of synchronization (from the natural frequency to the external force), and his work led him to the discovery of frequency demultiplication in relaxation systems (from Mr. van der Pol's terminology) (Krylov and Bogolyubov 1933, 10)

At the beginning of the 1930s, Van der Pol gave many lectures in Europe and specifically in France to present the concept of relaxation oscillations, the French scientific community therefore went through an in-depth analysis of his work that did not seem to address Andronov's fundamental results, including the connection with Poincaré's concept of limit cycles. In spite of this disregard, in January 1933, the first *Conférence Internationale de Non linéaire* (*International Conference on Nonlinear Oscillations*) was held in Paris, through the collaborative effort of both Van der Pol and Papalexi.

[15]The Luzin affair that begins in 1936 will lead to the arrest of Witt in 1937 and his conviction to five years in the gulag where he died June 26, 1938. See Bendrikov and Sidorova (1981), Bissell (1994) and Dahan Dalmedico (2004a,b).

[16]This problem is the study of the characteristics on the surface of the torus. See Poincaré (1885a, 220) and Denjoy (1932a,b). See *infra* Part III.

[17]The city of Gorki is become in the beginning of 1990s, Nijni-Novgorod. About Andronov's School see Dahan Dalmedico (2004a,b).

Fig. 5.1 Aleksandr Aleksandrovich Andronov, from Neimark (2001, 231)

The reception of Van der Pol's and Andronov's work as well as the impact of this conference, will be studied in the following paragraphs. During this time, some, such as Le Corbeiller, Denjoy, but also Krylov and Bogolyubov, called for the creation of a new branch of physics: Nonlinear Mechanics, whereas some others, such as Van der Pol, started what can only be called as a "hunt for the relaxation effect". Soon after, Andronov and Khaikin's (1937), as well as Krylov and Bogolyubov's (1937) books were notable publications in Russia in 1937, although they were not distributed in English before the 1940s. While the nonlinear oscillation theory was being formalised, the concept of relaxation oscillation needed to find a place in this structure. The issue of defining this new type of oscillation caused the classification of oscillations to be challenged, this will be analysed in the last paragraph.

The question of Henri Poincaré's scientific legacy, and especially the role played by Jacques Hadamard in this legacy, will be studied in Part III of this work.

5.4 The General Assembly of the I.U.R.S in 1934

During the General Assembly of the International Union of Radio Science (I.U.R.S.), which was held in London between the 12th and 18th of September 1934, Mandel'shtam, Papalexi, Andronov, Khaikin and Witt made an article entitled "Presentation of the recent research on nonlinear oscillations" (Mandel'shtam et al. 1935). This work, published in French in the *Journal de Physique Technique de l'U.R.S.S.* (*Journal of Theoretical Physics of U.S.S.R.*) contained all of the nonlinear oscillation theory's founding principles: the concept of limit cycle as a solution of a self-sustained dissipative system represented in Poincaré's phase plane, the concept of bifurcation (of codimension 1), and of course the concept of relaxation oscillation, were presented in the first paragraphs. The forced oscillation systems, the resonance, synchronization and frequency multiplication phenomena that were addressed, will be the subject of an analysis in Part III of this work. Mandel'shtam and his colleagues introduced Van der Pol's concept:

> We often see, in the field of radio-technics, the creation of conditions leading to relatively sinusoidal oscillations.* But during the last few years, it is partly thanks to van der Pol's work that there was an increase in interest shown to systems producing oscillations that strongly differ from the sinusoidal form, or "relaxation oscillations".
>
> ---
>
> *A very simple mechanical system allowing to obtain quasi-sinusoidal oscillations- the Froud pendulum- was studied by Strelkov.[18] (Mandel'shtam et al. 1935, 96)

This tribute given by Mandel'shtam to Van der Pol's work was only a courteous nod, since the Soviet delegation, eagerly awaited in Paris the previous year for the first *Conférence Internationale de Non linéaire* (see *infra*) had been invited by Van der Pol so he could participate in a congress where the theory of nonlinear oscillations was discussed. Mandel'shtam then recalled Van der Pol's second order differential equation (V_9) characterising the triode's oscillations, and offered to highlight the problem by supposing that the inductance in the circuit is null. This reduces Van der Pol's equation to an integrable first-order equation. He then added:

> After highlighting the problem, we must, in order to describe the physical phenomenon, introduce a new condition, in our case consisting in the current being subjected, at a certain moment, to a discontinuity, while the tension at the terminals of the capacitor stays constant. This postulate, of "discontinuity condition" is physically justified by the fact that the energy cannot vary in a discontinuous manner. We can give it another form, by explicitly asking for the continuity of the energy variation. This "discontinuous" theory, adjoined to the discontinuity condition, allows to evidence the periodical "discontinuous" movements, and to find their "amplitude" and period. This approach to relaxation oscillations, applicable to electrical and mechanical systems, is analogous, without being identical, to the processes used in mechanics to analyze elastic collisions. (Mandel'shtam et al. 1935, 96)

From this moment on, at the instigation of Mandel'shtam, Van der Pol's concept of relaxation oscillations was replaced by the concept of "discontinuous movement" in Andronov's publications (see *infra*). This can be perceived as a "battle of concepts". It was previously established that the concept of self-oscillation covers, on the one part the concept of limit cycles according to Andronov's own statement (1929a, 560) and on the other, the concept of relaxation oscillations, since a relaxation oscillator is always a self-oscillator (the reverse is false, see *supra*), for which the periodic solution is a limit cycle. Nevertheless, at this time, the imagery of this representation (see Fig. 5.2) was still vague.

---

[18]Mandel'shtam refers to Strelkov : Das Froudsche Pendel., Журнал технической Физики, III, p. 563, 1933.

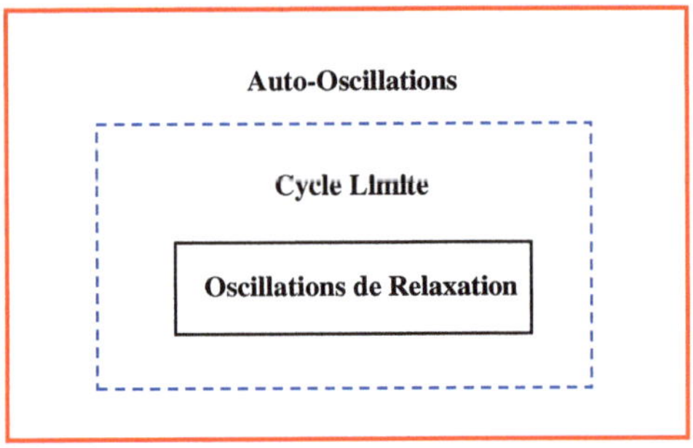

Fig. 5.2 Imagery of the concepts of self-oscillation, limit cycle and relaxation oscillation

Mandel'shtam and his student Andronov seem to admit that Van der Pol's concept can possibly describe, with one and the same prototype equation, an array of apparently very different phenomena. However, they reproached him for his inability to explain the inner nature of relaxation oscillations, in other words, how a system suddenly "jumps" from a slow phase to a fast phase. Moreover, the way in which Van der Pol tried to force his "theory of relaxation oscillations", without it having genuine mathematical foundations, in parallel to the one for (nonlinear) oscillations which was being developed by Andronov, Khaikin and Witt, caused a coexistence issue. In order to provide a rational explanation for the phenomenon and also to bring it back under the rule of the "Theory of Oscillations", Andronov, Khaikin and Witt chose to use another denomination.

Chapter 6
Response to Van der Pol's and Andronov's Work in France

At the end of the 1920s, and the beginning of the 1930s, Van der Pol was invited to France numerous times in order to present his work. On the 24th of May 1928, he was invited by the *Société des Amis de la T.S.F.* and the *Société Française des Électriciens* to give a lecture in Paris, in the *Salle de la Société de Géographie*, 184 boulevard Saint Germain, under the direction of General Gustave Ferrié. On the 10th and 11th of March 1930, he gave a series of lectures at the *École Supérieure d'Électricité*. He came back two years later, for the fiftieth *Congrès International d'Électricité* which was held in Paris[1] in July 1932. It was also at his and Nikolai D. Papalexi's (1880–1947) initiative that the very first *Conférence Internationale de Non linéaire*[2] was held in Paris on 29 and 30 January 1933. Lastly, he attended the Réunion Internationale de Physique-Chimie-Biologie at the *Palais de la Découverte* in Paris[3] in October 1937. It seems that Philipe Le Corbeiller was his main interlocutor, as shown in the footnote of one of his presentations, in which Van der Pol (1930, 32) thanked Le Corbeiller "for his devotion to the writing in French of his lectures". This implies that Le Corbeiller was in the audience at his presentations.

[1]The very first *Congrès International d'Électricité* was held in Paris in 1881. It was again held in this town for its 50th.

[2]Regarding this lecture, see *infra* and Ginoux et al. (2010).

[3]The *Palais de la Découverte* was inaugurated in May 1937 at the initiative of Jean Perrin (1870–1942).

J.-M. Ginoux, *History of Nonlinear Oscillations Theory in France (1880–1940)*,
Archimedes 49, DOI 10.1007/978-3-319-55239-2_6

6.1 Philippe Le Corbeiller's Work: *Towards History of Oscillations Theory…*

After graduating 36th of his class at the *École Polytechnique*, Philippe Le Corbeiller joined the *École Supérieure des Postes et Télégraphes*[4] (nowadays called *Télécom ParisTech*) and in 1914, became chief telegraph engineer at the state Electricity laboratory. He was promoted *sous-lieutenant de réserve* (reservist second lieutenant), and as a reward for his great activity in the field of transmissions during the *First World War*, he was decorated with the *Croix de Guerre*. In 1926, he defended at the *Université de Paris* his doctoral thesis, entitled *Contribution à l'étude des formes quadratiques à indéterminées conjuguées* (*Contribution to the study of quadratic forms with conjugated indeterminable conjugates*) under the direction of Émile Picard (1856–1941). He then became an expert in electronics for communication, electromechanical systems, and acoustics. He worked for the Ministry of Communication from 1929 to 1939, before being appointed as head of the technical program and national radio broadcasting network in France, he also taught at the *École Supérieure d'Électricité*. After fleeing from France during the German occupation, he settled in the United States in 1941, and during the Second World War, he taught electronics to the staff of the Navy and American army at Harvard, before joining this college in 1945 as a lecturer in applied physics, then in 1949 as a *maître de conférence*, professor, and finally, in 1959 as an emeritus professor. After Van der Pol (1889–1959), who had retired in Wassenaar (Netherlands), passed away, Le Corbeiller[5] grew closer with his widow Pietronetta Posthuma-Van der Pol, and married her in New York on the 7th of May 1964. In 1968, he became resident of the Netherlands, until his death in Wassenaar on the 26th of July 1980, at 89 years old.[6]

Le Corbeiller was talented when it came to vulgarization, and in the same manner as a historian of Sciences, presented, in a large number of lectures, remarkable synthesis of the various results obtained in the field of *relaxation oscillations*. His contribution to the understanding and reconstruction of the development process of the theory of nonlinear oscillations is fundamental, as he made an inventory of the crucial steps which would have fallen into oblivion without his intervention.

From 1930, Le Corbeiller gave a series of lectures in France and Europe:

- on the 6th and 7th of May 1931 at the *Conservatoire National des Arts et Métiers* in Paris,
- in September 1931 at the meeting of the *Société d'Économétrie* in Lausanne,
- on the 22nd of April 1932 at the l'*École Supérieure des Postes et Télégraphes* in Paris,

[4] *Telecom ParisTech*, Alumni, Directory 2009 (internal document).

[5] He had married Dorothy Leeming on 3 April 1924.

[6] For a biography of Le Corbeiller, see Lindsay (1981) and *The New York Times* of 27 June 1980.

- on the 30th of April 1932 at the *Société des Amis de la T.S.F.* in Paris,
- on the 3rd of April 1935 in front of the *Wireless Section* in London.

He also attended, along with Alfred Liénard and Balthazar Van der Pol, the Third International for Applied Mechanics in Stockholm between the 24th and 29th of August 1930, which will be specifically discussed (see *infra*).

6.1.1 *The Third International Congress for Applied Mechanics in Stockholm*

Le Corbeiller's presentation (1931b, 205), entitled: "Sur les oscillations des régulateurs[7]" ("On the oscillations of governors"), began with a note on the existence of analogies between various electrical and mechanical systems:

> The aim of this note is to attract attention to the analogies for these oscillations, which are well-known due to the work of numerous authors, with the analogies found in radioelectricity by using triode lamps. These analogies can easily cover other systems, either governors, or generators of oscillations, and as an example we will describe a continuous temperature regulating system. We ought to mention the general cause of the limitation of the amplitude of these oscillations, as well as their form variations, these two points being the result of Mr. VAN DER POL's work on relaxation oscillations. (Le Corbeiller 1931b, 205)

He therefore established the third order differential equation[8] (see *infra*) allowing the representation of the operating of a governor-machine.

At the same time, he described an electric assembly comprised of a triode exactly analogous to this system, and introduced a stability analysis. He then explained:

> The differences between the regulator-machine system and the oscillating grid-circuit triode-lamp generator must be noted, as well as their similarities. The most obvious one is that the oscillations known as long-period oscillations are the phenomenon we seek to avoid in the first case, and to produce in the second case. (Le Corbeiller 1931b, 209)

This excerpt presents two points of interest. Firstly, the mentioning of "long-period oscillations", observed by Henry Léauté (1885) in a hydraulic vanning-control device, which were soon after viewed as an example of relaxation oscillations. Secondly, the radical change in point of view, namely, that these deleterious oscillations are not sought to be eliminated anymore, but on the contrary, produced

[7]This title is reminiscent of J. C. Maxwell's (1867) "On governors".

[8]The equation (V_9) of the oscillating triode established by Van der Pol (1926d) is a second order differential equation, which has a solution that can be represented in the plane, i.e. in a phase space of dimension two.

in order to be studied.[9] In the conclusion of his presentation, Le Corbeiller made the very first reference to[10] Andronov's work (1929a):

> If we know that the regulator-machine system indeed presents periodic oscillations, this means that among the integral curves drawn on the characteristic surface, there will be at least *one* which is a closed curve. But since the system is not linear with constant coefficients anymore, the infinitely neighboring solutions will not be homothetic to this curve anymore, but will approach it asymptotically, in other words, that the periodic solution will correspond to a *limit cycle* of POINCARÉ's, as noted by Mr. ANDRONOW. Its amplitude will thus indeed be determined. (Le Corbeiller Le Corbeiller 1931b, 211)

Here, Le Corbeiller implicitly recalled Liénard's result (1928) regarding the existence and uniqueness of the periodic solution for this type of oscillator, and explicitly underlined the connection established by Andronov (1929a,b). He finished his article with this sentence:

> I can only refer to the remarkable works of this author,[11] for which Mr. LIÉNARD and Mr. ANDRONOW[12] have provided notably interesting additions. (Le Corbeiller 1931b, 212)

which should be compared to Van der Pol's sentence (1931, 179) (see *supra*). It is surprising to note that in both cases, the results of Mr. Liénard and Mr. Andronov were only viewed as "interesting", when they proved to be fundamental in the development of the theory of oscillations.

6.1.2 Presentations on 6–7 May 1931 at the Conservatoire National des Arts et Métiers

During this first lecture at the Conservatoire National des Arts et Métiers (C.N.A.M.) in France entitled "Les systèmes auto-entretenus et les oscillations de relaxation" ("Self-sustained systems and relaxation oscillations"), Le Corbeiller (1931a) presented Van der Pol's work, and overestimated their impact, hence going from Andronov's (1929a) "theory of self-sustained oscillations" to Van der Pol's (1926d) theory of relaxation oscillations.

> But it was a Dutch physicist, Mr. Balth. van der Pol, who, with his theory of *relaxation oscillations* (1926), furthered the question in a decisive manner. Scientists from various countries are now working on expanding the path he traced. Out of these contributions, we recognize that Mr. Liénard's (1928) seems to be the most important. Remarkably interesting mathematical researches are being conducted by M. Andronow, in Moscow. (Le Corbeiller 1931a, 4)

---

[9]This same behaviors in electrical devices were observed at the end of the 1960s.

[10]He then mentioned it for a second and last time in his presentation to the C.N.A.M. in 1932. See *infra*.

[11]Van der Pol.

[12]Le Corbeiller referred to Liénard (1928), Andronov (1929a), and Andronov and Witt (1930a).

As it is implied by the title of his presentation, he associated the terminology "introduced[13]" by Andronov (1929a), not to a type of oscillation, but to a type of system able to produce relaxation oscillations. Le Corbeiller's presentation (1931a) then starts with mechanical examples, such as the *Tantalus vase*[14] and the *culbuteur* (see Fig. 6.1), but also electrical, such as the oscillation of a neon tube, and the overvoltage circuit breaker in order to illustrate the definition of the self-sustained system. It must be noted that at the time in France, the only people to use the expression "self-sustained'" with this definition were, Blondel (1919d, 118) (see *supra* Part I), Pomey (1921, 120) (see *infra*) and Andronov (1929a). Moreover, the first two examples chosen by Le Corbeiller were described in great detail by Blondel (1919d, 124) and Van der Pol (1930, 307):

> Let (fig. 2) a *Tantalus vase*, *i.e.* a tank in which a tap continuously lets water trickle, with a large bent pipe comes out of the lower part

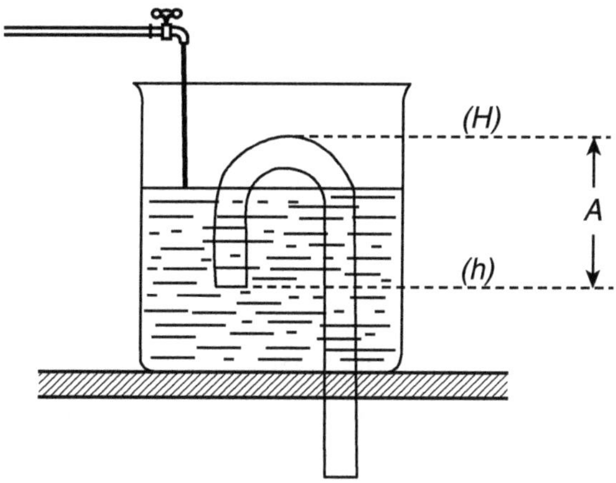

> When the water reaches a certain level (H), the siphon starts (h), and the phenomenon starts over. (Le Corbeiller 1931a, 6)

He also introduced a device corresponding exactly to the description Van der Pol made (1930, 307) during his lectures at the *École Supérieure d'Électricité* (see *supra*):

He then recalled of the principle of Liénard's construction for the graphical integration of the differential equation for a self-sustained oscillator, and explained that to the integral curve "will correspond to one, and only one *limit cycle*." (Le Corbeiller 1931a, 18). It is surprising to observe that Poincaré's name was not cited at all in this article, in which there are no references to his works either. Then, he clarified that it was by following Curie's approach (1891)-which allows, by

[13]See *supra*.

[14]The French denomination, *vase de tantale*, seems to have been introduced by Le Corbeiller. Blondel (1919d, 118) used the term *siphon auto-amorceur* (*self-started siphon*).

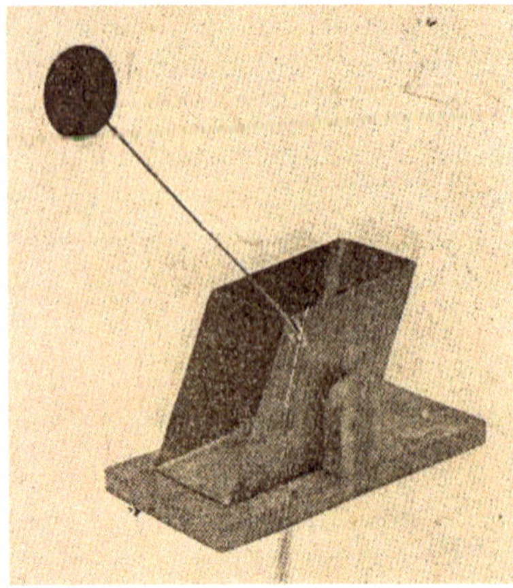

Fig. 6.1 *Culbuteur*, from Le Corbeiller (1931a, 42)

using changes of variables, to find "reduced-form equations" depending on only one dimensionless parameter- that Van der Pol (1926c) managed to integrate his equation (V_9) graphically, by using the "the classic but tiresome *isocline curves* method". Le Corbeiller (1931a, 19) (see *supra* Part I). He then formulated a proposition, which he attributed to Van der Pol, although it seems the latter did not formulate it explicitly:

> We therefore have this proposition (van der Pol): *when* ε *increases from a value very close to zero to a very large value, the steady state satisfying the equation* (V_9) *moves in a continuous manner from a sinusoidal oscillation (for* ε *very small) to a relaxation oscillation (for* ε *very large)*. Le Corbeiller (1931a, 21)

The continuous moving from one type of oscillation to the other is absolutely not trivial, and will be later researched more in-depth. Curiously, Le Corbeiller did not address this fundamental property again until his last presentation in Europe (see *infra*). Then, he recalled Van der Pol's results regarding the amplitude and period of the oscillations (see *supra* Part I) and added that the calculations for the amplitude carried out based on the fundamental component of the Fourier expansion of the differential equation (V_9) had already been carried out by Strutt and Rayleigh (1883) for a differential equation corresponding to Van der Pol's (see *supra* Part I). Finally, he summarized, clearly and objectively, Van der Pol's results (1926d, 1930):

> One of Mr. van der Pol's most substantial contributions has consisted in clearly recognizing this fact,[15] giving a *name* to these non-sinusoidal oscillations, and turning them into a tool for research in physics, in the same way as the sinusoidal equations, of which they actually are the counterpart. Le Corbeiller (1931a, 22)

[15]It alludes to the fact that if the parameter ε is very large, the number of harmonics increases significantly and the calculation of the amplitude and period of the Fourier transform becomes very difficult.

Then, Le Corbeiller described what he called "Mr. Janet's experiment", when it actually is Gérard-Lescuyer's experiment (see *supra* Part I), insisting more on the form of the *oscillation characteristic*:

> (...) the series-dynamo followed by a sufficiently weak ohmic resistance (the one in the coil carrying the inductance L) will have a characteristic symmetrical about the origin, analogous to the one on figure 25.

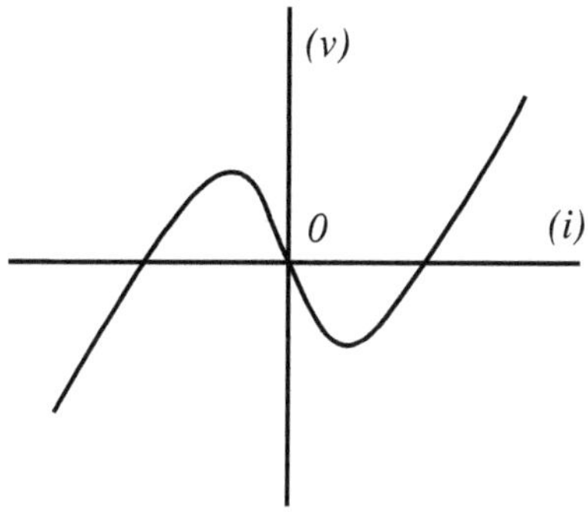

> It must therefore sustain the oscillations in a circuit comprised of the series-dynamo itself, N, an inductance L and a capacitance C. (Le Corbeiller 1931a, 28)

The figure 25 produced by Le Corbeiller transcribes the fact that the *oscillation characteristic* of the series-dynamo is represented by a cubic function, which implies that "such a device presents a negative resistance." Le Corbeiller (1931a, 14). However, even if he admitted that this experiment was "familiar to any engineer who followed Mr. Janet's teachings", he did not refer to Janet's demonstration (1919) which he obviously drew inspiration from in order to establish the equation for the phenomenon (see *supra* Part I).

> Let in effect i be the current in such an motor (with direct current and separate excitation), ω its angular velocity, Φ the permanent flux, ν the number of conductors per radian ($\nu = \frac{N}{2\pi}$), L, R the inductance and resistance of the armature, $\Im$ its moment of inertia, e the potential difference at the terminals, and $\mathcal{C}$ the shaft torque. We find the equations:
>
> $$\begin{cases} e = L\frac{di}{dt} + Ri - \nu\Phi\omega \\ \mathcal{C} = \nu\Phi i + \Im\frac{d\omega}{dt} \end{cases}$$
>
> If the motor runs idle, we find $\mathcal{C} = 0$, which yields:
>
> $$e = L\frac{di}{dt} + Ri + \frac{(\nu\Phi)^2}{\Im}\int idt$$
>
> The motor running idle is therefore equivalent to an inductance L, a resistance R and a capacitance C in series, the capacitance having a value $\frac{\Im}{(\nu\Phi)^2}$. (Le Corbeiller 1931a, 29)

Indeed, by writing $k = -\nu\Phi$ and $K = \Im$, the exact same equation can be found in Janet's article (1919). Finally, Le Corbeiller (1931a, 30) nevertheless recalled that this experiment was originally "conducted by Gérard-Lescuyer in 1880".

Having reached the end of his presentation, he returned to some moderation in regards to Van der Pol's merits:

> The mathematical theory of relaxation oscillations has just started. Several of the phenomena which we have discussed are only known empirically and consequently might prove surprising (...). In short, in front of us is the immense field of research of *nonlinear systems*, in which we have only just begun to advance. I hope that these lectures have shown the interest that comes with its exploration. (Le Corbeiller 1931a, 45)

This falls in step with what was established in the previous section, that is to say, that Van der Pol indeed authored a concept of relaxation oscillations, but not a mathematical theory of relaxation oscillations.

6.1.3 Presentation of September 1931 at the Société d'Économétrie in Lausanne

While the title of this new presentation is identical to the previous one, the contents and audience were completely different. Le Corbeiller first starts by recalling the fundamental role played by the mechanical analogy during this time, which marked the start of the modeling era.

> The research on mechanical model of electrical phenomena was one of the characteristics of the English school of physics during the previous century. Le Corbeiller (1933b, 328)

Here, he hinted at the work of Georges Darrieus (1888–1979) entitled: "Les modèles mécaniques en électrotechnique, leur application aux problèmes de stabilité" (Darrieus 1929) ("Mechanical models in electrical engineering, and their application to stability problems") to which Le Corbeiller (1931a, 148) soon referred to (see *infra*). He then explained that this method has known considerable development in many industries.

> The acoustics industries (phonograph, sound film,[16] ...) received over the span of a few years, an extraordinary level of technical and economic development. The mechanics industries, which have a long history behind them, put up more inertia, but have nonetheless started to evolve (shown by the complete works of Mr. Constantinescu). (Le Corbeiller 1933b, 328)

George Constantinescu (1881–1965) was a Rumanian engineer and physicist, who created the solid mechanics branch and sonics, i.e. the transmission of energy through vibrations. After migrating to London in 1910, he worked in various

[16]The first speaking movie: *The Jazz Singer* (with Al Jolson) was produced in 1927 by Warner Bros.

branches of activity, registering over 130 patents. One of his most famous inventions was the machine-gun synchronization gear allowing the shooting through the spinning blades of the propeller used aboard the planes during the First World War. Then, Le Corbeiller called for the creation of an *Oscillation Theory*[17]:

> The idea is starting to spread that methods born of luck among engineers should be established as a body of doctrines, the *Oscillation Theory*, for which the theorems would in principle be independent from mechanical, thermodynamic, electrical, chemical, physiological, biological or economical applications they might have, but already illustrated by examples taken from these branches. (Le Corbeiller 1933b, 328)

It is essential to note that he stopped using the term "mathematical theory of relaxation oscillations", and instead spoke of methods "which should be established as a body of doctrines, the *Oscillation Theory*". In these methods, Le Corbeiller appears to include Curie's (1891), Poincaré's (1892, 1893, 1899), Van der Pol's (1926a,b,c,d), Andronov's (1929a)...Moreover, the body of doctrines's name seems somewhat prophetic if we consider the one used a few years later by Andronov (1937) as the title of his book. Le Corbeiller then added:

> (...) I wanted to present a short report on recent improvement, which I believe is important, on the theory of oscillations: the one brought to the problem of the self-sustained systems, with the discovery of the relaxation oscillations, owing to the Dutch scientist, Dr. Balth. van der Pol. (Le Corbeiller 1933b, 329)

Le Corbeiller therefore does not consider van der Pol to be the author of the theory any more, but rather as its discoverer, as the inventor of a new type of oscillations. He henceforth considered a general theory of oscillations aiming, amongst other things, at studying self-sustained systems for which the discovery of relaxation oscillations allows to describe the oscillatory phenomenon which can occur in them (see *supra*). This presentation by Le Corbeiller marked a turn in his views on Van der Pol's works, and a returned to a better objectivity in this regard. This is even more visible in his following reports in Paris.

6.1.4 Presentation of 22 April 1932 at the **École Supérieure des Postes et Télégraphes**

This presentation was set in a venue of great importance.[18] As a matter of fact, Le Corbeiller studied in this school, and graduated there as an engineer in 1914. Le Corbeiller (1932) incidentally still called it the *École Professionnelle Supérieure des Postes et Télégraphes*, although it lost the denomination "Professionnelle" in 1912 (Atten et al. 1999, 52). It was also the venue for Poincaré's lectures (1908) (see

[17]A short time later, Krylov and Bogolyubov (1932c, 1933) called for it (see *infra*).

[18]The title and text of this presentation are absolutely identical to the one Le Corbeiller (1933a) presented on the 30th of April 1932 in front of the *Société des Amis de la T.S.F.*

supra Part I). This raises the following question: how was he not aware of this work? Indeed, any engineering student at the E.S.P.T. is required to attend these lectures dispensed by contributors from outside the school during the second year of their studies. Although Le Corbeiller's graduation and Poincaré's lectures are six years apart, one would have thought that traces of them would remain. Unfortunately, it seems this is not the case. Once again, his presentation started by a reminder the analogy's significance:

> The analogy between electrical and mechanical oscillations is now well-known, and it is barely necessary to say that the study of purely electric systems which follows is applicable, *mutatis mutandis*, to purely mechanical systems. (Le Corbeiller 1932, 698)

He then gave the definition of a self-sustained system, and his use of terminology (self-oscillator) shows a perfect knowledge of Andronov's work (1929a) although he paradoxically did not refer to it.

> Such a system, sustaining its own oscillations, is called a *self-sustained system*, or self-oscillator, and it is essentially a *nonlinear system*. (Le Corbeiller 1932, 698)

However, this article contains only one occurrence of the "relaxation oscillation" term introduced by Van der Pol, which appears on the tenth page, and to which Le Corbeiller (1932, 704) ostensibly preferred the term "self-sustained oscillations". He then recalled that "the necessity of writing a *nonlinear* differential equation in order to describe the periodic regime of a self-sustained system was clearly recognized by Strutt and Rayleigh (1883)" (Le Corbeiller 1932, 704). But in this article, Le Corbeiller goes much further by demonstrating a bijective correspondence between Rayleigh's equation (R_5) and Van der Pol's equation (V_9) (see *supra* Part I).

Previously, he drew attention to the grandfathering of Blondel's work (1919b) in relation to Van der Pol's (1920), in regards to the writing of an equation for the triode, the calculation of the amplitude and period of oscillations (see *supra* Part I). He then studied Van der Pol's equation (V_9), clarifying that his periodic solution "corresponds"[19] to a closed integral curve, and added:

> The general theory of the closed integral curves, or *limit cycles*, was established by H. Poincaré (^{2}) [1]. The demonstration of the existence of a unique limit cycle, in the present case, is owed to Mr. Liénard [16].
>
> ---
>
> (^{2}) To put it simply, the equation $\dfrac{dx}{X(x,y)} = \dfrac{dy}{Y(x,y)}$ is equivalent to the two following: $\dfrac{d^2x}{dt^2} - y\left(\dfrac{Y}{X} + x\right) + x = 0,\ y = \dfrac{dx}{dt}$. Poincaré's mentioned work is therefore equivalent to the studying of the conservative oscillating system $\dfrac{d^2x}{dt^2} + x = 0$, subject to dissipation and maintenance forces for which the resistance is *some* force of x and $\dfrac{dx}{dt}$:

[19] Le Corbeiller (1932, 708). This is exactly the term which Andronov used (1929a). See *supra*

$$F\left(x, \frac{dx}{dt}\right) = y\frac{Y}{X} + x.$$

> [1] H. POINCARÉ, Mémoire sur les courbes définies par une équation différentielle, Deuxième partie, *Journal de Mathém. Pures et app.* 8, 251, 1882 ; et *Œuvres*, T. 1, p. 44.
> [16] A. LIÉNARD, Étude des oscillations entretenues, Rev. gén. d'électr., 901 et 946, 1926.
> (Le Corbeiller 1932,708–709)

This excerpt is especially interesting as it shows the evolution of Le Corbeiller's point of view, which therefore went from Van der Pol's "mathematical theory of relaxation oscillations" to a "*Theory of oscillations*", and resulted here in a "general theory" which was established by Poincaré. After "promoting" Van der Pol's relaxation oscillation theory, he re-established the merits of some of the discoveries constituting fundamental steps towards the constitution of a *Theory of Oscillations*. This is notably the case with Liénard's work, of which he stressed the importance, but it is also especially the case for Poincaré's work on *limit cycles* which had, until then, never been mentioned. This footnote (2) provides information on said subject, showing precious indications on Le Corbeiller's level of understanding of this problem, for which he visibly grasped the main aspects, as indicated by the next part of the paragraph.

> Indeed, the thorough qualitative study of the solutions to a second order differential equation (to which the one for a self-oscillator would coincide) has yet to be carried out, despite Henri Poincaré's best effort. This allows us to appreciate how far we are from addressing the study of an equation of a higher order (Le Corbeiller 1932, 710)

It is on the one hand striking to see Andronov's neologism "self-oscillator" (1929b) (see *supra*) instead of "self-sustained systems" which Le Corbeiller used until then. On the other hand, he shows incredible perspicacity in regards to a result which would be the analogue in three dimensions of the Poincaré-Bendixson theorem,[20] and we nowadays know that it can never be established.

He continues by recalling that in order to create a self-sustained system, the presence of a dipole whose *nonlinear characteristic*[21] (i, v) presents a "falling" part is required. This implies that the directing coefficient of the tangent to the representative curve of this characteristic $v = f(i)$ is negative at any point of the falling part, which is written as: $\frac{dv}{di} < 0$.

> This quantity $\dfrac{dv}{di}$ possesses the physical dimensions of a resistance, and it is therefore often called a *negative resistance*. Although confusions, even mistakes, have been provoked by the use of this inaccurate expression. We will give the name *variance* to what follows this quantity $\dfrac{dv}{di}$ as proposed by Mr. Blondel in 1919 in an important paper ([1]).
>
> ---
>
> 1. The term negative resistance is, on the contrary, correct in the case of a series dynamo or a device with analogous properties. (Le Corbeiller 1932, 717)

---

[20]This theorem allows the demonstration a periodic solution's existence, or *limit cycle*, for the self-sustained two-dimension systems, as demonstrated in Van der Pol's system.

[21]See *supra* Part I.

Here, Le Corbeiller demonstrates his perfect knowledge of Blondel's (1919d, 124) and most probably Janet's work (1919), the footnote seems to indicate a reference being made to the triode and series-dynamo machine analogy. In the last part of his presentation, Le Corbeiller (1932, 723) accurately describes the principle of Liénard's construction (1928) (see *supra* Part I).

6.1.5 Presentation on the 3rd of April 1935, in Front of the **Wireless Section** *in London*

On the 3rd of April 1935, Le Corbeiller (1936) gave, in front of the Wireless Section of the Institution of Electrical Engineers what seems to be his last lecture in Europe,[22] in which he went over the same themes he used in previous presentations, with some modifications. Indeed, he again described the principle of Liénard's construction (1928) enabling the possibility to obtain the periodic solution to Van der Pol[23] equation (V_9) in the form of Poincaré's *limit cycles*, which from then on possesses the following properties:

> (...) the closed curve represents a periodic, sustained oscillation of our series electrical system. (Le Corbeiller 1936, p. 364)

It is interesting to note that even in Appendix 2, where he very succinctly summarized Poincaré's work, Le Corbeiller did not translate the expression *cycle limite* by *limit cycle*, but by *closed curve*. He then addressed the main characteristic that Andronov formulated (1929a, 560) (see *supra*), without referring to it:

> (...) there is only one such closed curve, no matter from what initial conditions we may have started (...) (Le Corbeiller 1936, 364)

Excepted for Léauté's work (1885, 76), he was one of the first French researchers to thus evidence the fact that the period and amplitude of such a system are "insensible to initial conditions". This remarkable property was used by Rocard (1941, 38) a few years later (see *infra*).

Another particular attribute of relaxation oscillators, already noted by Le Corbeiller (1931a, 21) (see *supra*), is again mentioned:

> We see, then, that when the parameter ε varies continuously from zero to infinity, the corresponding periodic oscillations vary continuously from the sinusoidal type to the relaxation type. (Le Corbeiller 1936, 366)

This "Van der Pol proposition" was the subject of a study by Jean Abelé in his thesis dissertation (see *infra*). Le Corbeiller then addressed Gérard-Lescuyer's experiment and the series-dynamo machine. At the very start of his presentation, he clarified:

---

[22]He emigrated to the United States during the Second World War.

[23]He wrote it with the form of Rayleigh's (R_5) by using the *duality principle* (see *supra* Part I).

> Therefore, a series dynamo should be able to sustained oscillations in series circuit. (see Appendix 1). (Le Corbeiller 1936, 362)

Unlike Van der Pol who established that this device is the seat of relaxation oscillations (see *supra* Part I), he did not view it as the oldest device allowing the observation of this oscillatory phenomenon anymore, and preferred the study of the oscillations of a neon lamp carried out by Professor Augusto Righi (1902).

> Oscillations of the relaxation type had been met with earlier by various physicists, and, to quote an early instance, Prof. Righi had in 1902 given a thorough study of oscillations of a neon lamp. (Le Corbeiller 1936, 366)

Le Corbeiller (1936, 374) however presents, in Appendix 1, Gérard-Lescuyer's experiment, which he had already called "Janet's experiment" (Le Corbeiller 1931a, 28) and now called "The Series Dynamo Oscillator".

> In the days of d.c. (direct current) power transmission it was found that if a d.c. motor, with separate excitation, were fed by a series dynamo, the motor would reverse periodically. (Le Corbeiller Le Corbeiller 1936, 374)

Le Corbeiller (1936, 374) again established the differential equation for oscillations occurring in Gérard-Lescuyer's experiment, drawing both from Janet's results (1919), to which he did not refer, and Van der Pol's results (1927c), which he explicitly quoted. He then evidenced the analogy between this equation and the one for relaxation oscillations, i.e. Van der Pol's (1926d). Moreover, he did not attribute to Janet (1900, 222) the merit of providing the explanation of the phenomenon anymore, crediting Van der Pol (1927c, 117) instead.

> The explanation given by van der Pol,[24] is that between the brushes of the dynamo, D, we have an e.m.f. v depending upon the current i according to equation (8) or Fig. 3. (Le Corbeiller 1936, 374)

Despite the fact that Le Corbeiller appeared quite enthusiastic during his first lecture (1931) in France regarding Van der Pol's work, leading him to crediting him as the author of a "mathematical theory of relaxation oscillations", his viewpoint quickly evolved. He considered that this concept had to be integrated into the broader frame of a theory of (nonlinear) oscillations, which he called for. He then stressed the importance of the contribution of French research such as Curie's work (1891) for the reduced-form writing of the differential equation governing the phenomenon, and the graphical integration, and Liénard's work (1928) for the existence and uniqueness of the periodic solution to the equation. During his two other presentations in Paris, Le Corbeiller (1932) recalled that the very first writing of the equation for the triode is owed to Blondel (1919b), and not to Van der Pol (1920). Then, he demonstrated that the equation (V_9) established by Van der Pol (1926d) to characterize relaxation oscillations can be reduced to Lord Rayleigh's (1883) with a simple change in variables. He then stressed the importance of correspondence between the periodic solution to Van der Pol's equation (V_9) and

[24] Le Corbeiller referred to Van der Pol (1927c).

Fig. 6.2 Philippe Le Corbeiller (left) and Nicolaas Bloembergen, from A.I.P.

Poincaré's concept of *limit cycle* (1882). At the same time, he precisely defined self-sustained or self-oscillating systems, and used Andronov's terminology (1929a), while almost never mentioning its reference. He nonetheless seems to have given a lot more credit to his work than it could have appeared, and it is indeed inside the conceptual frame of the theory of self-sustained oscillations that he included and introduced Van der Pol's research (Fig. 6.2).

6.2 Alfred Liénard's Work: *From Sustained Oscillations to Self-Sustained Oscillations*

Previous explanations lead us to expect that Van der Pol's (1926c) and Andronov's (1929a) articles had a notable impact on the French scientific community. As a matter of fact, less than six months after Andronov's note got published, Van der Pol (1930, 16) recognized during a presentation given in France (see *supra*) that the periodic solution to his equation (V_9) corresponded to Poincaré's *limit cycle*.

– A similar reaction could have been expected from Liénard. –

In 1928, Liénard, who demonstrated the existence and uniqueness of this periodic solution in an article entitled: "A study on sustained oscillations" (Liénard 1928), did not account for the problem of duality of the oscillatory phenomenon that Van der Pol had called relaxation oscillation.

Three years later, when he was invited to the Third International Congress for Applied Mechanics in Stockholm, his presentation was entitled: "Oscillations auto-entretenues" ("Self-sustained oscillations"). Liénard (1931) clearly shows his preference for Andronov's terminology. He then recalled the main results obtained in his article published in 1928, in these terms:

> If the function $F(x)$, negative between O and a certain value $X_0 > 0$ of x, then becomes positive and increasing, or at the least, never decreasing, there exists only one closed integral curve D. (Liénard 1931, 176)

However, it appears that Liénard had established no correspondence between this "closed integral curve" and Poincaré's *limit cycle* after Andronov's publication (1929a).

– Did he not know about it? –

In this article, Liénard (1931, 177) demonstrates the contrary, by addressing Andronov and Witt's demonstration (1930a) on the "Lyapunov stability" of the periodic solution, i.e. of the *limit cycle*. Before that, he used a clever change of variables to transform the differential form with a degree of one (3.22), to an exact differential form in order to calculate his "characteristic exponent" (see *supra* Part I). Starting with the equation (5.6), he proceeded by writing

$$xdx + (y - F(x))\,dy = 0 \tag{6.1}$$

he used the "calculus of variations" introduced by Poincaré (1892, 162), which consists in substituting in the equation (3.22) the ordered pair (x, y) with the ordered pair $(x + \delta x, y + \delta y)$, where the variations δx and δy are such that

$$\delta x = x\delta n \quad \text{and} \quad \delta y = (y - F(x))\,\delta n$$

represent the direction parameters of a normal with a speed vector field (L_4). We therefore obtain, by differentiation and taking into account the fact that $F'(x) = f(x)$:

$$d\delta x = xd\delta n + dx\delta n \quad \text{and} \quad d\delta y = (y - F(x))\,d\delta n + [dy - f(x)\,dx]\,\delta n$$

By writing: $u = \left[x^2 + (y - F(x))^2\right]\delta n$ the equation (3.22) yields

$$du - \frac{f(x)\,dy}{x}u = 0 \tag{6.2}$$

which is trivially integrated

$$u = Ke^{\int \frac{f(x)dy}{x}} \tag{6.3}$$

Liénard then explained that:

> If the primitive integral curve is a closed curve, then the factor $x^2 + (y - F(x))^2$ has the same value at the points 5.1 and 5.3 (fig. 3)[25] and the value relation of δn is the same as u, and it is again the same (to the sign) to the value relation of δy. (Liénard 1931, 176)

By substituting dy with its value $-\omega xdt$, he finally obtained:

$$Lg\frac{u_{A_2}}{u_{A_1}} = \int_{A_1}^{A_2} \frac{f(x)\,dy}{x} = -\omega \int_{A_1}^{A_2} f(x)\,dt$$

This led him to a stability condition for the "closed curve":

$$\int_{A_1}^{A_2} f(x)\,dt > 0 \tag{6.4}$$

It is essential to note that, contrary to the stability condition (5.6) (see *supra*), the condition (L19) is expressed inside polar coordinates, and not cartesian ones. It is therefore not with the condition (5.7) defined by Andronov (see *supra*) that a comparison must be established anymore, but with the condition (A_5):

$$\int_0^{2\pi} \left[f'_x\left(R\cos(\xi), -R\sin(\xi); 0\right) + g'_y\left(R\cos(\xi), -R\sin(\xi); 0\right)\right]d\xi < 0$$

By comparing Liénard's system with Andronov's (see *supra* Table 5.1), by writing: $k = 1, f(x, y; \mu) = -F(x)$ and $g(x, y; \mu) = 0$ and substituting in (A_5), it yields:

$$\int_0^{2\pi} -F'(x)dt < 0 \Leftrightarrow \int_0^{2\pi} f(x)dt > 0$$

We can deduce that the stability condition (6.4) established by Liénard (1931) is absolutely identical to Andronov's (A_5) (1929a) and to the condition (5.6) that Liénard (1928) had obtained previously (see *supra*) with a different process.

In the next part of his presentation, Liénard (1931) offers a generalization of the result provided a few months earlier by Andronov and Witt (1930a) on the "Lyapunov stability". In order to insure this, he modified the equation (3.20) he had been using until then (see *supra* Part I) and substituted it by the following:

$$\frac{d^2x}{dt^2} + \omega f\left(x, \frac{dx}{dt}\right) + \omega^2 x = 0 \tag{6.5}$$

[25] See *supra* Part I or Liénard (1931, 175).

He then wrote:

> If the equation (6.5) admits a periodic solution, with a period T, the condition for this solution to be stable is that the integral, for a period of $\frac{\partial f(x, x')}{\partial x'} dt$ must be positive. The proposition, established by Misters ANDRONOW and WITT ([26]) in the specific case where the function $f(x, x')$ is very small, is then immediately generalized. (Liénard 1931, 177)

This second stability condition established by Liénard is therefore written:

$$\int_0^T \frac{\partial f(x, x')}{\partial x'} dt > 0 \tag{6.6}$$

On January the 20th of 1930, Jacques Hadamard introduced a note by Russian mathematicians Andronov and Witt, entitled: "Sur la théorie mathématique des auto-oscillations" ("On the mathematical theory of self-oscillations") (Andronov and Witt, 1930a). Firstly it is important to note that the term "self-sustained oscillation" introduced (see *supra*) by Andronov (1929a) in his previous note, was replaced by "self-oscillation". It is interesting to compare Andronov and Witt's approach with Liénard's.

> In this Note, we will focus on the study of the periodic solution to the differential equations system
>
> $$\ddot{\xi} + \omega_1^2 \xi = \mu f\left(\xi, \dot{\xi}; \eta, \dot{\eta}; \mu\right), \quad \ddot{\eta} + \omega_2^2 \eta = \mu g\left(\xi, \dot{\xi}; \eta, \dot{\eta}; \mu\right) \tag{6.7}$$
>
> where μ is a parameter which can be chosen arbitrarily small, and where $\dot{\xi} = \frac{d\xi}{dt}, \ddot{\xi} = \frac{d^2\xi}{dt^2}$, etc. (Andronov and Witt 1930a, 256)

It is a system with two degrees of freedom, and for which the solution can be represented by a rolling of the characteristics on the surface of a torus.[27] In the case where the parameter μ is null, the solution is periodic, and Andronov and Witt (1930a, 237) recommend Poincaré's approach:

> According to Poincaré ([1]), it is easy to obtain a periodic solution for μ being sufficiently small, with the form of an ordered series following the potencies of μ, converging when μ is sufficiently small (...)
>
> ---
>
> ([1]) H. POINCARÉ, Les méthodes nouvelles de la Mécanique céleste, 1, p. 89, Paris, 1892. (Andronov and Witt 1930a, 237)

It was indeed by using Poincaré's work (1892, 1893, 1899) on "periodic solutions" that Andronov (1929a) managed to establish the *limit cycle*'s stability (see *supra*). Using the same method, Andronov and Witt (1930a, 237) established

---

[26] Liénard referred to Andronov and Witt (1930a).

[27] This problem had already been studied by Poincaré (1885a, 220) and was addressed two years later by Denjoy (1932a,b). See also Arnold (1974, 33 and 215)

Table 6.1 Differential equations found by Liénard (1931), Andronov and Witt (1930a)

Liénard (1931)	Andronov and Witt (1930a)
$\ddot{\xi} + \omega^2 \xi = -\mu F\left(\xi, \dot{\xi}\right)$	$\ddot{\xi} + \omega_1^2 \xi = \mu f\left(\xi, \dot{\xi}; \eta, \dot{\eta}; \mu\right)$

the "Lyapunov stability" of these self-oscillating systems. Indeed, by writing "in the usual way the equations at the variations ([28])" they obtained the "characteristic exponents" which are reduced, when μ is sufficiently small, to the following conditions:

$$\begin{cases} \int_0^{\frac{2\pi}{\omega_1}} f'_{\dot{\xi}}\,(R\cos\omega_1 q, -\omega_1 R\sin\omega_1 q; 0, 0; 0)\, dq < 0 \\ \int_0^{\frac{2\pi}{\omega_1}} g'_{\dot{\eta}}\,(R\cos\omega_1 q, -\omega_1 R\sin\omega_1 q; 0, 0; 0)\, dq < 0 \end{cases} \tag{6.8}$$

It was by using Poincaré's "calculus of variations" (1892, 162) that Andronov, Witt (1930a), and then Liénard (1931) found a term for a condition corresponding to the characteristic exponent of the studied differential equation,[29] allowing statements on the stability of the periodic solution i.e., of the *limit cycle*.

In order to establish a comparison with the condition submitted by Liénard (1931), we must put Andronov and Witt's equation (1930a) as well as Liénard's (1931) in a same form,[30] presented in the synoptic chart below (see Table 6.1).

We can immediately deduce that: $-F\left(\xi, \dot{\xi}\right) = f\left(\xi, \dot{\xi}; \eta, \dot{\eta}; \mu\right)$.

The integrand of the first condition (6.8) is then written

$$f'_{\dot{\xi}} = \frac{\partial f}{\partial \dot{\xi}} = -\frac{\partial F}{\partial \dot{\xi}} = -F'_{\dot{\xi}}$$

By substituting in the first condition (6.8), it yields

$$\int_0^{\frac{2\pi}{\omega_1}} f'_{\dot{\xi}}\left(\xi, \dot{\xi}; 0, 0; 0\right) dt = \int_0^{\frac{2\pi}{\omega_1}} -F'_{\dot{\xi}}\left(\xi, \dot{\xi}\right) dt < 0$$

Finally, we obtain the following inequality:

$$\int_0^{\frac{2\pi}{\omega_1}} F'_{\dot{\xi}}\left(\xi, \dot{\xi}\right) dt > 0$$

[28] Andronov and Witt again referred to Poincaré (1892, 162).

[29] The negativity of the characteristic exponent of the differential equation studied.

[30] Liénard's function is written in order to differentiate it from Andronov and Witt's (1930a). We wrote: $x = \xi$ and $\omega = \mu$ in the second member to allow for a comparison.

This stability condition is identical to the condition (6.6) introduced by Liénard (1931, 177) (see *supra*). Liénard's (1931, 177) demonstration therefore makes it certain that he had read Andronov and Witt's article (1930a).

– Why is there no mention of Poincaré's limit cycles in his article? –

However, there is a reference to limit cycles in a footnote on the very first page of the article:

> ([2]) For the definition of self-oscillations and the discussion on the case of one degree of freedom see A. ANDRONOW, *Les cycle limites de Poincaré et la théorie des oscillations auto-entretenues* (Comptes rendus, 189. 1929. p. 559). (Andronov 1929a,b, 559)

We can therefore assume that even if he had not read Andronov's article (1929a), Lienard must have at least seen the title, in which there is a relatively explicit correspondence between self-sustained oscillations and Poincaré's limit cycles. Moreover, Liénard, Van der Pol and Le Corbeiller were all participant to this congress. Assuming that they could have only sent a written contribution, Liénard's (1931, 173–177) preceded Van der Pol's (1931, 178–180) in which the latter explicitly referred to Poincaré and Andronov (see *supra*). It is consequently unlikely that he did not at least read Van der Pol's article.

– Why did Liénard not establish this connection? –

Out of all the mathematicians related to this problem, such as Élie and Henri Cartan, he was the most likely to set up this correspondence. Indeed, he not only demonstrated the existence and uniqueness of the periodic solution in a more general manner than Van der Pol's equation (V_9), but he also offered a graphical integration method allowing the representation of this solution (Liénard 1928, 902–906). Although several hypotheses may be considered, it is difficult to find an answer; especially as three years later, Liénard wrote and published a note on his scientific work, curiously enough the title of his 1928 article became: "Étude des oscillations auto-entretenues" ("A study of self-sustained oscillations") (Liénard 1934, 18), and once again no mention of Poincaré's limit cycle can be found in its summary.

Chapter 7
The First International Conference on Nonlinear Processes: Paris 1933

7.1 The First International Conference on Nonlinear Processes: *The Forgotten Conference?*

7.1.1 The "Three Sources" Enigma

In a famous article entitled "The nonlinear theory of electric oscillations", published in 1934 in the *Proceedings of the Institute of Radio Engineers*, Balthazar Van der Pol ended his introduction by saying:

> Although the first researches in connection with our subject date back to 1920 and although the development of this theory has gradually continued ever since, recent years have shown a considerable increase of activity in this field by many research workers scattered all over the world, and a special international conference dedicated solely to the problems arising in the nonlinear oscillation theory was recently held in Paris, on January 28–30, 1933. (Van der Pol 1934, 1051)

During the celebration of the hundredth anniversary of Papalexi's birth in 1981, Vladimir Vasilevich Migulin explained that during the first international conference on nonlinear oscillations, which was held in January 1933 in Paris, Nikolai Dimitrievich Papalexi presented two articles on the research carried out in U.S.S.R. in this field. Twenty-five years later, Russian academician Evgenu Lvovich Feinberg wrote in homage to Papalexi:

> It is not surprising that, when the first international conference on nonlinear oscillations was convened in Paris in 1932 (among its participants were such pioneers in this field as B. Van der Pol, L. Brillouin, and others), it was Papalexi who represented the Moscow school of Mandel'shtam and Papalexi, their closest disciples and colleagues Andronov, A. A. Vitt, Khaikin, and others, reporting on its achievements. (Feinberg 2006, 67)

Apart from these references, no literary documents concerning this conference can be found. There is no indication regarding the venue where it was held in Paris, the list of participants, or the program. It therefore seems reasonable to wonder if

J.-M. Ginoux, *History of Nonlinear Oscillations Theory in France (1880–1940)*,
Archimedes 49, DOI 10.1007/978-3-319-55239-2_7

this event actually happened, in other words, if it wasn't simply a "vague project" mentioned in several places, but which for unknown reasons would never take place. Moreover, according to Van der Pol (1934) and Migulin (1981), this conference, held in Paris in January 1933, created international echoes, it would at least be expected to find the announcement in the French scientific periodic publications and journals such as the *Comptes Rendus de l'Académie des Sciences de Paris* or the *Revue Générale des Sciences Pures et Appliquées*, or *Le Figaro* as it was the case for the *Congrès international d'Électricité* held in Paris in July 1932 (*C.R.A.S.* 1932, t. 194, 2192). Indeed, *Le Figaro* wrote on the 5th of July 1932 that the opening ceremony was held at the Grand Amphithéâtre of the Sorbonne, in the presence of French President Mr. Albert Lebrun as well as the minister of National Education. Interestingly enough, nothing can be found in the *C.R.A.S.*, or in the *R.G.S.P.A.*, or in this journal regarding the first *Conférence Internationale de Non linéaire*. This could imply that this conference could have not been organized by the French community. All the more so that the title of this conference seems to indicate that it could have been the first I.C.N.O. (International Conference on Nonlinear Oscillations). Nevertheless, almost all the I.C.N.O. presented in Table 7.1 were identified, excepted for the second one. According to Professor Mikhail Rabinovich (in a personal communication), even though it was a national event, the conference was entitled "First Soviet Conference on Oscillations", and was held in Moscow in November 1931. Since then, all the I.C.N.O. were held in Eastern Europe countries, every three years (see Table 7.1). It therefore appears that the first *Conférence Internationale de Non linéaire* cannot be the first I.C.N.O.

While Van der Pol (1934) and Migulin (1981) spoke of a conference held in 1933, Feinberg (2006) stated that the dates for this first conference were predetermined in 1932. Is it the same conference? Which circumstances led them to decide in 1932? Finally, Migulin and Feinberg stated that Van der Pol, Brillouin and Papalexi were present. The first two's biographies and bibliographies do no reveal, however, any trace of this conference.

Table 7.1 International Conference on Nonlinear Oscillations (I.C.N.O.)

International Conference on Nonlinear Oscillations (I.C.N.O.)		
Edition	Year	Town
Ist	1961	Kiev
IInd	1962	Varsovie
IIIrd	1964	Berlin
IVth	1967	Prague
Vth	1969	Kiev
VIth	1972	Poznan
VIIth	1975	Berlin
VIIIth	1978	Prague
IXth	1981	Kiev
Xth	1984	Varna

The only remaining possible proof is a publication by Papalexi relating this event.[1] In an article published in 1934 in the periodic *Zeitschrift für Technische Physik* and entitled: "Международная нелинейным конференция" which means, "*Conférence Internationale de Non linéaire*", Papalexi (1934) exposed the minutes of this assembly in great detail. Its translation allows a description of the way the first *Conférence Internationale de Non linéaire* proceeded, and solves, notably, the enigma posed by Feinberg's statement.

(...) The researchers working in this field had the idea during the International Congress of Electricity, which was held in Paris in July 1932, to try and establish international contact between the physicists and engineers on one side, and the mathematicians on the other. (Papalexi 1934, 209)

Discovering this reference led to the finding of other sources mentioning. Firstly, in Andronov's autobiographical note, we found a letter of recommendation written by Mandel'shtam, entitled "examination of Andronov's scientific works" dated from the 11th of June 1933, in which we can read:

Ценность и значение работ А. А. Андронова и А. А. Витта в области нелинейных колебаний вполне признаны за границей. На конференцию по нелинейным колебаниям, имевшую место в январе сего года в Париже, и А. А. Андронов, и А. А. Витт были приглашены для доклада о своих работах.[2] (Andronov 1943, 4)

Secondly, in the first volume of Mandel'shtam's collected work, there is a brief summary of his life and scientific activity written by Papalexi, which relates this event:

На первую Международную конференцию по нелинейным колебаниям в Париже в 1933 г. были приглашены, кроме Леонида Исааковича, одного из инициаторов этой конференции, также Н. Д. Папалеши, А. А. Андронов и А. А. Витт (принимал в ней участие лишь Н. Д. Папалеши), а в 1934 г., по приглашению оргбюро, подробный доклад об исследованиях в области нелинейных колебаний, проведенных у нас, и составленный Л. И. Мандельштамом, Н. Д. Папалеши, А. А. Андроновым, А. А. Виттом и С. Э Хайкиным, был представлен Конгрессу Радиофизического союза (URSI[3]) в Лондоне.[4] (Mandel'shtam 1947–1955, vol. 1, 30)

[1]This research made possible through the insight of M[lle] Gaelle Chapdelaine.

[2]The value and importance of work of A. A. Andronov and A. A. Witt in the field of nonlinear oscillations is also recognized abroad. Thus, during a conference on non-linear oscillations, which took place in January this year in Paris, A. A. Andronov and A. A. Witt, were invited to make a presentation of their work. (Andronov 1943, 4)

[3]Author referred to the Russian version of the article of Mandel'shtam et al. (1935)

[4]At the first International Conference on nonlinear oscillations which took place in Paris in 1933 were invited, except Leonid Isaakovich, one of the initiators of the conference, the other being Papalexi: A. A. Andronov and A. A. Witt (Papalexi was the only one to participate). In 1934, at the invitation of the organizing committee, a detailed report on the research in the field of nonlinear oscillations was conducted and summarized by Mandel'shtam, Papalexi, A. A. Andronov, A. A. Vitt and S. E. Khaikin and has been submitted to the general assembly of the International Union of Radio Science (U.R.S.I.) in London. (Mandel'shtam 1947–1955, vol. 1, 30)

Lastly, this conference is also mentioned in Aleksandr Adolfovich Witt's biography written by Bendrikov and Sidorova:

> В январе 1933 г. в Париже состоялась « Международная нелинейная конференция », на которой присутствовали : Ван-дер-Поль (Голландия), Льенар, Картан, Эксклагон, Абрагам, А. Бриллюэн, ле Корбелье (Франция) и другие видные математики и физики из различных стран [[5]]. От Советского Союза на конференцию были приглашены Л. И. Мандельштам, Н. М. Крылов, П. Д. Папалеши, А. А. Андронов и А. А. Витт. Однако принять участие в конференции смог лишь Папалеши.[6] (Bendrikov et al. 1981, 159)

As for the French researchers who attended the event, there is no mention of it in their publications, as if it did not occur. The impact of this event of the French scientific community, and it importance in the developing of the theory of nonlinear oscillations in France will therefore be the subject of an in-depth analysis.

7.1.2 *The Venue: The Henri Poincaré Institute*

As told by Papalexi (1934, 210) "The conference was held at the *Institut de Recherche Henri Poincaré* on the 28th, 29th and 30th January 1933". Unfortunately, research carried out at the Henri Poincaré Institute did not help in finding any trace of this conference. However, it was possible to establish that notorious scientists such as Nikolai Krylov and Nikolai Bogolyubov were invited in 1935.[7] Papalexi then explained that the last day was dedicated to the visit of Abraham's laboratory at the *École Normale Supérieure*, where the conference came to an end. Investigations were carried out at this school, with the kind support of Mrs. Dauphragne, but yielded no result (Fig. 7.1).

7.1.3 *The List of Participants*

Regarding the participants, this conference was supposed to bring together internationally renowned mathematicians and physicists, as recalled by Papalexi:

[5]Bendrikov and Sidorova referred to Papalexi (1934).

[6]In January 1933, held in Paris an international conference on nonlinear processes to which attended Van der Pol (Netherlands), Lienard, Cartan, Esclangon, Abraham L. Brillouin, The Corbeiller (France), and other eminent mathematicians and physicists from different countries []. Representatives of the Soviet Union invited to the conference were Mandel'shtam, N. M. Krylov, N. A. Papalexi, A. A. Andronov and A. A. Witt. However, the only one to take part in the conference was Papalexi. (Bendrikov et al. 1981, 159)

[7]Investigations were carried out at this school, with the kind support of Mrs. Brigitte Yvon-Deyme and M. Dominique Dartron.

Fig. 7.1 Institut Henri Poincaré, from Mosseri (1999, 121)

> A lot of individuals from various countries who are working in this field, and whose collaboration was wished for, were invited to this conference. Amongst them, we can mention Professor Volterra, who used mathematical analysis in order to answer the questions of fluctuations in animal species in the struggle for life, mathematicians such as Hadamard, Cartan, Esclangon, as well as the instigator of the conference, Van der Pol. As for the U.S.S.R., the academicians L. I. Mandel'shtam, N. M. Krylov and N. D. Papalexi, author of this article, had also been invited, as well as young students at Mandel'shtam's academy: Andronov and Vitt. (Papalexi 1934, 210)

The world's specialists in the field of nonlinear oscillations were invited on the initiative of Van der Pol, whose unifying role will be explained. This extraordinary line-up could have foreshadowed a laudable outcome, since it could have led to the "second birth" of the theory of nonlinear oscillations that Mandel'shtam and his "students" were developing. Unfortunately, as Papalexi explained, due to the influenza epidemic that hit Europe at the time, a lot of the guests did not attend. In the end, this conference was held in the presence of:

- Balthasar Van der Pol (Netherlands),
- Alfred Liénard (1869–1958), Élie Cartan (1869–1951) et Henri Cartan (1909–1008), Ernest
  Esclangon (1876–1954), Henri Abraham (1868–1943), Léon Brillouin (1889–1969), Philippe
  Le Corbeiller (1891–1980), Yves Rocard (1903–1992) and Camille Gutton (1872–1963) (France),
- Charles Manneback (Belgium) and
- Nikolaï Papalexi (U.S.S.R.).

Even though the list was considerably shortened, most of the members of the French scientific community, whose members were very involved in the field of nonlinear oscillations, were present. As for the Belgian researcher Charles Manneback (1894–1975), who was more specialized in the field of *Electromagnetism and Electrodynamics*,[8] his presence in such an assembly is startling. However, the "Note on Charles Manneback" written by Marc de Hemptine and Maurice A. Biot, and published in the 1978 *annuaire* of the Royal Academy of Belgium explains it:

> His educational skills, very well-liked in other countries, led him to being invited to a great number of events, whether in Europe or America. Hence, let us mention, amongst others between 1932 and 1933, an invitation to a series of lectures at the Poincaré Institute.[9]

The fact that it was organized at Van der Pol's initiative shows from the list of personalities who were invited.

> The aim of this International Conference on nonlinear oscillation (which is the subject of an article and was called for at the initiative of one of the pioneers of this branch, prof. Van der Pol), was to gather researchers from various countries who worked in this field, and to give them the opportunity to discuss and exchange opinions on the problems related to nonlinear oscillations, and to establish a common terminology, as well as define, at least partially, the direction future research would take. (Papalexi 1934, 210)

In the first part of his article, Papalexi (1934, 209–210) then stresses the importance of the nonlinear oscillation process in the branches of applied and theoretical sciences such as Physics, Mechanics, Acoustics, Biology and Radio Engineering. He then recalls a first conference of this kind being held in Moscow in November 1931, therefore clearing any possible confusion between the "First Soviet Conference on Oscillations" (I.C.N.O. see *supra*), and this "Conférence Internationale de Non linéaire". It is nevertheless important to clarify that neither the Moscow Conference nor the Paris Conference can be viewed as "International", as the first was held in Moscow and only gathered members of the Soviet scientific community, and the second one held in Paris could only gather a majority of French scientists, plus one Belgian, one Dutch, one Russian scientist, due to an influenza epidemic. Papalexi's article (1934) can however be seen as an accurate report on the discussions and debates held during this conference, since he describes the various presentations of each guest. As the organizer, Van der Pol delivered an opening address, in which he deplored the absence of a number of participants due to an influenza epidemic which hit Europe at the time. The first presentation was given by Philippe Le Corbeiller.

[8]This domain is called nowadays *Quantum Mechanics*.

[9]See *Annuaire de l'Académie Royale de Belgique* (1978, 14).

7.1.4 Proceedings of the Conference

7.1.4.1 Philipe Le Corbeiller's Presentation (Saturday 01/28/1933)

According to Papalexi (1934, 210), Le Corbeiller presented various types of self-oscillating (self-sustained) systems, such as the "metaphorical" *Tantalus Cup* example, which he first introduced during his presentation at the *C.N.A.M.*, as well as the neon tube example (Le Corbeiller 1931a, 6) to illustrate the *relaxation phenomenon*. He added that Le Corbeiller also included in this list of relaxation oscillators, "Volterra's fluctuations in animal species in the struggle for life" (Papalexi 1934, 210). This is quite surprising, as Volterra's predator-prey model is not the seat of relaxation oscillations, at least in its initial design (see *supra*). It therefore seems improbable that Le Corbeiller actually mentioned Volterra's work. However, Papalexi explicitly referred to the book entitled "Lessons of Mathematical Theory of Struggle for Life" written by Volterra, who was supposedly invited, and hence covered the example given by Andronov (1929a, 560) in his note to the *C.R.A.S.* (see *supra*). He then added that Le Corbeiller addressed and covered various problems found in the theoretical study of self-sustained oscillations, such as the existence condition of a stationary-amplitude regime, and that he recalled the period's value (see *supra* Part I), as the *relaxation oscillations's form* (see *supra*). He recounted that Le Corbeiller presented a historical summary (probably similar to his presentation to the *C.N.A.M.*):

> After recalling that linear methods could not be used anymore in order to analyze these problems, he made a short historical summary of the evolution of the theory of nonlinear mathematics. He briefly described Van der Pol's main work regarding the oscillations of a Thomson-type system, and then focused on the work of the Cartans and Liénard, and mentioned the connection between Van der Pol's graphical analysis, Mr. Liénard's geometrical construction, and Poincaré's theory of limit cycles, the latter's importance in regards to oscillation problems has been stressed by our young scientists: A. Andronov and A. Witt. The speaker then addressed relaxation oscillations, and mentioned the importance of Van der Pol's research in this field, noting the in-depth study of the theory of nonlinear oscillations carried out by the Soviet scientists. (Papalexi 1934, 210)

It is therefore established that from January 1933, the most prominent members of the French scientific community working, or having worked, in the field of research of nonlinear oscillations could not ignore the correspondence established by Andronov anymore, between the periodic solution of a Van der Pol type of oscillator, and Poincaré's theory of limit cycles. As for the Cartans, Liénard, Abraham and Gutton, none of them used this result. This attitude may be explained partly by the fact that for some of them, this type of research was not at the heart of their preoccupations, as is the case for the Cartans, and also in part by the fact that, aside from Henri Cartan, Brillouin, Le Corbeiller and Rocard , all the other attending French scientists were almost retired in 1933 (Liénard (64 years old), Élie Cartan (64 years old), Esclangon (57 years old), Abraham (65 years old), Gutton (61 years old)), and therefore they were not really involved in these investigations. Papalexi then explained:

> In the next part of his presentation, Le Corbeiller addressed the branch of quasi-periodic oscillations. He summarized Van der Pol's work on the processing of the problem arising from the periodic forcing of self-oscillating systems, and dwellt on the forced synchronization phenomena (or "Mitnehmen"). He stressed the importance of these phenomena from the fundamental and experimental point of view. By noting the importance of a rigorous analysis of the synchronization phenomena carried out by A. Andronov and A. Witt, he ended his speech with a reminder that new research conducted in the field of periodic forcing on nonlinear systems and "fractionary" phenomena would be presented by the U.S.S.R. representative. (Papalexi 1934, 210)

7.1.4.2 Balthazar Van der Pol's Presentation (Saturday 01/28/1933)

Papalexi said that Van der Pol made a historical summary of his work, without detailing them, simply indicating that the references could be found in the special issue of the *Zeitschrift für Technische Physik* of the U.S.S.R., Band 4, Heft 1. The subject of the discussion that followed his presentation was: the phenomenon of forced synchronization.

7.1.4.3 Alfred Liénard's Presentation (Saturday 01/28/1933)

According to Papalexi (1934, 211), Liénard's presentation was a reminder of the main results of his "Study on sustained oscillations" (Liénard 1928). Then, he added:

> Using his graphical approach as a basis for the construction of differential integral curves, he deduced that the condition that must be met by the nonlinear characteristic of the system in order to allow periodic oscillations to occur, i.e. for the integral curve to be a closed curve, i.e. a limit cycle. (Papalexi 1934, 211)

In regards to what was presented about Liénard (see *supra*), this statement must be regarded cautiously. Let us keep in mind that on the one hand, the narrator already possessed a remarkable knowledge of Andronov's work (1928, 1929a,b) and on the other, his "minutes" were also intended for the members of the Academy, to whom he must have justified his presence at this conference in France, and show how much the Soviet research have spread in Europe. He then recounted the discussion that followed:

> During the discussion that followed the presentation, the researchers have talked about two characteristics: the type N and S characteristics, based on Le Corbeiller's terminology. These two types correspond to two possible forms of representation of the characteristics as functions of the first or second variables ($i = f(v)$ or $v = \varphi(i)$), and illustrate two different ways of including a "negative resistance" in the circuit, as parallel or series connections. (Papalexi 1934, 211)

This result is crucial since it allows, via the *duality principle in electrical engineering*, on the one hand to demonstrate the analogy present in various self-oscillators (series-dynamo machine, singing arc, triode), and on the other, to establish that Van der Pol's equation, prototype of the relaxation oscillations, and Lord Rayleigh's, are completely identical (see *supra* Part I).

7.1.4.4 Henri Abraham's Presentation (Saturday 01/28/1933)

Henri Abraham presented the results of experiments regarding the behavior of self-oscillating systems at the operating limit. He showed that under some known conditions, the existence of very slow oscillations can be noted, and they correspond to the passing from synchronization to the "beat", and vice-versa. Papalexi explained that these experiments were presented on Monday the 30th of January 1933 during a visit of Abraham's laboratory.

7.1.4.5 Ernest Esclangon's Presentation (Sunday 01/29/1933)

Ernest Esclangon (which will hold an important part in Part III of this work), presented his research on the quasi-periodic functions, and stressed their importance in different uses, especially the processing of statistical data. Papalexi then explained:

> During the discussions that followed the speech, the researchers mentioned the possibility of observing states of quasi-periodic oscillations, in non-conservative and conservative oscillating systems. (Papalexi 1934, 211)

7.1.4.6 Nikolai Papalexi's Presentation (Sunday 01/29/1933)

Papalexi (1934, 211) then presented his own presentation, in which he explained the two methods used for processing of self-oscillations. The first is the qualitative theory of nonlinear differential equations based on the research on limit cycles which would correspond to periodic solutions to these equations. He then summarized Andronov's work (1928, 1929a,b) and Andronov and Witt's (1930b). Then, he clarified that the second method "consists in finding the periodic solutions to differential equations depending on a small parameter" (Papalexi 1934, 211) by using the "methods of Lyapunov and Poincaré", hence allowing the establishment of the solution's stability. He then explained that the discussions that followed his presentation were focused on questioning the legitimacy of the idealization of relaxation oscillations with the form of "discontinuous motions" called "jump hypothesis", or "Mandel'shtam condition", which will constitute the root of "Discontinuous theory" (see *supra*). The next morning, Papalexi introduced the resonance and forced synchronization phenomena.

7.1.4.7 Visit at Abraham's Laboratory (Monday 01/30/1933)

Papalexi (1934, 212) recounts that the last day of the conference was dedicated to visiting Abraham's laboratory, at the *École Normale Supérieure*.[10] The various experiments that Abraham talked about during his presentation were thus presented, as well as the work carried out in collaboration with Eugène Bloch. In particular, the calibration of the frequency of a valve oscillator stabilized by a quartz. Papalexi then noted:

> One of the most interesting results of this study is the discovery of the influence of sudden air pressure variation on the oscillations's regularity in the piezoelectric quartz, when the latter is in a thermally insulated room in the basement, in which a constant temperature is maintained. Professor Abraham's experiment therefore emphasize that these changes are not directly linked to variations in temperature of the quartz produced by the adiabatic expansion and compression of the air surrounding the quartz. These changes disappear when the space of the air surrounding the quartz is made hermetic. (Papalexi 1934, 212)

Here, Papalexi implicitly referred to Abraham's latest work (1931) entitled: "Peut-on maintenir une salle à température constante ?" ("Can a room be maintained at a constant temperature?"). This note may seem out of place in a conference on nonlinear oscillations. If we keep in mind that according to Mandel'shtam, the frequency (or period) of a relaxation oscillator depends on *extrinsic* quantities, it appears that the problem questioned by Abraham (1931) actually regards the variability of the oscillation's frequency depending on the temperature during the experiments. Hence the necessity of "maintaining the room at a constant temperature".

7.1.4.8 Van der Pol's Closing Speech (Monday 01/30/1933)

After this visit, Van der Pol gave a final speech to conclude the first *Conférence Internationale de Non linéaire* by recalling that in spite of the absence of numerous scientific personalities working in this field, this conference reached its main aim, which was to gather physicists from various countries working on nonlinear oscillations, and mathematicians. He then added:

> Since this first experience was successful, we can hope that such meetings will occur regularly. The main presentations of the Conference shall be published in the collection "*Actualités Scientifiques et Industrielles*" (édition Hermann, Paris).[11] (Papalexi 1934, 212)

[10] See "À la mémoire de Henri Abraham, Eugène Bloch, Georges Bruhat : Créateurs et Directeurs de ce Laboratoire Morts pour la France", École Normale supérieure, Laboratoire de physique, École Normale Supérieure, imprimerie Lahure, Paris, 1948. Voir *Les trois physiciens Henri Abraham, Eugène Bloch, Georges Bruhat*, éd. Rue d'Ulm 2009.

[11] The proceedings was never published indeed.

Having so many people be absent, it appears for the one part that this conference did not have the international impact that it should have had, and for the other, that it did not lead to the constitution of a sort of "nonlinear dynamics community" that Van der Pol and other such as Le Corbeiller called for. Indeed, the absence of mathematicians Hadamard and Volterra, who were probably the most likely to appreciate Poincaré's work, was certainly very detrimental, and so was the absence of Mandel'shtam and his "students" Andronov and Witt, as well as Krylov and Bogolyubov,[12] in spite of the fact that Papalexi had quite accurately summarized their work.

This meeting of physicists and mathematicians therefore seems to have no impact on the French scientific community. Notably, in regards to Le Corbeiller and Rocard , this conference did not lead to publications, nor any form of collaboration with members of the Soviet scientific community (and incidentally, nor with Van der Pol). Their work in this field do not appear to have been significantly influenced by the various presentations given or by the discussions that followed.

The next year newly vice-president of the International Union of Radio Science (U.R.S.I.) Van der Pol again invited the most eminent members of the Soviet scientific community to present their work During the General Assembly of the International Union of Radio Science (U.R.S.I.), which was held in London between the 12th and 18th of September 1934. This gathering led to two publications, which mark a breakthrough in the developing process of the theory of nonlinear oscillations. The first was Van der Pol's (1934), entitled: "The nonlinear theory of electric oscillations", in which he synthesized his main results. The first part of this article is dedicated to the presentation of his "slowly-varying amplitude method" to which Mandel'shtam et al. (1935, 89) gave the name "Van der Pol method" (see *infra* Part III). In the second part, he synthesized his previous conferences on relaxation oscillations. Van der Pol visibly tried to crystallize the focus of the community on his work, in order to have them be considered as a starting point of a theory of oscillations.

> The above paragraphs give a connected account of many oscillation phenomena which present themselves to the radio worker and which can only be understood on the basis of a nonlinear theory. (Van der Pol 1934, 1082)

This seems to be corroborated on one hand by the fact that a majority of his references are to his articles, and on the other, by his way of mentioning numerous fundamental results, such as the correspondence established by Andronov (1929a) between the periodic solution of a relaxation oscillator, and Poincaré's concept of limit cycles, which was not featured in this article. Although in his conclusion, Van der Pol established a brief "state of the art" of various researches carried out in this field, notably Mandel'shtam and Papalexi's (1932), Krylov and Bogolyubov's

[12]Let's recall that two weeks before this conference, Krylov and Bogolyubov (1933) published in the *R.G.S.P.A.* an article entitled: "Problèmes fondamentaux de la Mécanique Non linéaire" ("Fundamental Problems of Nonlinear Mechanics"). However, it was not possible to establish a link between these two events.

(1932a,b,c), and Le Corbeiller's (1931a, 1932), to which was added a very thorough bibliography of international publications, this does not change the subjective way he presented the theory of nonlinear oscillations.

The existence of this first *Conférence Internationale de Non linéaire* held in Paris in 1933, is in some ways quite unexpected, almost unhoped for. Indeed, it could have been believed that such a gathering would have produced an "electroshock" on the French scientific community, inciting it to globally get invested in this emerging field, to which France had previously contributed. Unfortunately, this was not the case, and this conference now looks like a missed opportunity. It seems that the failure is linked to the fact that the French scientific personalities who attended (excepted for Henri Cartan, Brillouin, Le Corbeiller and Rocard) were all concerned by the limit of age, and therefore did not feel involved in the problem. Secondly, it must be succinctly reminded that at the time, the research in France was mostly focused on *Wave Mechanics*, *Radioactivity* and *General Relativity* (see Taton (1957–1964) (1964, Second part)). Lastly, the influenza epidemic deprived this conference of numerous participants. While it is difficult to imagine what could have happened, it is nevertheless very likely that the situation would have been the same, considering that this conference may have not necessarily triggered research in this field.

It was only during the *Réunion Internationale de Physique - Chimie - Biologie*, organized on the year of the inauguration of the *Palais de la Découverte* in Paris in October 1937 that Van der Pol (1938a) apparently had a new opportunity to make a presentation in France. Nevertheless, it was not a part of a *Conférence Internationale de Non linéaire*.

It was not before the 18th of September 1951 that the "second" *Conférence Internationale de Non linéaire* was held, on the île de Porquerolles (Var) in the presence of Misters Van der Pol, Haag, Minorsky, Vogel, Stoker, Wasow, Lasalle, Graffi, … under the supervision of Théodore Von Karman.

Chapter 8
The Paradigm of Relaxation Oscillations in France

8.1 Haag and Rocard's Works from a Mathematical Viewpoint

8.1.1 Jules Haag: From Self-Sustained Oscillations to Relaxation Oscillations

After studying at the *École Normale Supérieure*, Jules Haag (1882–1953) was, in 1906, a successful candidate for the *agrégation* of mathematics. In 1910 he received his doctorate in sciences under the supervision of Gaston Darboux (1842–1917). The claim of his thesis was: "Familles de Lamé de surfaces égales. *Généralisations et applications*". He then taught Rational Mechanics at the science faculty of Clermont-Ferrand. In 1927 he became head of the Chronometry Institute in Besançon, which later became the *École Nationale Supérieure de Mécanique et Microtechniques* (E.N.S.M.M.). At the beginning of the 1930s, he wrote numerous publications in the fields of synchronized oscillations, sustained oscillations, and relaxation oscillations. While some, such as Andronov (1929a), Liénard (1928, 1931) or Le Corbeiller (1931a,b) tried to broaden the frame of relaxation oscillations to one of the self-sustained oscillations, in a movement starting at the specific, towards the general, Haag chose the opposite, and proceeded from general to specific (Fig. 8.1).

8.1.1.1 Note to the *C.R.A.S.* of 1934 *on Self-Sustained Oscillations*

Therefore, in his note to the *C.R.A.S.* entitled "Sur les oscillations auto-entretenues," Haag (1934b) began with:

> The mathematical theory of these oscillations has been studied in-depth by various authors, especially by Van der Pol and Liénard. The research on periodic oscillations is essentially

J.-M. Ginoux, *History of Nonlinear Oscillations Theory in France (1880–1940)*,
Archimedes 49, DOI 10.1007/978-3-319-55239-2_8

Fig. 8.1 Jules Haag (Internet source)

the determination of *cycles* of a specific first order differential equation, of which Mr. Liénard gave a very simple geometric interpretation. (Haag 1934b, 906)

Remarkably, Haag spoke of a mathematical theory of "self-sustained", rather than "relaxation" oscillations, which he credited to Van der Pol and Liénard, rather than Andronov. Moreover, while he seemed to have perfectly grasped the correspondence between periodic solution and Poincaré's *limit cycle*, he did not use this terminology, and Poincaré's name was not featured in this note. This is due to the fact that here, Haag's problematic is the exact opposite of his predecessors'. He did not try to solve a differential equation for which the periodic solution is a *limit cycle*, but claimed that "*algebraic cycles*[1] can easily be found" (Haag 1934b, 907), which correspond to relaxation oscillations, that he apparently considered as a specific case of self-sustained oscillations (see *infra*).

I decided to *determine the more general cycle and the curve*[2] Γ *associated to it*. This is a small geometry problem, which presents no difficulty. (Haag 1934b, 906)

For Jules Haag, student of Gaston Darboux, it is therefore indeed a problem of *Differential Geometry* and not of *Mathematical Analysis* as implied by the name of the section[3] in which Andronov's note was published (1929a). However, all algebraic cycles do not correspond to relaxation oscillations. If we imagine a moving point made to move on such a cycle in a way that this point is animated by two different speeds (one slow, one fast), the algebraic cycle cannot be a circle in any case. By using an analogy with the representation of Van der Pol's graphical integration of the equation (3.8) (1926d, 983) (see *supra* Part I), Haag explains that "such a circle must be elongated in the direction Oy." (Haag 1934b, 907). He implies

[1]Actually, the non-algebricity of the limit cycle of Van der Pol's equation (3.8) considered by Mitropolski (1964, 1966) was demonstrated by Odani (1995).

[2]The curve (Γ) corresponds to the *oscillation characteristic* of the triode modeled by Van der Pol (1920, 703) with a cubic function (see *supra* Part I).

[3]Nevertheless, it was in the *Mécanique* section that all of Haag's notes will be published in the *C.R.A.S.*

that the (limit) cycle representing the periodic solution to Van der Pol's equation (3.8) is not algebraic. He then provides the "means of *asymptotically determining the cycle*, from the curve Γ." (Haag 1934b, 908). According to Minorsky (1962, 87), the origin of the use of *asymptotic methods* could be traced back to Poincaré (1892, 1893, 1899), and one of the first applications to the theory of oscillations is owed to Haag. These methods are still used nowadays under the term *theory of singular perturbations*. In the case of relaxation oscillations, it consists in "cutting" the limit cycle (see *infra*) solution to Van der Pol's equation (3.8) into *successive arcs*, and to approximate each part of the cycle by using an asymptotic expansion formed as a series. Difficulties then arise at the "junctions" which are the points linking one part of the circle to the other. Haag first considered that the cycle is not necessarily symmetrical. He then cut it in a *superior arc* (OA) *inferior arc* (AB), before dividing the latter into two subsections which he called soon after (see *infra* Haag 1943, 1944) *arc of the first kind* (AF) and *arc of the second kind* (FB) where A is named the terminal point, i.e. the rightmost point, and F the frontier point, i.e. the minimum of the function $y = f(x)$ representing the curve (Γ). This "cutting" allowed Haag to determine the value of the amplitude and period of relaxation oscillations, with a far better approximation than the one provided by Van der Pol (1927b, 114–115) (see *supra* Part I).

8.1.1.2 Note to the *C.R.A.S.* of 1936 *on Relaxation Oscillations*

His note to the C.R.A.S., entitled "*Sur l'étude asymptotique des oscillations de relaxation*" ("*On the asymptotic study of relaxation oscillations*") (Haag 1936), allowed him to further clarify his method, his notation and his objectives:

> We know that the determination of self-sustained oscillations is essentially the research on cycles of the differential equation
>
> $$xdx + (y - \lambda)\,dy = 0$$
>
> where λ designates a function[4] determined by x. (Haag 1936, 102)

Haag therefore demonstrated that he had perfectly integrated the correspondence established a few years earlier by Andronov (1929a), in other words, that a condition necessary and sufficient to the possibility of self-oscillations of a device is the existence of a limit cycle in its phase portrait. He then polished the definition of the concept of relaxation oscillations:

> To obtain *relaxation oscillations*, we must suppose that we multiply the function λ by a very large factor that we will call $1/\varepsilon$, ε being infinitely small. (Haag 1936, 102)

[4]The function λ is a graphical representation of the curve (Γ), i.e. the cube root Van der Pol (1920, 703) used to schematize the triode's property (see *supra* Part I).

The presence of a "very large factor" in the equation transcribes the existence of two characteristic time scales and stresses the slow-fast aspect of the oscillatory phenomenon. Haag then clarified:

> The *approximate determination of the cycles* and the *approximate calculus of the corresponding periods* stem from this. (Haag 1936, 102)

The term "of *corresponding periods*" indicates the time travel of the fictitious moving point on each section (arc) of the cycle. He then clarified that in a previous note,[5] he had "already tried to conduct this study, by starting from general equations of a cycle" (Haag 1936, 102), without finding a result. This second note on the subject is a new step allowing him to suggest a first correction of an amplitude's value, even though it is quite insignificant. He then posed the main problem inherent to the *successive arc* method:

> The main difficulty of the problem resides in the *junction between two different kinds of arcs*. (Haag 1936, 103)

He then added:

> Much more complex specificities can occur, if the function λ possesses several minima and maxima, especially if the curve Γ possesses angular points or horizontal tangents. (Haag 1936, 104)

Here, Haag referred to Liénard's work (1928), which considered that the function possesses a more general aspect than the cubic root used by Van der Pol (1920, 703). He therefore presented a method's principal and the obstacles which may occur. As for the results, they can only be found in his later work (see *infra*).

8.1.1.3 Note to the *C.R.A.S.* in 1937: *On the Theory of Relaxation Oscillations*

The title of this third contribution to the *C.R.A.S.*: "*Sur la théorie des oscillations de relaxation*" ("*On the theory of relaxation oscillations*") Haag (1937) is startling when compared to previous ones. At first glance, Haag seems to have gone from *self-sustained oscillations* to the *theory of relaxation oscillations*. However, it seems that the use of this term in this note written by Haag aimed at stressing the difficulty posed by the calculation of the amplitude and period of oscillations which he planned on carrying out, rather than giving Van der Pol's concept the worth of a theory. This was also confirmed by Haag's objective vision of his previous work:

> I indicated ([6]) a method allowing to calculate the main section of the amplitude of the period of a relaxation oscillation. In order to understand the practical value of the formulae I found, I carried out a numerical verification on an example. This verification showed me that the approximation of said formulae was quite mediocre, and I decided to improve them. This led me to significantly deepening my first theory. (Haag 1937, 932)

---

[5] See Haag (1934b).

[6] Haag referred to his previous note to the *C.R.A.S.* See Haag (1936).

The term "main section" clearly shows that Haag was going to use series expansions in his calculations. The example which he considered was Van der Pol's (1926d), although the method he suggested can be applied to Liénard's equation (1928), in other words, for an *oscillation characteristic* which is not necessarily a cubic root. This note, however, was only a declaration of intent, since no tangible result was presented, as shown in the conclusion:

> As a consequence, we can finally calculate the *amplitude of the periodic oscillation* depending on ε. As for the period, I have not yet addressed it, but it seems likely that the present theory must also allow its evaluation, with an error of arbitrarily large value. (Haag 1937, 934)

We can therefore note that Haag did not give the value of a theory to Van der Pol's results, but actually to his own work, that Minorsky (1967, 86) later called "asymptotic theory" (see *infra*).

8.1.1.4 Note to the *C.R.A.S.* of 1938: *On Asymptotic Formulae*

This fourth note to the *C.R.A.S.*, entitled: "*Formules asymptotiques concernant les oscillations de relaxation*" ("*Asymptotic formulae regarding relaxation oscillations*") (Haag 1938) gave him the opportunity to present and apply the *successive arc method*. Before that, it must be noted that this method, nowadays called: *shooting method*, originated in external ballistics problems.[7] As early as the beginning of the First World War, Haag (1915a,b,c) took interest in problems faced by the artillerymen. He participated in the *Commission d'expériences d'artillerie navale de Gâvre*[8] (*Commission for experiments at the Gâvre naval artillery*), under the supervision of the General Prosper-Jules Charbonnier. A few years later, Haag summarized his work carried out at the time:

> At the end of 1915, there was no new progress, at least to our knowledge. We had not been concerned with making a theory of calculation *errors* by successive arcs,[9] and we had not considered the calculations for the *differential alterations produced* by given perturbations along a trajectory either. The aim of the present work is to show the results found, on this double path, by the author at the end of 1915 and at the beginning of 1916, using the calculation method used at the *Commission de Gâvre* as a starting point ([1]).
>
> ---
>
> ([1]) (1) A summary of these results was presented to the Academy of Sciences in February 1917, but has not been published, as of interest to the *Défense Nationale*. (Haag 1921, 2)

[7]External ballistics is the branch that studies the free flight of the projectiles without internal propulsion, as it was the case for the artillery during WWI.

[8]Nowadays, Gâvres: a peninsula near Lorient in *Loire Atlantique*.

[9]According to Robert D'Adhémar (1934, 24) it was Félix Hélie (1795–1885) and Pierre Henri Hugoniot (1851–1887) who recommended successive arc calculations, derived from the Euler method, in ballistics.

This excerpt is interesting on several levels. Firstly, it shows the importance of the "military secret", implied in the first sentence: "at least to our knowledge", and in the footnote. Secondly, because it appears that it was during this time of the war that Haag developed the *successive arc method*. At the time, it was applied to external ballistics, in other words, the open trajectories (mainly parabola), and Haag transposed it to relaxation oscillations characterized by limit cycles, i.e. by closed trajectories. The principle of the method is as follows. Considering an initial condition, Haag made the flow act on the system, which generates a trajectory, a "shot". When the parameter ε goes from the value zero to a non-null, but infinitely small value, the cycle distorts and the trajectory is subjected to a deviation, which modifies the value of the amplitude of the oscillation's period. By then expressing, using asymptotic expansions, the amplitude and period as functions of this deviation, Haag (1938) managed to establish a better approximation of their value than the one provided by Van der Pol (1927a, 114–115) (see *supra* Part I). After making explicit the formulae relating to the deviations at several points of the cycle, he concludes his note by saying:

> From these formulae, we can easily deduce the *amplitude of the periodic oscillation*, as an expansion comprising the powers 2, 3 and 4 of $\varepsilon^{\frac{2}{3}}$. (Haag 1938, 1236)

Halfway through the Second World War, Haag (1943, 1944) explicitly provided the expansions allowing the finding of a better approximation of the period and amplitude. In two striking essays, which deserve a more in-depth study, Haag (1943, 1944) wrote the foundations for *asymptotic theory*.

8.1.1.5 The Memoir of 1943: *Towards a Definition of Relaxation Oscillations*

The first memoir: "Étude asymptotique des oscillations de relaxation" ("An asymptotic study on relaxation oscillations") (Haag 1943) began with a paragraph entitled *self-sustained oscillations*, in which Haag (1943, 35) established a differential equation characterizing this type of equation.

$$m\frac{d^2x}{dt^2} + \frac{f(x)}{\varepsilon}\frac{dx}{dt} - kx = 0 \tag{8.1}$$

It is interesting to note that his equation transcribed the motion of a mass m subjected to moving under the action of an elastic force $-kx$ and subject to a friction force $\frac{f(x)}{\varepsilon}\frac{dx}{dt}$, mechanical analogue of a "negative resistance" (see *supra* Part I). Therefore, by comparing the *successive arc* method developed by Haag (1915a,b,c, 1921) for external ballistics, and the one he developed for relaxation oscillations, and by assimilating the "alterations" of the air's friction to the variations of the parameter ε, we find that a trajectory deviation still takes place. It then becomes possible to consider that it is again the analogy (between the force of the air's friction and the force of "negative friction") which led him to coming up with the transposition from one branch to the other. At first it can seem surprising that it is not through Van der Pol's equation (3.8) (1926d, 979), which is a prototype of the relaxation oscillations, that Haag focused on, but on Liénard's equation (3.20) (Liénard 1928, 901), which represents the oscillations(self) sustained by a series-dynamo machine. Nevertheless, this choice seems to be justified by his formation as a geometrician, and his position as head of the Chronometric Institute of Besançon, which probably led him to considering a problem of a mechanical matter rather than a "thermionic valve". Haag (1943, 35) then moves to Liénard's phase plane and thus reduces the equation to a first degree differential form:

$$\varepsilon^2 x dx + (y - \lambda)\, dy = 0 \tag{8.2}$$

where $\lambda = \frac{1}{m\omega} \int_0^x f(x)\, dx$, $\omega = \sqrt{\frac{k}{m}}$, $y = \lambda + \frac{\varepsilon}{\omega} \frac{dx}{dt}$ and ε is a positive constant.

Haag then clarified the designation of relaxation oscillations that he had outlined until this point (see *supra*):

> When the constant ε is very small,[10] self-sustained oscillations exhibit very specific characteristics and are named *relaxation oscillations*. (Haag 1943, 36)

Consequently, Jules Haag considers *relaxation oscillations* as inscribed in the theoretical frame of self-sustained oscillations, as later defined by Andronov and Khaikin (1937, 201) in their book (see *infra*). He then designated the "cutting" (see Fig. 8.2) of the cycle, which he incidentally did not call *limit cycle*.

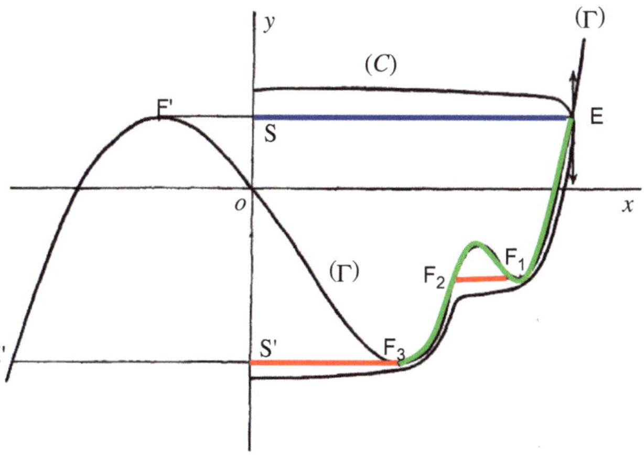

Fig. 8.2 "Cutting" of the cycle of the equation (8.2), by Haag (1943, 36)

[10]Haag has chosen a parameter ε which is exactly the inverse of that used by Van der Pol (1926d) in order to allow asymptotic expansions. So, he obtained relaxtion oscillations for $\varepsilon \ll 1$ and not for $\varepsilon \gg 1$ as Van der Pol (1926d, 981).

Haag (1943, 36–37) called:

- (Γ) the *fundamental curve* representing the function[11] λ,
- ES the *superior arc* (in blue on Fig. 8.2),
- ES' the *inferior arc*,
- $[S'F_3]$ and $[F_2F_1]$ the *arcs of the first kind* (in red on Fig. 8.2),
- F_3F_2 and F_1E the *arcs of the second kind* (in green on Fig. 8.2),
- E : the *terminal point*,
- F_3 : the *frontier point of the first kind* (because it limits, on the right, an arc of the first kind),
- F_2 : the *frontier point of the second kind* (because it limits, on the right, an arc of the second kind),
- F_1 : the *frontier point of the first kind* (because it limits, on the right, an arc of the first kind),
- S: the *superior peak*,
- S' : the *inferior peak*.

He considered that the "true trajectory", i.e. the integral of the equation (8.2), represented by (C) is placed below (Γ) on the *inferior arc* ES' and above (Γ) on the *superior arc* ES. The *terminal point* E, admitting a vertical tangent, is therefore its rightmost point. He then named "limit trajectory" ($S'F_3F_2F_1E$) the curve "comprised of horizontal segments and arcs of (Γ), which are respectively *arcs of the first kind* and *arcs of the second kind*" (Haag 1943, 37). The hypotheses that he formulated on the fundamental curve (Γ) clearly show that he considered an *oscillation characteristic* for which the representation is not necessarily a cubic as it was in Van der Pol's case (1926d), but "can present some angular points" (Haag 1943, 38) as it was the case for the generalization proposed by Liénard (1928). Then, Haag (1943) presented various possible expansions of the deviation in an asymptotic series. It is however unfortunate that the complexity of his calculations, as well as the lack of clarity in his notations, have made his results almost hermetic to numerous his contemporary peers, excepted for Flanders and Stoker (1946), Dorodnitsyn (1947), ... (see *infra*).

8.1.1.6 The Memoir of 1944: *The "Tangible" Example of Relaxation Oscillations*

In this essay entitled: "*Exemples concrets d'étude asymptotique des oscillations de relaxation*" ("*Tangible examples of asymptotic study on relaxation oscillations*") (Haag 1944) he considered two noteworthy cases relating to Van der Pol's equation. Firstly, Haag (1944, 82) hypothesized that the *oscillation characteristic* is

[11] In the case of van der Pol's equation (3.8) (1926d, 979), Haag chose $\lambda = \frac{x^3}{3} - x$ (see *infra*).

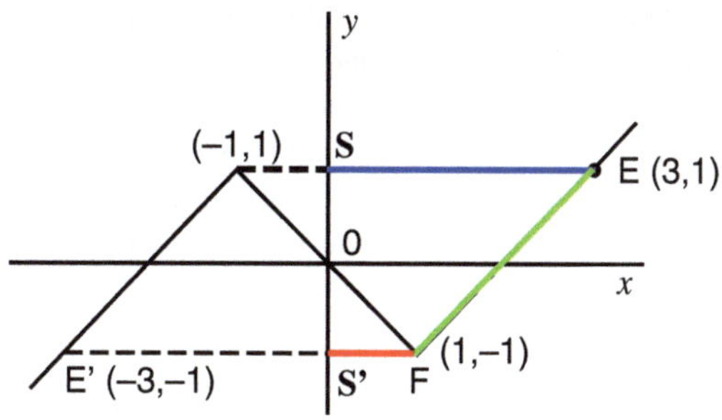

Fig. 8.3 Piecewise linear model, from Haag (1944, 93)

represented by a *piecewise linear model.*[12] Van der Pol's equation hence modified can be integrated analytically. Secondly, he focused on Van der Pol's classic model (1926d). In both cases, Haag provided an approximation for the amplitude and period of oscillations. The example of the piecewise linear model demonstrates Haag's approach. In this case the limit cycle and oscillation characteristic have the appearance schematized on Fig. 8.3.

By using the previous framework, (see *supra*), that is:

- ES the *superior arc* (in blue on Fig. 8.3),
- ES' the *inferior arc*,
- $[S'F]$ the *arc of the first kind* (in red on Fig. 8.3),
- FE the *arc of the second kind* (in green on Fig. 8.3),
- E the *terminal point*,
- F the *frontier point*,
- S the *superior peak*,
- S' the *inferior peak*.

Haag (1944, 96) provided, through asymptotic expansions, the value of the period for each arc. He therefore found the travel time:

- for the *arc of the second kind*:

$$t_1 = \frac{1}{\varepsilon} Log\,(3) + \varepsilon\,[1 - Log\,(3)] + \varepsilon^3\,[1.5 - Log\,(3)] + \ldots$$

- for the *arc of the first kind*:

$$t_2 = -\varepsilon Log\,(2\varepsilon^2) - \varepsilon^3\left[1 + 2Log\,(2\varepsilon^2)\right] + \ldots$$

- for the *superior arc*:

$$t_S = [Log\,(2) - Log\,(3) - 2Log\,(\varepsilon)]\,\varepsilon + \varepsilon^3\,[-2 + 3Log\,(2) - Log\,(3) - 2Log\,(\varepsilon)] + \ldots$$

[12] This type of representation was previously used by Rocard (1937a). Later, it was used by Stoker (1950, 141) who used Van der Pol's "piecewise linear model". See *infra*.

The significance of the piecewise linear model resides in the fact that its *direct integration*, by using quadrature, allows for a numerical comparison of the times found for each arc by using asymptotic formulae with the value found using *direct calculations*. This was Haag's approach, which led him to rectify his result. Therefore, the half-period of the limit cycle of this model is not exactly equal to: $t_1 + t_2 + t_S$ as could have been expected. Indeed, after Haag made rectifications (1944, 96), he suggested the following formula for the half-period:

$$\frac{T}{2} = \frac{Log\,(3)}{\varepsilon} + \varepsilon\left[\frac{4 - 10Log\,(\varepsilon)}{3} - Log\,(3)\right] + \varepsilon^3\left[\frac{4}{9} - 4Log\,(2) + 3Log\,(3)\right.$$
$$\left. - Log\,(3)\,Log\,(6) - \frac{2}{9}Log\,(\varepsilon) - \frac{2}{3}Log\,(6)\,Log\,(\varepsilon) - \frac{2}{9}Log^2\,(\varepsilon)\right] + \ldots$$

Once it was verified, he concluded for this case:

As we can see, the *precision of our asymptotic formulae is excellent*. (Haag 1944, 99)

Then, Haag focused on the study of Van der Pol's model (1926d) and gave the *oscillation characteristic* the appearance of a cubic function: $\lambda = x^3/3 - x$.

Using the same framework as he did previously (see *supra*), which is:

- ES the *superior arc* (in blue on Fig. 8.4),
- ES' the *inferior arc*,
- $[S'F]$ the *arc of the first kind* (in red on Fig. 8.4),
- FE the *arc of the second kind* (in green on Fig. 8.4),
- E the *terminal point*,
- F the *frontier point*,
- S the *superior peak*,
- S' the *inferior peak*.

Using asymptotic expansions of the value of the period for each arc, Haag (1944, 96) obtained the travel time:

- for the *arc of the second kind*:

$$t_F = \frac{3 - 2Log\,(2)}{2\varepsilon} + 1.0189\varepsilon^{1/3} + 0.024\varepsilon - 0.531\varepsilon^{5/3} + \frac{1}{3}\varepsilon Log\,(\varepsilon)$$

- for the *arc of the first kind*:

$$t_{S'} = 1.31920\varepsilon^{1/3} - 1.15\varepsilon - 0.382\varepsilon^{5/3} - \frac{2}{9}\varepsilon Log\,(\varepsilon)$$

- for the *superior arc*:

$$t_S = 1.76528\varepsilon - 1.38055\varepsilon^3 - \frac{2}{3}\varepsilon Log\,(\varepsilon) - 1.72104\varepsilon^3 Log\,(\varepsilon) - \frac{16}{81}\varepsilon^3 Log^2\,(\varepsilon)$$

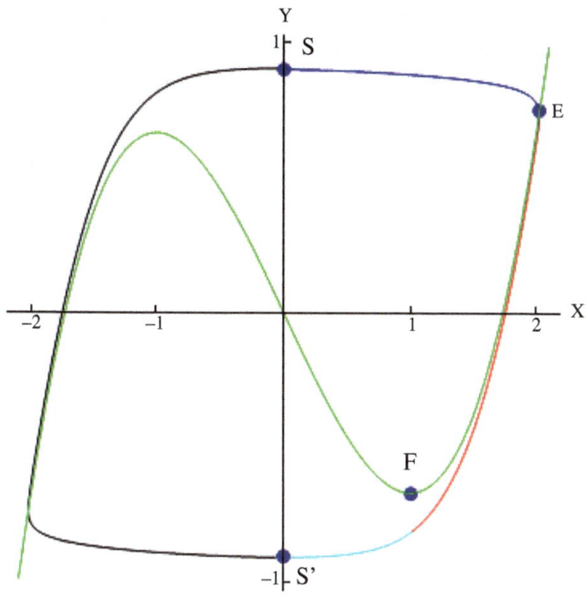

Fig. 8.4 "Cutting" of the (symmetrical) limit cycle of Van der Pol's equation (3.8)

Since in this case, the direct integration is not possible anymore, Haag (1944, 103 used the *successive arcs method* in order to verify his results. After making adjustments he suggested, for the half-period, the following formula:

$$\frac{T}{2} = \frac{3 - 2Log\,(2)}{2\varepsilon} + 6.445\varepsilon^{1/3} + \left[-3.31 + \frac{19}{9}Log\,(\varepsilon)\right]\varepsilon - 2\varepsilon^{5/3} + \cdots$$

In order to compare this result with Van der Pol's formula (3.17) (1927b, 114–115) (see *supra* Part I) or Liénard's (3.34) (1928, 953) (see *supra* Part I), ε must be replaced by its opposite (see *supra*) and be multiplied by two in order to have the whole period. We then have:

$$T = [3 - 2Log\,(2)]\,\varepsilon + \frac{12.89}{\varepsilon^{1/3}} + \frac{2}{\varepsilon}\left[-3.31 + \frac{19}{9}Log\,(\varepsilon)\right] - \frac{4}{\varepsilon^{5/3}} + \cdots \qquad (8.3)$$

We can deduce that at the first order in ε the formulae (3.17), (3.32) and (8.3) coincide perfectly. Moreover, if we choose $\varepsilon = 10$ (case of the *relaxation oscillations*, see *supra* Part I) the period $T \approx 20.4$, or nearly 2ε, we therefore find the value estimated by Van der Pol (1926d, 990). Hence, less than 20 years later,[13] Haag's work (1944, 115) finally led to a much better approximation than

[13]Actually, if we compare the development in Haag's essays (1943, 1944) with the results of his notes published in the *C.R.A.S.* (Haag 1934b, 1936, 1937, 1938) we see that as early as 1938 he had already calculated the amplitude, if not the period of relaxation oscillations.

the one suggested by Van der Pol. Nevertheless, Haag stressed that his asymptotic expansions were, by nature, tainted by some errors, and therefore perfectible. It was Russian mathematician Anatoly Alekseevich Dorodnitsyn who improved Haag's result (1944). In an article published in Russian, Dorodnitsyn (1947, 326) provided the following expansion:

$$T = [3 - 2Log\,(2)]\,\varepsilon + \frac{7.01432}{\varepsilon^{1/3}} - \frac{22}{9}\frac{Log\,(\varepsilon)}{\varepsilon} + \frac{0.0087}{\varepsilon} + O\left(\varepsilon^{-4/3}\right) \tag{8.4}$$

With this formula (8.4), the period of the relaxation is $T \approx 18.83$ for $\varepsilon = 10$, hence a values still close to 2ε but this time slightly inferior. When in 1952, the first[14] of the two volumes of Haag's work entitled: "Les mouvements vibratoires" ("Vibratory movements") (Haag 1952–1955) was published. While he recalled the correspondence established by Andronov (1929a) by referring[15] to Solomon Lefschetz's translation of Andronov and Khaikin's work (1937), he did not integrate Dorodnitsyn's expansion (8.4) but reproduced his own. It is possible that Haag was not aware of it, since Lefschetz, in the American version of 's book Andronov and Khaikin (1949, 347), addressed Flanders and Stoker's demonstration (1946), which was limited to the first term of the expansion, but did not address Dorodnitsyn's (1947).

8.1.2 *Yves Rocard:* Relaxation Oscillations and Self-Oscillations

Having joined the *École Normale Supérieure* in 1922, Yves Rocard was a successful candidate at the physical sciences *agrégation* in 1925. Two years later, he obtained his doctorate in mathematical sciences at the science faculty of Paris, his dissertation was entitled "*L'hydrodynamique et la théorie cinétique des gaz*" ("*Hydrodynamics and the kinetic theory of gases*"). After preparing a *thèse* at the *laboratoire d'enseignement* of Physics supervised by Charles Fabry, he obtained in 1928, a doctorate in physics, for which the dissertation focused on the "*Théorie moléculaire de la diffusion de la lumière par les fluides*" ("*Molecular theory of light diffusion by fluids*"). He was then appointed lecturer at the *Fondation Peccot* within the *Collège de France*, and was named *maître de recherches* in 1932. He became maître de conférences in physics at Clermont-Ferrand's science faculty in 1939, then on 1 October 1939, maître de conférences in experimental fluid mechanics (and then physics) at the faculty of science in Paris. At the beginning of the 1930s, Rocard was already interested in electric oscillations and mechanical vibrations (Fig. 8.5).

[14]The second volume was published after his death in 1955 and written by Jean Chazy (1882–1955).

[15]See Haag (1952–1955, 204).

Fig. 8.5 Yves Rocard (Internet source)

8.1.2.1 The Note to the *C.R.A.S.* of 1932: *Oscillations Near the Operating Limit*

In this note to the *C.R.A.S.* entitled: "Sur les oscillateurs réglés près de la limite d'entretien" ("On oscillators set near the operating limit") (Rocard 1932) he established[16] in a different manner the differential equation for the oscillations of the triode (3.8), by using the operating point, which consists in "linearizing" the problem, by expanding the voltage of the plate current in the form of a series. Rocard's approach is perceptibly similar to the one previously used by Blondel (1919b) (see *supra* Part I). However, contrary to Blondel, who formed a series with odd terms, Rocard kept all the terms (even and odd). After some simplifications, he obtained the following equation:

$$\omega^2 v - \varepsilon \left[1 - \alpha v - \beta v^2 - \gamma v^3 - \delta v^4 - \ldots\right] v' + v'' = 0 \tag{8.5}$$

He then added:

> This equation is known, it is the one from which Mr. Van der Pol obtained his remarkable results regarding relaxation oscillations. (Rocard 1932, 1326)

He therefore appeared to fully adhere to Van der Pol's terminology, as well as his work. He then proceeded to what he called "solving through successive approximations" (Rocard 1932, 1326), although it actually is the search for a solution by way of a Fourier-series expansion, and calculated the amplitude of the third harmonic (H3) and the fundamental (H1) depending on all the coefficients of the expansion (($\varepsilon, \alpha, \beta$ and γ ; δ having been neglected). He then demonstrated that the pulsating varies with the harmonic proportion H_3/H_1 (see *infra* Part III).

[16]Even though he did not mention them, he found Blondel's results (1919b) for the amplitude and period of the oscillations (see *supra* Part I).

8.1.2.2 Marcel Brillouin's Jubilee in 1935: *Towards a New Type of Oscillation*

Three years later, Rocard started research "Sur la stabilité de route des locomotives" ("On the directional stability of locomotives") (1935b,c), and Marcel Brillouin's jubilee was for him the opportunity to present part of his results regarding the relaxation oscillations. His presentation, entitled: "Sur certains nouveaux types d'oscillations mécaniques" ("On new types of mechanical oscillations"), aimed at studying the oscillations of the axles of a bogie by representing their lateral motions with a differential equation. He planned to thus evidence new types of oscillations, which would supposedly include relaxation oscillations.

> We leave the "theory of directional stability of locomotives", that can be developed from this work, for other publications. I would like to bring attention to these equations' form, on new types of mechanical oscillations that they reveal, showing very intriguing specificities. Broad generalizations notably ensue, on the notion of relaxation oscillation, for which Van der Pol's simple second order equation gives a too specific conception. (Rocard 1935a, 402)

His attitude towards Van der Pol's work apparently changed somewhat, as shown by this excerpt. After he established the (fourth order) differential equation for the axles' motion, Rocard (1935a, 403) studied the stability of the solution based on his characteristic equation. He then demonstrated that some are unstable, and explained the "snaking" motion of a bogie isolated on a railroad. He then suggested an "electromechanical assembly comprised of triode lamps and electric motors" (Rocard 1935a, 403) aiming at representing the oscillations produced by the differential equation he established. This allowed him to evidence these "very curious specificities":

> Contrary to what happens in usual mechanical systems (a combination of systems with springs and coupled masses), for the electric circuits (combination of coupled resonating circuits), the introduction of simultaneous friction forces acting on each degree of freedom proves unable to restore the stability (...) (Rocard 1935a, 402)

Here, Rocard tackles the "dogma" on which rested the concept of relaxation oscillations: the existence of a nonlinear component analogous to a "negative resistance" able to sustain the oscillations instead of damping them. A few years later, Rocard (1943, 194) demonstrated that the presence of a "negative resistance" is not a necessary or sufficient condition to establish self-sustained oscillations (see *infra*). Moreover, it is possible that the four degrees of freedom system that he envisioned presents solutions for which the behavior could be chaotic in nature (or even hyperchaotic). This would explain the instability observed by Rocard. While he was not able to perceive this type of behavior for the solution to his system, he nevertheless seems to be aware of the fundamental differences existing between the differential equation representing it and Van der Pol's .

> However, this equation strongly diverts from the type given by Mr. Van der Pol, where the relaxation character is obtained by the use of a variable resistance depending on the amplitude of the oscillating quantity (this resistance is negative for small amplitudes

> and positive for the large ones). This is it justified to say that the concept of relaxation oscillations as considered until now can be notably broadened and generalized unless we would prefer to introduce a new terminology to characterize more general oscillations, for which the equations are obtained by substituting, in a system of simultaneous linear differential equations, the variable of some terms with a univalent function of this variable, which would not be a simple proportionality. (Rocard 1935a, 402)

Rocard considered Van der Pol's denomination as being too restrictive and seemed to be willing to suggest a new one. This is also what Andronov did (1937, 201) in his book (see infra*infra*). Therefore, in this article, Rocard countered one of the foundations of the concept of relaxation oscillations, which he tried to broaden, or rather integrate into a more general theory. This could explain the title he chose for the study he started two years later on "Les phénomènes d'auto-oscillations dans les installations hydrauliques" ("Self-oscillation phenomena in hydraulic installations") (Rocard 1937b). Although this is clearly an appropriation of Andronov's terminology, there is no reference to this author in Rocard's book, nor any definition of this terminology. Moreover, on page 41, he used the "theory of relaxation oscillations" (Rocard 1937b) in order to justify the introduction of nonlinear frictions analogous to a "negative resistance".

8.1.2.3 The *Société Française des Électriciens* in 1937: *Piecewise Linear Model*

On the 10th of February 1937, having been invited to present a communication in front of the Administrative committee of the *Société française des Électriciens* and the Study committee of the *Société française des Radioélectriciens*, his presentation entitled: "*Relaxation, synchronisation et démultiplication de fréquence*" (Rocard 1937a), thus started with:

> Since *Van der Pol*'s fundamental essays, relaxation oscillations have led to remarkable research, some of which are extremely important in regards to the generality and results, and mathematical rigor of the deductions. But the clear and numerical description of the phenomena is still nowadays reduced to minor details, at least this is what we believe, of the results produced by *Van der Pol* regarding his simple equation.
>
> $$\omega^2 v - \varepsilon \left(1 - v^2\right) \dot{v} + \ddot{v} = 0 \tag{8.6}$$
>
> (...) (Rocard 1937a, 396)

As always, Rocard's viewpoint consists in insisting on the importance of Van der Pol's results, while evidencing the particular characteristics of the oscillatory phenomenon's analysis, which is reduced, according to him, to the study of the equation (3.8), of which he questions the "genericity" and relevance:

> The choice of model for the equation is obviously arbitrary. (Rocard 1937a, 396)

He then explains that it was for its mathematical aspect that Van der Pol chose a cubic function to represent the oscillation characteristic of the triode. Then, he opposed, very legitimately, that outside the areas where this device behaves as

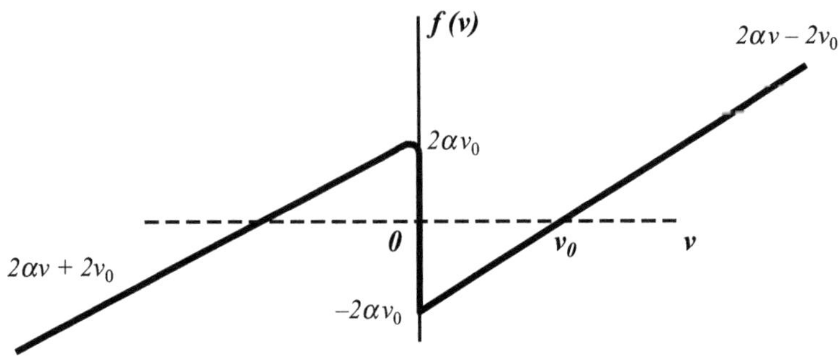

Fig. 8.6 Piecewise linear oscillation characteristic, by Rocard (1937a, 397)

a "negative resistance[17]" Van der Pol's model "will give to the phenomena an appearance very different from their reality." (Rocard 1937a, 396). He then offered to go further in the simplification by presenting a piecewise linear model. It must be noted that except for Andronov and Khaikin (1937, 159), he appears to be the only French scientist to have considered such a process. Indeed, rather than representing the oscillating characteristic with a cubic function $f(v) = \varepsilon\left(\frac{v^3}{3} - v\right)$, he used the following function:

$$f(v) = \begin{cases} 2\alpha v + 2v_0 \text{ if } v < 0 \\ 2\alpha v - 2v_0 \text{ if } v > 0 \end{cases}$$

for which he provided the following graphical representation (see Fig. 8.6).

The equation (8.6) then yields:

$$\omega^2 v + 2\alpha \dot{v} + \ddot{v} = 0 \tag{8.7}$$

When v goes through zero, the amplitude is subjected to a sudden variation of $-4\alpha v_0$ and the appearance of this piecewise linear function reproduces the behavior of the *oscillation characteristic* . The significance of this process therefore resides in the fact that the equation (8.7) is now easily integrable. His characteristic equation is written:

$$r^2 + 2\alpha r + \omega^2 = 0 \tag{8.8}$$

From this we deduce that the following conjugated complex roots[18]: $r_{1,2} = -\alpha \pm \sqrt{\omega^2 - \alpha^2}$ from which Rocard (1935a, 398) explicates the solution to the equation (8.7)

[17] See *supra*, Rocard (1935a, 402).

[18] The complex case is the only one likely to produce undamped oscillations.

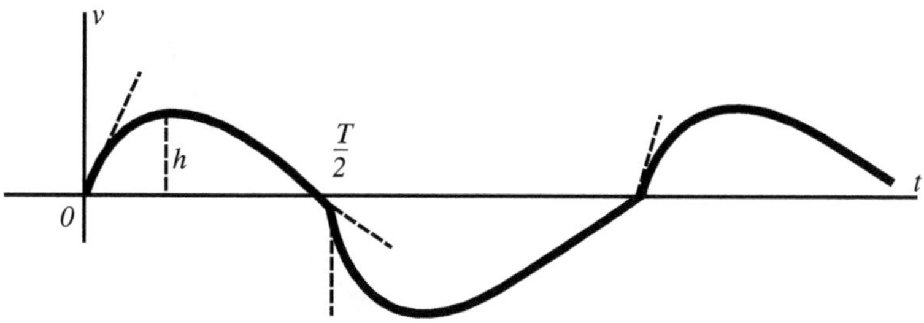

Fig. 8.7 Graphical representation of the solution (8.9), by Rocard (1937a, 398)

$$v(t) = Ae^{-\alpha t} sin\left(\sqrt{\omega^2 - \alpha^2 t}\right) \tag{8.9}$$

where A represents the amplitude of the oscillations and $\sqrt{\omega^2 - \alpha^2}$ the angular frequency or pulsation. By using the slope of the tangents at the times $t = 0, \frac{T}{2}, T$ where he provides the graphical representation of the solution (8.9) (see Fig. 8.7) and calculates the maximum value of the amplitude v_m and the period of oscillations.

It is noteworthy that these oscillations' form corresponds exactly to the one for Van der Pol's relaxation oscillation (1930, 303) (see *supra*, Fig. 4.8). But instead of bringing his result closer to Van der Pol's by demonstrating that the period is expressed as a relaxation time, meaning, depending on the parameters R and C and not L and C, Rocard viewed Van der Pol's work from a critical standpoint:

> We see that, of course, α must be limited to the value giving the critical damping, otherwise the first half-sine (Fig. 8.7) never goes back to the axis: such a limitation does not exist in Mr. Van der Pol's model. We also see that the amplitude v_m constantly decreases when α increases, whereas in Mr. Van der Pol's model, we find a constant amplitude for both large and small values of ε very similar to this constant for the intermediate values of ε (result owed to Mr. Liénard). We want to insist on these points, in order to avoid that an excessive generalization be attributed to results obtained from specific models of oscillators. (Rocard 1937a, 399)

The last sentence is particularly interesting since it shows a strong response to the "hunt for the relaxation effec" (see *supra*) started by Van der Pol during his numerous conferences given in France, which Rocard possibly attendedt.[19] He then concluded this first example:

> Our model defined by the equation (8.7) or Fig. 8.6, is essentially the concentration around $v = 0$ of all the self-oscillation capacity of the oscillator, if we may put it that way. (Rocard 1937a, 399)

The fact that Rocard used the neologism "self-oscillation" to describe the *relaxation oscillations* of his *piecewise linear model* seems to indicate that he

[19]It was established, however, that Rocard attended the first *Conférence Internationale de Non-linéaire* in Paris in 1933 under the supervision of Van der Pol. See *infra*.

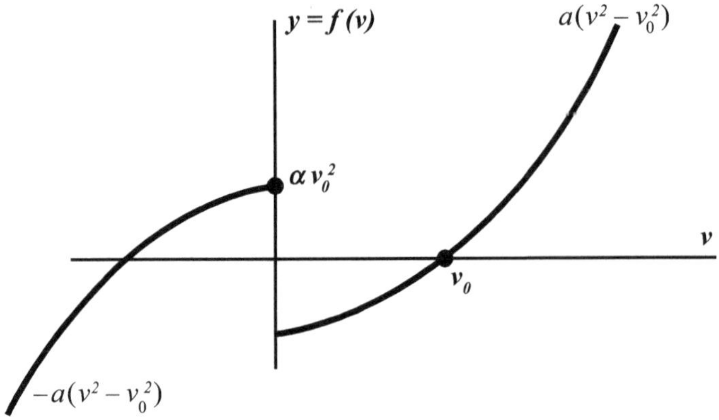

Fig. 8.8 Parabolic oscillation characteristic, by Rocard (1937a, 401)

counted this type of oscillations in the broader frame of a *theory of oscillations* started a few years before by Andronov (1929a), which reached its apex with Andronov and Khaikin's book being pusblished (1937). In the second example he presented, Rocard seems to have a more nuanced vision regarding Van der Pol. This was again an oscillator that is completely integral, in which he substitutes this time, the cubic function by the following function:

$$f(v) = \begin{cases} -a\left(v^2 - v_0^2\right) \text{ if } v < 0 \\ +a\left(v^2 - v_0^2\right) \text{ if } v > 0 \end{cases}$$

for which he also provides the graphical representation (voir Fig. 8.8).

But instead of integrating, as he did before, Van der Pol's equation (8.6), Rocard (1937a, 401) integrates Liénard's equation (see *supra* Part I, Eq. (3.21) in which he substitutes $F(x)$ with the piecewise linear function $f(v)$. By writing $v^2 = u$ he obtains:

$$v(t) = \begin{cases} -\sqrt{C_1 e^{-2ay} - \dfrac{y}{a} + \dfrac{1}{2a^2} + v_0^2} \text{ if } v < 0 \\ +\sqrt{Ce^{-2ay} + \dfrac{y}{a} + \dfrac{1}{2a^2} + v_0^2} \text{ if } v < 0 \end{cases}$$

He then provides a joining condition in order to have the two branches of the solution form a "limit cycle" (see Fig. 8.9).

Andronov and Khaikin (1937, 159) excepted, it seems he was once again the first French scientist to present the limit cycle of an integrable model. He then demonstrates that the amplitude of oscillations almost does not vary with a whereas the period slowly increases with a. He deducts that his results "are somewhat close to Mr. *Van der Pol's*" (Rocard 1937a, 402). Rocard considers, in the second part of his article, the issue of synchronization and demultiplication which will be the

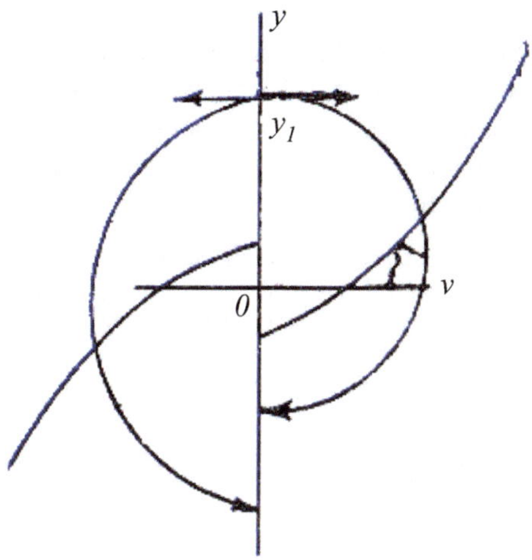

Fig. 8.9 Limit cycle of the piecewise linear model, by Rocard (1937a, 402)

subject of a more in-depth study in Part III (see *infra*). After his presentation, there was a discussion between the people present, amongst which was Alfred Liénard. His presence at such a gathering is startling, since after the publication of his works in this field, Liénard (1928, 1931) seemed to not have given upsurge to Andronov's results (see *supra*) and had taken his research in a new direction. He took part in the discussion and clarified that he had "considered, as Mr. Rocard had, the use of an angular characteristic in order to simplify calculations" (Rocard 1937a, 413). Nevertheless, after this statement, Liénard did not publish anything on the matter again. Le Corbeiller, who participated in Rocard's presentation, made a note in which he recalled:

> Mr. van der Pol, in his fundamental essays of 1927, made a considerable breakthrough on the question by studying a continuous system depending on an arbitrary parameter ε, the discontinuities only appearing as a border-line case when ε is infinitely big (relaxation oscillation). On the other extremity of the scale, the sinusoidal oscillations are presented as a border-line case when ε is infinitely small (result owed to Lord Rayleigh). (Rocard 1937a, 417)

This formation of relaxation oscillations already presented by Le Corbeiller (1931a, 21) a few years before (see *supra*) was used some time later by Rocard (1941, 1943) (see *infra*), by Jules Haag (1943, 36) (see *supra*) then by Jean Abelé (1943, 22) in his thesis dissertation (see *infra*).

8.1.2.4 The Self-Oscillation Phenomena in 1937

The year 1937 was marked by numerous publications (Andronov and Khaikin 1937; Krylov and Bogolyubov 1937; Timoshenko 1928) in the field of oscillations (see *infra*). In France, Yves Rocard published a research entitled "*Sur les phénomènes d'auto-oscillations dans les installations hydrauliques*" ("*On the self-oscillatory phenomena in hydraulic installations*") (Rocard 1937b). This title instantly shows his view on the concept of relaxation oscillation and the place it occupies in the theory of self-sustained oscillations, or rather self-oscillations, developed by Andronov between 1929 and 1937. In this book Rocard studied a problem of the hydraulic valves control, which is a reminder of the one addressed half a century earlier by Léauté (1885):

> The present work is destined to broach the study of even more complex phenomena, where a mechanical system able to move is coupled to a duct, and for which the displacements cause in return a reaction on the fluid's flowing. Such a system can, under specific conditions, start an oscillation *by itself*, as it is well known, and the classic theory of the battering ram can in no way explain the generating of such oscillations, nor what will occur with them. (Rocard 1937b, 15)

In the first page of his manuscript, Rocard (1937b, 17) considers a centrifugal pump for which the output can be set by guide valves, which have an opening angle determined by the angular variable α.

By calling:

- i: the inertia coefficient,
- K: the hardness coefficient,
- S: the surface of the vanes,
- x: the arm lever with a force corresponding to the rotation axis of the vanes,
- u: the speed of the water,

and writing $\beta = (\gamma Sx)\,\eta$ Rocard (1937b, 20) obtains the following differential equation:

$$\left[i\frac{d^2}{dt^2} - f\frac{d}{dt} + K\right]\delta\alpha - \beta u = 0$$

He then adds:

> We note that the coefficient f with a negative sign yields a negative damping in the motion of α: it is generally what will be responsible for the oscillations.[20] (Rocard 1937b, 20)

A few pages later, he explains that the coefficients of this type of equation vary with the amplitude and that:

> All these self-oscillating phenomena have been solved in a perfectly clear manner in the problem of the oscillation of triode lamps with curving characteristic. (Rocard 1937a, 24)

In the second part of his work, in the paragraph entitled "La courbure des caractéristiques entretient et limite l'amplitude des oscillations" ("the curve of self-sustained properties and the amplitude limit of oscillations", Rocard 1937a, 40), he then modifies the friction coefficient to $\frac{d\alpha}{dt}$.

[20]Self-oscillations.

We obtain a law of this type by giving to the term the following expression (¹) :

$$-\varepsilon\left(1-\frac{\alpha^2}{\alpha_0^2}\right)\frac{d\alpha}{dt}$$

α_0 being a constant characterizing a degree of opening of the valve.

(¹) *Note.* – Here, we use the representation of a resistance with a variable sign depending on the amplitude corresponding to the one used in the theory of relaxation oscillations.[21] (Rocard 1937a, 41)

Rocard (1937a, 42) then establishes a differential equation completely similar to the prototype of the relaxation oscillations proposed by Van der Pol (1926d), which allowed him to evidence the existence of self-oscillations in hydraulic systems. It therefore appears that Rocard may have considered relaxation oscillations as a particular type of oscillation belonging to the more general phenomenon of self-oscillations. This is unexpected considering that the problem of the *hydraulic ram* that he studied is quite close to the examples of relaxation oscillations described by Van der Pol (1930, 307), Van der Pol and Van der Mark (1928a, 371) and Le Corbeiller (1931a, 42) (see *supra*). During the following years Rocard kept doing research on the theory of oscillations.

8.1.2.5 The Theory of Oscillators of 1941: *On the Usefulness of a Negative Resistance*

At the beginning of the Second World War, Rocard (1941) published a book entitled "Théorie des Oscillateurs" ("The theory of Oscillators") which reminds of Andronov and Khaikin's (1937). Its first version entitled "теория колебаний[22]" was published in 1937. Chapter II of Rocard's book which contains nine of them, is fully dedicated to relaxation oscillations. In it, Rocard (1941, 31) suggest a synthesis of the various research carried out in this field, a summary of Van der Pol's article (1926d), as well as a definition of relaxation oscillations:

> For the small values of ε, the oscillation tends towards a pure sinusoidal regime, with pulsation ω. For the large values of ε, $\varepsilon \gg 1$, the oscillation tends towards a regime oscillating with sudden periodic starts, which corresponds quite well to the physical case of our figure 2,[23] which occurs with a pulsation $\omega_1 = 2\pi n = 2\pi/T$ such that the period T is to the order of ε/ω. (Rocard 1941, 38)

A few pages later, Rocard establishes that $T = \varepsilon/\omega = RC$, then writes

> There does not exist, to our knowledge, a direct mathematical demonstration of this result which is however extremely important. (Rocard 1941, 48)

[21] It seems it was the only occurrence of this expression in Rocard's whole book (1937a).

[22] "Theory of oscillations".

[23] Rocard's Figure 2 (1941, 32) represents a device analogous to the one presented by Van der Pol (1930, 307) (see *supra*) and implemented by Le Corbeiller (1931a, 42) (see *supra*).

Contrary to what he stated, this demonstration based on dimensional analysis had already been provided by Van der Pol (1926d, 987) (see *supra* Part I). Nevertheless, Rocard insists on a point on which Van der Pol might have not insisted enough, i.e., the origin of this terminology. Indeed, it was only two years after his first publication that Van der Pol (1928a, 36) gave a precise definition of the concept he had introduced (see *supra*). It is likely that the level of generalization and abstraction that he envisioned for this new type of oscillations able to describe apparently very different phenomena (electrical, biological, economical, hydrodynamic ...) had pushed to the background this "result which suggested the generic name of relaxation oscillation" (Rocard 1941, 49)

Rocard then reproduces the figures from Van der Pol's article (1926d, 983–986) representing the periodic solution to the equation (3.8) which he calls limit cycle (Rocard 1941, 49) without referring to Poincaré's (1882) nor Andronov's work (1929a). He then recalls that the main results regarding the calculation of the period and amplitude of the oscillations (see *supra* Part I) and adds:

> Some researchers have tried to give a greater mathematical rigour to deductions, to demonstrate the existence of solutions, etc. Important research has been made by Krylov and Bogoliuboff, as well as Mandel'shtamm and Papalexi: they do not bring new material. (Rocard 1941, 49)

Rocard's stance concerning the "Russian school of thought" is, to say the least, condescending and just as surprising as his sudden disdain for mathematical rigor, also showing that he did not fully take measure of Andronov's results (1929a), Andronov and Witt (1930a), Krylov and Bogolyubov (1932a,b,c, 1933, 1934a,b,c,d,e, 1935a,b, 1936a,b, 1937) as well as Mandel'shtam and Papalexi (1934). Nevertheless, it seems unlikely that this was due to ignorance (see *infra*).

He then introduces the principle of Liénard's construction (1928) for the graphical integration of the equation (3.8) and exposes more in depth the integration of piecewise linear models which he presented during a lecture at the *Société Française des Électriciens*. He then starts the third chapter of his work dedicated to the oscillator with no inductance or inertia and explains:

> In Mr Van de Pol's description of relaxation oscillators, we still have the same situation, the notion of a positive or negative damping depending on whether the variable is large or small is introduced, and he has done his best to explain facts which had been cautiously un-researched until now. However, the resistance of a force of inertia (or of self-inductance) with an elastic force (or capacity) still subsists, and Mr. Van der Pol believes it to be so crucial that when he gives a theory of the multivibrator, an assembly which comprises triode lamps, capacities, resistances, but no self-inductance, he claims that connection wires are the cause of a certain self-inductance, which is necessary to find an oscillating attribute in the system. (Rocard 1941)

Here Rocard implies on one hand that the presence of a device analogous to a "negative resistance" is not a necessary and sufficient condition for the establishing of a self-sustained oscillation regime, and on the other that Van der Pol's interpretation (1926d, 988) regarding Abraham and Bloch's multivibrator, is erroneous. He then addresses other subjects such as the synchronization and demultiplication of frequencies, which will be studied in depth in Part III (see *infra*).

8.1.2.6 General Dynamics of Vibrations in 1943

Two years later, his now famous treatise on the "*Dynamique générale des vibrations*" ("*General Dynamics of vibrations*") (Rocard 1943) is published. Its first part: "vibrating systems" has no less than 16 chapters. Nevertheless, if chapters XIV, XV et XVI focus more specifically of the field of nonlinear oscillations, Rocard (1943, 188, 216 and 235) preferred the terminology "self-oscillations" to Van der Pol's, and only briefly dealt with "relation oscillations" in this research (Rocard 1943, 219–227).

Paragraph 114 of chapter XIV entitled "Duality of the modes of oscillations" covers *in extenso* Rocard's note (1942) published in the *C.R.A.S.* the previous year.

> It is generally believed that, in order to have self-oscillations in a resonant circuit, it is necessary and sufficient to introduce in it a properly measured out negative resistance. The aim of the present note is to show that this is not the case. If a sufficient negative resistance will always start self-oscillations, it is in no way necessary. As soon as we consider a system with more than one degree of freedom, we find another mechanism able to start the self-oscillation and consisting in straining two natural frequencies of the system towards one another. (Rocard 1942, 601)

He therefore demonstrated that in an oscillator with more than one degree of freedom (which is not the case for the triode, but it is for the multivibrator) the presence of a component analogous to a "negative resistance" is not a necessary and sufficient condition for the existence of self-oscillations. In this case the additional variable, or variables, play the part of a factor limiting the amplitude of oscillations and generating in the phase space[24] a limit cycle or a chaotic attractor. In 128 of chapter XV, Rocard (1943, 220) recalls once again Liénard's method (1928) for the graphical integration of the equation (3.8) and then explains:

> (...) we note (as H. Poincaré demonstrated before) that the integral curve coiled up numerous times and tends towards a closed curve called a *limit cycle*, which in this representation, corresponds to the permanent oscillation regime. (Rocard 1943, 220)

This book contains no bibliography and there are no references associated to this, Poincaré is only mentioned. Rocard's stance regarding the subject is somewhat ambiguous. In the preface of his "Théorie des oscillateurs" ("Oscillators Theory") he states:

> This volume does not contain a bibliography, and we apologize for this. It was written during an inconvenient time (July 1940). (Rocard 1941, viii)

This is startling since Rocard, who was appointed *maître de conférences* in physics at the faculty of science in Clermont-Ferrand in 1939, was also appointed on the 1st of October 1939 as *maître de conférences* in Experimental Fluid Mechanics (and later, Physics) at the faculty of science in Paris. It is therefore expected that during the time he wrote the manuscript, he had access to various books to refer to.

---

[24] With dimensions superior or equal to three.

Especially as he writes about Jean Cavaillès[25] in his scientific work pertained to the 1940–1945 period:

> He sometimes found himself in pain from seeing his philosophical mediations, or even his technical work on mathematical logic, made impossible, and at times I would give alms to him by fetching for him at the Sorbonne, forbidden to him - there was a price on his head - some books were for him a nutrient as precious as a parcel to a starving prisoner. (Rocard 1962, 14)

Moreover, in the preface of his "General Dynamic of Vibrations", he writes:

> We only gave few bibliographical references mainly because, too often, we had to completely rework questions on which there only existed rudimentary presentations, and which had not been fully exploited on a theoretical level. (Rocard 1943, vi)

Analyzing his work shows that the work which Rocard referred to with great accuracy is generally his or written by his students.[26] Moreover, in the field of nonlinear oscillations, Rocard's books (1941, 1943) are only a synthesis of numerous older results obtained by scientists such as Blondel, Van der Pol or himself, as he recalls in the preface:

> Mr. Van der Pol inspired the whole beginning of chapter XV, which only transcribes his works. We also had to borrow from our volume *Théorie des oscillateurs*, published at *Revue Scientifique* in 1941, because we could not let some of the questions addressed there fall into oblivion in an educational work, we even consider that we considerably improved their presentation. (Rocard 1943, vii)

Paragraph 130 of chapter XV entitled "The relaxation", corresponds to chapter II of Rocard's *Théorie des Oscillateurs* (1941, 31). He then recalls one of the fundamental aspects of the oscillatory phenomenon evidenced by Van der Pol and recalled by Le Corbeiller (1931a, 21, 1936, 366) (see *supra*), i.e. the continuous passing from sinusoidal oscillations to relaxation oscillations:

> This is Mr. Van der Pol's greatest merit, having recognized that it is possible to have a gradual passing from the pendulum type of oscillations to the relaxation type, by considering one equation where one parameter ε varies. (Rocard 1943, 224)

Rocard's stance on Van der Pol's work is characteristic of French scientists' perception, such as Misters Le Corbeiller, Liénard, and Haag. Indeed, even though they unanimously granted him the merit of having evidenced an oscillatory phenomenon of a new kind to which he gave the name, relaxation oscillation, they however considered that it must be included in the frame of theory of self-sustained oscillations, or self-oscillations, which is being constituted. Moreover, it seems that the lack of mathematical rigor Van der Pol showed in the calculations for the value

[25] Jean Cavaillès (1903–1944) was a French mathematics philosopher, *résistant* from the start, he was arrested twice. During his imprisonments in 1942 and 1943 he wrote a treaty "Sur la logique et la théorie de la science" (Cavaillès 1946) and was shot to death by German soldiers on 17 February 1944.

[26] Similarly to Jean Abelé, see Rocard's example (1943, 218).

of the period and amplitude of oscillations, as well as his desire to impose relaxation oscillations as a new paradigm led to general scepticism regarding his work.

8.2 Phenomenological Aspect: "*Hunt for the Relaxation Effect*"

At the same time relaxation oscillations generated a systematic search for this phenomenon in numerous and varied fields. As recalled by A. V. Gaponov-Grekhov and M. I. Rabinovich who mentioned the "hunt for the butterfly effect"[27] which occurred a few years after the publication of Edward Lorenz's article (1963):

> The same thing happened thirty years ago with limit cycles: after Andronov announced a "hunt" for them, they were discovered in chemistry, biology, ecology, and other often unexpected fields, and "there was no getting away from them". (Gaponov-Grekhov 1979, 600)

Therefore, Van der Pol's discovery (1926d) of a new type of oscillations, susceptible of characterizing a periodic phenomenon for which Andronov (1929a) (see *supra*) established that the mathematical representation corresponds in the phase space to a stable limit cycle, started a true "hunt for the relaxation effect"[28], as shown by a long series of publications in France (Bedeau 1928; Kao 1930; Fessard 1931; Hamburger 1931; Hochard 1933; Gause 1935; Kostitzin 1936, 1937; Eck 1936; Bourrières (1932, 1937a,b, 1939); Auger 1938; Parodi 1942a,b,c; Parodi and Parodi 1943) and other countries (Hamburger 1931; Gause 1934, 1935; Gause et al. 1936; Kolmogorov 1936; Richardson 1937, ...).

8.2.1 François Bedeau: Relaxation Oscillations in the C.R.A.S.

Oddly enough, Balthazar Van der Pol never published in the *Comptes Rendus de l'Académie des Sciences de Paris* (*C.R.A.S.*),[29] and references to his work can therefore only be found through other authors. It seems that the first ones to mention relaxation oscillations were François Bedeau and Jehan-Georges de Mare (1928) in a note introduced on the 17th of July 1928 by Charles Fabry. The latter had been, in 1921, appointed as holder of the chair of Physics of the science faculty of science at the University of Paris, previously occupied by Edmond Bouty, then

[27] See *supra*.

[28] See *supra*.

[29] He was however elected correspondent with the Academy of science (*division des académiciens libres et des applications de la science à l'industrie*) somewhat late: on the 6th of May 1957.

as director of the *laboratoire d'enseignement* of Physics, with Eugène Darmois, Louis Décombe and François Bedeau as collaborators. Bedeau and Mare's note, entitled "Stabilisation des oscillations de relaxation[30]" ("Stabilization of relaxation oscillations") provided one of the first occurrences within the *C.R.A.S.* of this terminology. Considering a circuit comprised of a resistance, a capacitor and a neon tube, they explain:

> Van der Pol gave the name relaxation oscillations to oscillations occurring in the circuit and he noted in a recent lecture that the period of such oscillations was not very constant. We established an objective consisting in stabilizing these oscillations by using a tuning fork, and measuring the frequency of the tuning fork afterwards. (Bedeau and Mare 1928, 209)

Bedeau must have been present at Van der Pol's conference (1928a) on the 24th of May 1928 at the *Salle de la Société de Géographie* in Paris, as the minutes of this presentation were only published in September 1928, while Bedeau's note was presented on the 17th of July. It is also possible that Van der Pol gave him an offprint of his presentation, or that Bedeau (1928), who published as Van der Pol did in the periodic *Onde Électrique*, could have had access to this document before its publication. Bedeau and Mare's statement regarding the period of relaxation oscillations faithfully transcribed the conclusion formulated by Van der Pol and Van der Mark (1928a) during this conference, regarding the period:

> The fact that these periodical phenomena's frequency is not rigorously constant is due to a relaxation time being determined, in particular, by a resistance of some type, and it is a well-known fact that external circumstances can modify a resistance much more easily than a mass or elasticity. (Van der Pol and Van der Mark 1928a, 370)

They therefore seemed to be aware of Van der Pol's recent work on relaxation oscillations and the concerns of their time, namely the determination of radio waves frequency:

> We then stroboscopically observe tuning forks vibrating freely with periods equal to 10^{-2} and 10^{-1} second. We can thus control the frequency of the tuning fork sustained electrically. In all likelihood, we will be able to, by using the last neon tube turning on at frequency 1, compare this frequency to the one in a known clock, and as it was said in the Note previously mentioned, deduce the frequencies used in radiotelegraphy. (Bedeau and Mare 1928, 210)

Nevertheless, even though the calibration principle of the frequencies that they used is quite close to the one suggested by Abraham and Bloch (see *supra* Part I) during war time, the method seems more empirical and imprecise. In his research, Bedeau (1931, 237) recalls the various methods he used, such as the multivibrator and the modulating tuning fork. In the next paragraph, he presents the principle of "calibration of the frequency of the tuning fork using Van der Pol and Van der Mark's "relaxation" oscillations". Bedeau (1931, 240) then summarized his note's results to the *C.R.A.S.* as well as Van der Pol's (1926d), and both Van der Pol and Van der Mark's (1928a), from which he visibly drew inspiration. Even if it does not

[30]A printing error occurred in the first original version entitled "Stabilisation des oscillations de relation" ("stabilization of relation oscillations").

truly constitute a "hunt for the relaxation effect" as the ones which will be presented later, this research is a first account of the way the concept of relaxation oscillation was perceived in France right after Van der Pol evidenced it.

8.2.2 *Panc-Tcheng Kao:* Oscillation Relaxations in a Piezoelectric Quartz

In regards to his biography, it has been established thanks to the kind cooperation of Mrs. Anne Vidal (Librarian) and Misters Jean-Alain Hernandez (Scientific Director) as well as Renaud Mounier (Archives Department) of the *Institut Telecom ParisTech* formerly *École Supérieure des Postes et Télégraphes*, that Panc-Tcheng Kao had been engineering studen in this *École*, class of 1929.[31] On the 17th of November 1930, it was again Charles Fabry who presented his only contribution to the *C.R.A.S.*, entitled: "Oscillations de relaxation produites par un oscillateur à quartz piézoélectrique" ("Relaxation oscillations produced by a piezoelectric quartz oscillator") (Panc-Tcheng Kao 1930). He constituted an assembly comprised of a triode, with a piezoelectric quartz blade[32] fit between the grid and the coil. In his plate circuit, an antiresonant circuit and a milliammeter are inserted serially. He then explains:

> In some cases, while we maintain the capacitance of the antiresonant circuit as fix and equal to the value corresponding to the stalling of the oscillations of quartz, we observe a periodic fluctuation, in the milliammeter, of the plate current. High-frequency oscillations are modulated by beats of somewhat variable period, of around half a second. We identified these beats as being relaxation oscillations. (Panc-Tcheng Kao 1930, 933)

He then explains that he evidenced these "relaxation oscillations, with very long periods" (Panc-Tcheng Kao 1930, 933). He uses an analogy to demonstrate it, but before that, he describes a device able to produce this kind of oscillation:

> Let us consider the typical case of a relaxation oscillator ([33]) comprised of a neon tube discharging periodically while illuminating a capacitor continually charged by a battery, through a strong resistance. We note in this example that the existence of a device C able to store energy, and of a phenomenon starting the motion towards the initial state, as soon as the energy stored in C reaches a certain value. The device storing the energy is, in our oscillator, that quartz itself, which explains the length of the observed period. (Panc-Tcheng Kao 1930, 934)

After isolating the *nonlinear characteristic* component, i.e. the device which may play the part of the neon tube or of the triode, he refers to Van der Pol's work (1930) and shows that the observed oscillatory phenomenon is "analogous to the disruptive discharge of a neon tube" (Panc-Tcheng Kao 1930, 934). This study, which provides

---

[31] Telecom ParisTech, Alumni, Directory 2009 (internal document).

[32] Quartz is a mineral component presenting the particularity of oscillating with a stable frequency when it is traversed by an electric current. Piezoelectric properties of the quartz allow the possibility to obtain very accurate oscillation frequencies.

[33] Panc-Tcheng Kao referred to Van der Pol (1930).

yet another example to the long list of devices identified by Van der Pol (see *supra*) as being the seat of relaxation oscillations (triode, series-dynamo machine, singing arc, Wehnelt interrupter, heartbeats, ...) is however still very much qualitative. He did not explain anything regarding the form of the *oscillation characteristic* of the quartz (cubic? quintic?) and did not propose any mathematical representation of the phenomenon. Of course, he did identify the relaxation oscillation, but it was only thanks to the analogy that he was able to add this example which "opened" the "hunt for the relaxation effect".

8.2.3 *Alfred Fessard:* Relaxation Oscillations in the Nerve Rhythms

A very detailed biography of neurophysiologist Alfred Fessard, published in the *Encyclopédie Universalis* by Misters Buser, Y. Galifret and Y. Laporte allowed to establish that Fessard had obtained a *Licence* in Physics in 1920 at the faculty of science in Paris (Sorbonne),and that he had then worked at the *Laboratoire de Psychologie Appliquée* at the *École Pratique des Hautes Études*, supervised at the time by Jean-Maurice Lahy (1872–1943). There, he met Henri Piéron (1881–1964), founder of experimental psychology in France, he occupied the senses physiology chair at the *Collège de France*. Fessard worked in the latter's laboratory, first as a chemist's assistant, then, from 1927, as vice-director of the *École Pratique des Hautes Études*. Fessard then founded at the *Collège de France* his own laboratory of electrophysiology, thanks to the Singer-Polignac foundation's support. He might have been the first researcher in France to record the action potential in nerve and muscle fibers. In 1931, Fessard (1931) published a extensive article entitled "Les rythmes nerveux et les oscillations de relaxation" ("Nerve rhythms and relaxation oscillations") in the periodic *L'Année psychologique*. Five years later, Fessard (1936) defended his doctorate dissertation in sciences which was dedicated to the analysis of autorhythmicity phenomena that the nerve fibers can produce. In 1947, he was appointed by the *Conseil National de la Recherche Scientifique* (C.N.R.S.), founded on the 19th of October 1939, as head of the *Centre d'Études de Physiologie Nerveuse* et d'*Électrophysiologie*. In 1949, he was elected professor at the *Collège de France*, in the chair of General Neurophysiology, which he directed in 1971.

Barely three years after Van der Pol (1928) proved that heartbeats can be represented by relaxation oscillations, Fessard (1931) established a new correspondence between the concept introduced by Van der Pol and oscillations observed in the domain of biology. In the first two parts of his article, entitled: natural and artificial activities, he presents the theoretical founding principles regarding the nerve rhythms. He then suggests "general interpretations" in which he considers "rhythms imposed by an external periodicity" of the "circular phenomena" (return of a phenomenon to its starting point), of "antagonistic phenomena", before studying the possibility of representing them using harmonic and relaxation oscillators.

After having demonstrated "the harmonic theory's lack of success in linking the elementary nerve rhythms with a defined class of physical phenomena" (Fessard 1931, 83), he addresses relaxation oscillations, and explains:

> There is no shortage of examples of periodic machines which have an oscillation form much removed from the sine-wave (in other words, developable in a very-slowly converging Fourier series, which would deprive this representation of its interest), which presents in its general characteristics, striking analogies with the probable processes occurring in a "pulsating" neuron. These machines, which we will address again later, are a part of the more general category of self-sustained systems, their main characteristic being their ability to give an intermittent form to a continuous flow of energy. (Fessard 1931, 84)

In the first part of the first sentence, Fessard covers *in extenso* one of Le Corbeiller's conclusions, which he refers to:

> *Case of ε being very large.*[34] - Here the oscillation curve has numerous harmonics. In mathematical language, the corresponding Fourier series converges very slowly. It is therefore absolutely unrealistic in this case to laboriously calculate the first, second or third terms of the series. (Le Corbeiller 1931a, 22)

This crucial point of relaxation oscillations apparently did not escape this physicist by training. In the second part he speaks as a neurophysiologist, by once again using an analogy between the form of relaxation oscillations (see *supra*) and the oscillogram representing an action potential.[35] He used the term "pulsating" which seems to have been introduced by Henri Bouasse (1926, 327) and has been replaced nowadays by "burst" or "spike". In the last sentence, Fessard agrees with Le Corbeiller's point of view (1931a, 45), regarding relaxation oscillations, which he seems to place inside the more general frame of self-sustained systems. He then explains:

> In order to narrow down the analogy, we must address a special type of self-sustained system, by first choosing from simple cases (our usual motors are often complicated combinations of machines belonging to several categories), and in particular from the ones which present a specific kind of oscillation, which were recently called relaxation oscillations. (Fessard 1931, 84)

It seems that Fessard indeed wanted to specify the relaxation oscillation phenomenon, and to include it in a more general theory, as Le Corbeiller later did (1933b, 328) (see *supra*). He then reproduced, in a very educational manner, the metaphorical example characterizing the concept of relaxation oscillations: the *Tantalus cup*, explained by Le Corbeiller (1931a, 6) (see *supra*) and describes the two phases of filling and emptying of this device, then adds:

> These are two antagonistic processes, the existence of one favouring the production of the other. Except that the change, instead of happening in a continuous manner, as is the case

[34]This corresponds to the case of relaxation oscillations. The case of $\varepsilon \ll 1$ is the one for sinusoidal oscillations.

[35]The action potential consists in a sudden modification of the resting potential. This is an electrical phenomenon produced by an electrical stimulation of a nerve cell.

> for the elastic or pendular oscillations, happens suddenly. The system works by "jolts". (Fessard 1931, 85)

With this example, Fessard shows that he perfectly integrated the concept of relaxation oscillations possessing a slow phase and a fast phase, and which is consequently able to characterize the "pulsating" that an action potential is. It must be noted that the concept of self-sustained oscillations is not addressed here. Indeed, it is the slow-fast phenomenological aspect that Fessard was interested in, rather than the mathematical angle, connected to the schematizing which he nevertheless considered a bit later (see *infra*). He then describes the oscillations of a neon lamp studied by Le Corbeiller (1931a, 8), and establishes a comparison with the nerve's rhythm:

> As for nerves, what can we see? We know that when the discharge happens, the nerve enters a phase of complete inexcitability, called absolute refractory period, followed by a phase where it slowly returns to normal excitability, or relative refractory period: this is the repair phase, which can be compared to a kind of "filling" (or "charge"). At one point, coinciding with the start of the action potential, something analogous to an "emptying" (or "discharge") occurs, which leads to a new filling, and so on. Adrian, as soon as he discovered the first nerve rhythms of the isolated fiber ([36]), thought to link them with the existence of the refractory period (absolute and relative), and based on this, proposed a clever explanation of the relation between the frequency and intensity of the stimulation. There is here what seems to be a solid starting concept, which only needs furthering by considering the necessary variations, for the application to the various cases (receptors, centers, effectors or nerves). The linking of these nerve phenomena to the general class of relaxation oscillations can only improve the original hypothesis, and gives a direction to the research aiming to verify its pertinence. (Fessard 1931, 86–87)

Again, this analogy led Fessard to link the oscillatory phenomenon observed in the nerve rhythms to the concept of relaxation oscillations. This was once again a qualitative study, which aimed to be first and foremost descriptive, and denied any ambition to create a model.

> The term *relaxation oscillation* is quite recent, and little known. We might think that a new label, in conjunction with the presentation of a few crude models - which are certainly not the firsts of their kind to be proposed - would not make for a great advancement. Indeed, without an accurate general theory, the labels might be arbitrary, and the models1 are useless- they might even lead to erroneous ideas - because the knowledge of one's mechanism does not inform on the mechanism of another (...). Yet, until the last few years, self-oscillating systems had indeed received from physicists the attention they deserved, but their general theory was yet to be developed, even if the problem had been written in a mathematical form long ago (Lord Rayleigh[37]). In 1926, at the same time Adrian discovered elementary nerve rhythms, Dutch physicist, B. van der Pol,[38] suggested his general theory of *relaxation oscillations*, and gave them a name. Since then, other scientists have tackled the question, which still holds many obstacles to overcome. But a great breakthrough was accomplished, since phenomena that had been viewed until now as disparate, have been

[36]Fessard refers to Adrian (1930). Edgar Douglas Adrian (1889–1977) was an English physiologist, who obtained the Nobel Prize in Medicine in 1932.

[37]Fessard refers to Strutt and Rayleigh (1883).

[38]Fessard refers to Van der Pol (1926d, 1930).

grouped in a same natural family. Consequently, the study of a specific system will from now on benefit, in part - the one they have in common- to the understanding of any other. Saying now that the elementary nerve rhythms are determined by a relaxation oscillation is not, according to us, an overly bold statement. While this is not strictly speaking an explanation, it is at least a natural classification, - which does not seem really debatable - with a still-new label belonging to a promising theory.

1. Let us also note that we did not intend to build a nerve "model", but rather to only give a concrete idea with extremely simple example of what constitutes a relaxation system. A true model of the systems we study should be more reminiscent of the characteristics of the nerve. (Fessard 1931, 87–88)

Fessard describes with great perspicacity the real value of the concept of relaxation oscillation, and the importance of Van der Pol's research. But he first dismisses the possibility of a "trend effect" or rather, a "relaxation effect", although it was already quite present, by explaining that Van der Pol's merit which consisted, on one hand, in evidencing this new type of oscillations, and, on the other, in naming it, would be a meaningless new label, if it did not have a budding underlying theory. He then recalls the precedence of Rayleigh's work (1883) over Van der Pol's (1926d), agreeing with Le Corbeiller (1931a, 22) (see *supra* Part I). Finally, he explains in detail what gave its value as a concept to the relaxation oscillation phenomenon, that is to say, its ability to describe apparently very different phenomena, but which were from then on gathered under a same type of oscillations. By analogy, he states that the nerve rhythms can be described in terms of relaxation oscillations and then concludes by quoting Le Corbeiller (1931a, 45) (see *supra*). In the following paragraphs, Fessard analyzes Van der Pol and Van der Mark's work (1928a) on heartbeats, and explains that "if the theory can be successfully applied to the complex case of the heart, it is endorsed to use it for elementary nerve rhythms." (Fessard 1931, 90). Fessard therefore provides on the one hand a new neurobiological example of relaxation oscillations which can be added to Van der Pol's already exhaustive list, and on the other hand, an analysis of Van der Pol's results, admittedly marked by Le Corbeiller's (1931a), but with an outsider's point of view, that is of a neurophysiologist's. He also gives a clear definition of the concept of relaxation oscillations, and confers the scope of Van der Pol's work (Fig. 8.10).

8.2.4 Étienne Hochard: Relaxation Oscillations in Photoelectric Cells

No biographical elements regarding Étienne Hochard[39] were found. On the 27th of March 1933, Paul Villard presented a note written by Étienne Hochard (1933)

[39]However it was established that he participated in 1932 along with Bernard Champion, to the translation of Werner Heisenberg's work: "Les Principes physiques de la théorie des quanta", and that in 1935, he defended a thesis entitled "Mesure de petits courants thermoélectriques application à l'étude d'un faisceau de rayons infrarouges polarisé elliptiquement" at the Faculty of Science of the University of Paris to obtain the rank of *docteur ès sciences physiques*.

Fig. 8.10 Alfred Fessard and Denise Albe-Fessard (Internet source)

entitled: "Sur les oscillations entretenues" ("On sustained oscillations"). Hochard considered a circuit comprised of a photoelectric cell serially assembled with a high-tension battery, a great resistance R, and a capacitor C. He then noted on the one side, that when the cell is lit up, the permanent regime is not instantaneously reached, and on the other side, that if the luminous flux is properly oriented:

> (...) we can produce a series of charges and discharges of the capacitor, and sustain the oscillations. (Hochard 1933, 905)

In order to do this, he placed a light-spot galvanometer[40] in the circuit, and had the light spot act on the photoelectric cell. He thus proved that the period of the phenomenon is a function of C and R, and explained:

> By recording the galvanometer's deviations, we obtain very varied forms of curves. We can especially recreate the various types of curves described by Van der Pol ([41]). We go from a quasi-sinusoidal form to the form characteristic of relaxation oscillations by increasing the spot's lighting. (Hochard 1933, 905)

[40]It is an ammeter with a mobile frame, for which the pointer was replaced with a mirror reflecting a beam, indicating the value of the electric current's voltage.

[41]Hochard referred to Van der Pol (1930).

He also used the analogy between the curves he found and the oscillograms that Van der Pol produced (1930, 303) (See *supra*) to show that the photoelectric cell is the seat of relaxation oscillations.

8.2.5 *Jean-Louis Eck:* Relaxation Oscillations in the Gas Triodes

Regarding Jean-Louis Eck's civil status, the dates (1908–1984) can be found, but they might be a homonymous person's.[42] In 1936, he published an article in the *Journal de Physique et du Radium*, entitled "Oscillations de relaxation à période stable obtenues avec une triode à gaz" ("Stable-period relaxation oscillations obtained with a gas triode"). At the time, he was attached to Armand Gramont's laboratory in Levallois-Perret. In 1944, he defended his doctorate thesis at the Faculty of Science in Montpellier (Eck 1947). In his 1936 article, Eck demonstrates that:

> The gas-discharge triodes allow for the production of relaxation oscillations, for which the amplitude depends on the negative current applied to the grid. (Eck 1936, 227)

He carried out the same assembly as Blondel (1919b) or Van der Pol (1920) (see *supra* Part I) but used a thyratron[43] (see Fig. 8.11) instead of a triode.

He drew the thyratron's oscillation characteristic (see Fig. 8.12).

It is interesting to compare Fig. 8.12 which represents the oscillation characteristic of the thyratron with the dynamic characteristic of the arc (see *supra* Part I, Fig. 1.8). We see that it is an observation of the same phenomenon. He then explains that the thyratron operates between the voltages E_0 and V as shown on Fig. 8.13:

From this, he deduces that the time t is expressed by the relation:

$$t = RC\, Log \frac{E - E_0}{E - V}.$$

He then uses various calculations in order to have the period be independent from the voltage, and found the following expression:

$$t = RC\, Log\,(1 + k)$$

[42] However, it was established that he presented in 1929 an essay entitled "Étude expérimentale de l'effet Peltier et de l'effet Thomson" at the science faculty of the University of Paris to obtain his *Diplôme d'Études Supérieures Sciences Physiques*.

[43] It is a discharge chamber filled with gas, containing a coil at the cathode, a plate at the anode, and one or more grids. An inert gas or metallic vapor fills the discharge chamber. In the case of the RCA 885, the gas used is Xenon. The grid only controls the current's start, and thus provides a triggering effect.

Fig. 8.11 Thyratron RCA 885 used by Eck (1936) (Internet source)

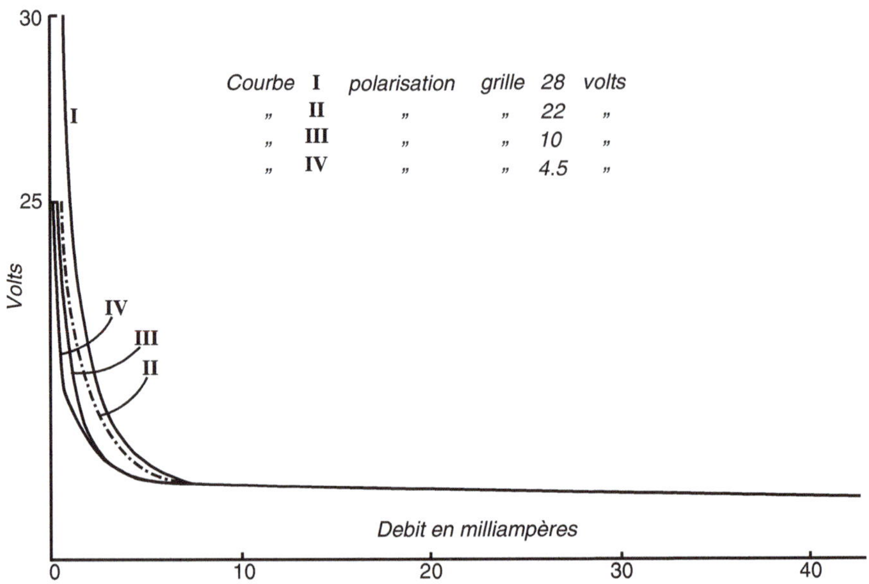

Fig. 8.12 Thyratron RCA 885 characteristic, by Eck (1936, 227)

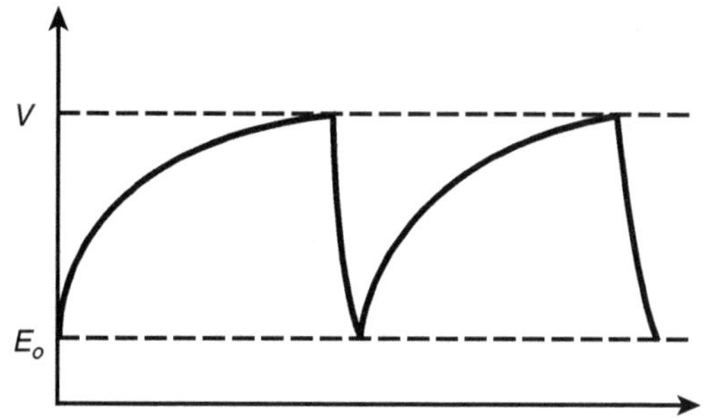

Fig. 8.13 Charge and discharge of the capacitor, by Eck (1936, 228)

where k represents an amplification factor. He then demonstrates the oscillations' amplitude increases linearly with the applied voltage. This led him to the following conclusion

> Thus, we obtain a fix-period and variable-amplitude relaxation oscillation system. (...) In short, the gas-discharge triodes allow, among other uses, to produce especially stable relaxation oscillations. (Eck 1936, 229)

Contrary to his predecessors, Eck's work was mostly experimental, although he succeeded in the feat of "demonstrating", without using the analogy, and without referring to Van der Pol (or anyone else), that relaxation oscillations occur in the thyraton.

8.2.6 *François-Joseph Bourrières:* Relaxation Oscillations in Garden Hoses

A biography of François-Joseph Bourrières[44] can be found in Mutabazi et al.'s book (2005, 30). It explains that he studied for his bachelor's degree in Bordeaux with Henri Bénard (1874–1939) and Pierre Duhem (1861–1916), before working at Bénard's laboratory at the *Institut de Mécanique des Fluides in Issy-les-Moulineaux.* In 1932, he published a first note in the *C.R.A.S.* entitled: "Sur les oscillations libres des extrémités de tubes élastiques parcourus par un courant uniforme de fluide" ("On free oscillations at the extremities of elastic hoses traversed by an unvarying fluid current") (Bourrières 1932). As astonishing as it may seem, he considered demonstrating the existence of self-oscillations using a garden hose subjected to strong water current. In order to experiment on the phenomenon, he attached a flashlight to the free extremity of a hose, and took pictures of the trajectory's motion.

[44] It was Mr. Emmanuel Delangre, Professor at the *Laboratoire de Mécanique des Fluides de l'École Polytechnique, who informed me of M. F. J. Bourrières's work. This paragraph is dedicated to him.*

In this first qualitative rather than quantitative study, he describes the observed phenomenon and claims it was governed by laws that he intends on clarifying:

> The motions of the extremities of tubes where a fluid circulates (air or water) are subjected to simple and precise laws. (Bourrières 1932, 50)

He then indicates the reason why he conducting this study, a surprising one to say the least.

> The phenomenon is periodic when it occurs in fixed conditions. It seemed interesting to study it, as it is one of the cases where a continuous excitation results in an oscillation. As it can occur on a large scale and with moderate speeds, it could be expected that its sustaining would be easier to analyse than with other cases of continuous-excitation periodicity (Bénard-Karman vortex streets, reed beats, etc.). (Bourrières 1932, 50)

It therefore appears that his work was part of a set of studies carried out in the field of hydrodynamics (see Camichel 1927; Camichel et al. 1928 and later Camichel and Teissié-solier 1935; Rocard 1937b; Teissié-Solier et al. 1937; Teissié-Solier and Castagnetto 1938 then Castagnetto 1939) as well as in acoustics (see Carrière 1924; Foch 1935, Auger 1937, 1938, 1939, 1945, ...). He then explains the three laws (summarized below) from which the oscillations originate. Beforehand, he made sure to clarify that the motions study was carried out in-plane.

I. Law of amplitudes: a minimum length of hose is necessary for the oscillations to "start".
II. Law of periods: the duration of oscillations depends on the length of the hose and the pressure.
III. Law of motion sustaining: the sustaining is explained by the asymmetry between the initial position of the hose and its position after its horizontal crossing (see Fig. 8.14).

We can see on this chronophotography that the initial position of the hose (first picture on the left) does not present a symmetric value (in red on Fig. 8.14) horizontal to the amplitude observed on the fourth picture.

According to the *Annales de l'Université* de Paris (1934, 432), Bourrières finished writing his thesis dissertation in 1934, and defended it at the science faculty of the University of Paris. It was entitled "Sur un phénomène d'oscillation auto-entretenue en mécanique des fluides" ("On the phenomenon of self-sustained oscillation in fluid mechanics') and it was only published five years later with the support of the *Ministère de l'Air* (Bourrières 1939). In 1937 he submitted two other notes to the *Académie des Sciences* which were presented by Marcel Brillouin, as it was the case for the previous one. In the first, evocatively entitled: "Sur les oscillations auto-entretenues des extrémités de tubes élastiques déversant un courant continu de fluides et celles d'anches libres encastrées" ("On the self-sustained oscillations of the extremities of elastic tubes discharging a continuous current of fluids and the self-sustained oscillations of framed free reeds") presented on the 28th of February 1937, he went from a phenomenological study to a mathematical analysis. By equating the liquid jet to an "infinitely flexible string", he managed to establish the equation partial derivative equation for the motion. He

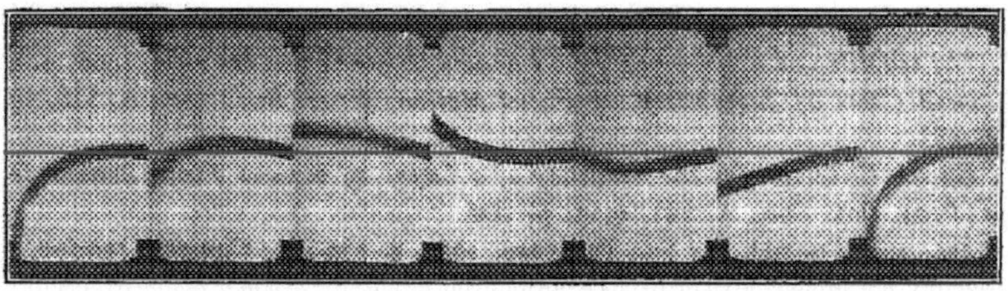

N. B. — Le film ci-dessus montre une période d'oscillation. On y remarque deux positions rectilignes du tube différemment placées, une pour l'aller, l'autre pour le retour. On remarquera sur le film les 7 événements principaux d'une période en allant de gauche à droite : maximum d'élongation en bas, proéminence vers le haut, position rectiligne en haut, maximum d'élongation en haut, proéminence vers le bas, position rectiligne en bas, maximum d'élongation en bas. Le tube est encastré à droite. On n'en voit sur la figure que les deux tiers à partir de l'encastrement.

N.B. – The film above shows an oscillation period. We can see two differently placed rectilinear positions of the hose, one to go and one to return. It is noteworthy from what is seen on the film that there are seven main events in a period, from left to right: maximum elongation at the bottom, prominence towards the top, rectilinear position at the top, maximum elongation at the top, prominence towards the bottom, rectilinear position at the bottom, maximum elongation at the bottom. The hose is framed on the right. Only two-thirds are visible on the figure due to the framing.

Fig. 8.14 Chronophotography of the oscillations produced by the hose, by Bourrières (1932, 51)

then deduced an expression of the oscillation period, which he confronted with the stroboscopic observation of the phenomenon. In the second note presented on the 5th of May 1937: "Sur les auto-oscillations de jets réels en milieu identique par autoplissements" ("On the self-oscillations of real jets in an identical environment by "self-wrinkling""), he describes a new experiment aiming at evidencing the self-oscillations of liquid jets depending on the nozzle[45] adapted to the hose's opening. It is interesting to note the way in which Bourrières used Andronov's terminology (1928, 1929a) (see *supra*) and the ease with which he went from using the term self-sustained oscillation to self-oscillations in just a few months. It seems that until now Bourrières had a keen interest for Andronov's work, evidencing new examples of self-oscillators without actually having demonstrated or shown the existence of a limit cycle in the phase space corresponding to the mathematical representation of the phenomenon. In a work he published in 1939, he synthesized his thesis dissertation and the two notes published in the *C.R.A.S.* In it, he carried out lengthy mathematical developments based on partial derivative equations found in Fluid Mechanics, in order to establish equations for the motion in the studied cases. However, while keeping his perspicuity regarding the interpretation of the phenomenon, he wrote in a footnote:

[45]Hose adapted an opening for the flow of a fluid under pressure in order to modify the output or form of a jet.

> One must face the fact that no matter how interesting the determination of the critical period of self-oscillations by considering the radiant may be, it does not analyze what happens during the phenomenon's period itself.
> Studying what occurs inside this period might be as difficult in self-perturbation as it is in astronomy or electricity. Interrogations on cycles and stability might have to be considered. This would therefore make the comparison closer to the *relaxation* phenomenon (See bibliography).
> The problem would moreover be more difficult, as it would be a question of stability and instability, not around the static balance position, but around the *dynamic balance* trajectory. (Bourrières 1939, 44)

The analogy he proposed with Astronomy and Electricity does not seem to be a mere coincidence, as confirmed by the following sentence and the bibliography which refers to Le Corbeiller's (1931a), Andronov's (1929a), Van der Pol's (1926d) and Liénard's (1928) work. The "stability around dynamic trajectories" mentioned by Bourrières is none other than the limit cycle stability (see *supra* Part I). His sentence perfectly summarizes the previous half-century, during which the scientists went from studying "linear stability" to studying "Lyapunov stability". Looking back on his experimental study consisting in attaching a flashlight to the free extremity of a hose, he observes that the motion of its trajectory draws an eight-shaped curve, and explains:

> Later, we wondered if the eight drawn at the second order of approximation by the extremity of the hose, *was an essential part of the self-sustaining mechanism* and if it should be explained by a non-uniform motion, caused by static or dynamic instability phenomena, as is the case for the fascinating general phenomena of *self-relaxation* ([46]). (Bourrières 1939, 60)

In this excerpt, he questioned the inherent nature of the self-sustained phenomenon, which could also be the seat of relaxation oscillations, which would be shown by the fact that the extremity of the hose indeed draws an eight shape, but presenting slow parts and fast parts. This idea led him to the creation of another neologism, *self-relaxation*, which incidentally could have been used to describe a self-sustained system, in which relaxation oscillations may occur. The many examples suggested by Van der Pol could have been called: the triode, the series-dynamo machine, the singing arc, ...

At pages 79 and 80 of his memoir, he presents the plates V (reproduced below). On them, we can see the trajectory of the hose's extremity drawing an eight. It must be noted that the trajectory cannot in any case intersect itself.[47] Moreover, it appears that the trajectory Bourrières observed presents all the characteristics of a *limit cycle* (Figs. 8.15 and 8.16).

Bourrières took these shots by using a camera with an open shutter. This allowed him to empirically evidence these "limit cycles" characterizing the self-sustained oscillations of a hose.

---

[46] Bourrières refers to Le Corbeiller (1931a).

[47] From a theory of dynamical systems point of view, this would mean that for a same given initial condition, a trajectory may have two different futures. See Bergé et al. (1988, 120).

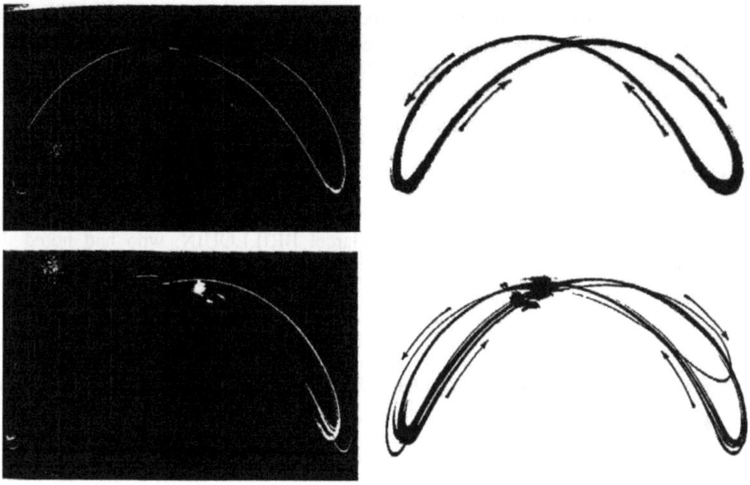

Fig. 8.15 Self-oscillations of a garden hose, by Bourrières (1939, 79)

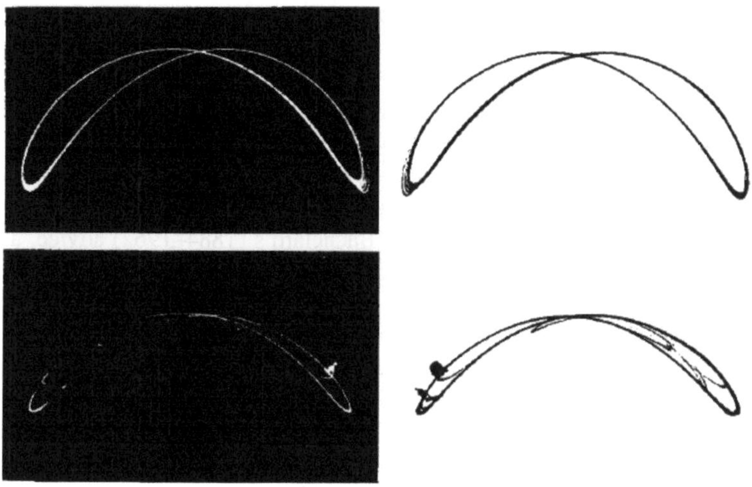

Fig. 8.16 Self-oscillations of a garden hose, by Bourrières (1939, 79)

Unlike his predecessors, he thus provided another example of a self-oscillator potentially likely to produce relaxation oscillations. This could imply that he actually observed "limit cycles" rather than sinusoidal oscillations, which would produce elliptical trajectories close to the eight-shaped ones that he managed to photograph, meaning that no matter what the initial conditions are, the amplitude and trajectory remains the same. This was also confirmed by Henri Bénard in the preface of Bourrières's book:

But we did not look for the non-periodic shape that the motion would have with any initial conditions (we would probably reach this by letting ourselves be guided by an analogy with the solution given by H. RESAL for beams). Nor did we look for the reason which, in case of self-oscillations, causes the system to strain more or less fast towards a *limit cycle* which is identical whatever the initial conditions are. (Bourrières 1939)

Bénard also recalls in this preface that this was not a new concept:

The self-oscillation of rubber hoses discharging a continuous current of fluid was noticed as early as the year 1885 by my teacher Marcel BRILLOUIN, who had noted their particularities in laboratory notes that were never published. (Bourrières 1939)

8.2.7 *Léon Auger:* Relaxation Oscillations in Percussion-Reed Pipes

There are numerous biographical elements regarding Léon Auger in the obituary notice written by Rochot (1965). Former student at the *École Normale* in Auteuil, he taught Physics in Saint-Maur, then at the Arago, Turgot and Lavoisier Schools, before starting research in the field of *Acoustics*. During the 1930s, he published a series of notes on said subject in the *C.R.A.S.*, which were presented by Villat (Auger 1932, 1937, 1938, 1939). In 1943, he defended his thesis dissertation at the Faculty of Science of the University of Paris, which was entitled "Oscillations auto-entretenues dans les tuyaux sonores à anches battantes" ("Sustained oscillations in percussion-reed acoustic pipes"), and was published in 1945 by Gauthier-Villars under the supervision of the *Ministère de l'Air*. He then turned to humanity studies and presented, following Gaston Bachelard's (1884–1962) advice, a thesis dissertation on Roberval. He then focused his research on the History of Science.

While the first two notes written by Auger (1932, 1937) in the *C.R.A.S.* were very empirical, the third, entitled: "Sur l'accord des tuyaux sonores à anche battante considéré comme phénomène de relaxation" ("On the tuning of percussion-reed acoustic pipes considered as a relaxation phenomenon") (Auger 1938) allowed him to provide another example of relaxation oscillation. In order to carry out his study, Auger attached a reed[48] to a pipe linked to a resonator. He then measured the frequency of the emitted sound with varied reed lengths. Afterwards, he represented on a diagram the frequency depending on the reed length, from which he deduced:

All in all the self-sustained oscillation of the reed is adjusted along AB for the inherent period of the resonator closed at the smaller end. The relaxation character of the phenomenon is thus confirmed ([49]). (Auger 1938, 218)

He went back over his study the following year, and published a fourth note entitled "Sur la stabilité des sons émis par un tuyau à anche soumis à une pression

[48] Thin reed or metal strip placed on various wind instruments which are made to vibrate from the passage of the air produces the sound.

[49] Auger referred to chapter VI of Foch's work (1935, 55) dedicated to relaxation oscillations.

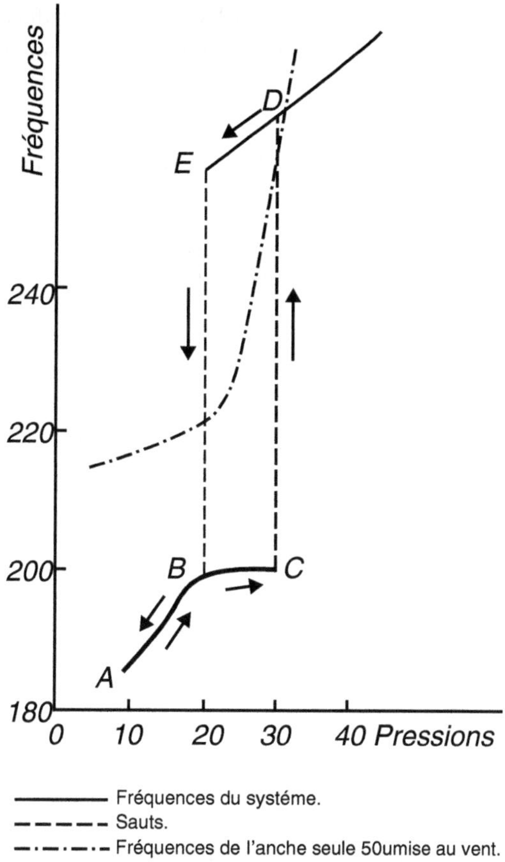

Fig. 8.17 Hysteresis cycle of a sound percussion-reed pipe, by Auger (1939, 509)

constante" ("On the stability of sounds emitted by a reed pipe subjected to a constant pressure") (Auger 1939). While the title is less explicit than his previous note's, his demonstration is more convincing. Auger presents the graph reproduced below (Fig. 8.17).

Auger explains that when the pressure increases from zero, the representative point of the frequency draws the arc ABC. Then at the frequency 30 the representative point "jumps" from C to D. If the pressure decreases, the point follows along the curve DE and "suddenly falls" to B when the frequency reaches the value 19. The "sudden jump" phenomenon described by Auger accurately characterizes Van der Pol's concept of relaxation oscillation with its slow phase and fast phase. He furthers his experimental study in his thesis dissertation[50] with a more in-depth analysis of

[50]Defended on the 17th of June 1943 before Misters Bruhat (President), Foch and Rocard (Examiners).

this self-sustained system (Auger 1945) which resulted in the evidencing of another device in which relaxation oscillations occur, although he was not aware of it.

8.2.8 *Hippolyte Parodi:* Relaxation Oscillations in Running of Trains

Many biographies on Hippolyte Parodi can be found, such as the ones written by Darrieus (1968), Peychés (1970) or Merger (1997). He was the second son of Dominique Alexandre Parodi, a literary man and writer who lived in Paris. After successful studies at the *lycée* Condorcet Hippolyte, he prepared for the competitive entrance examination for the *École Normale Supérieure* and the *École Polytechnique*, as his brother did. Being a successful candidate for the two, he chose the *Polytechnique*, which he joined in 1893. During these years, he met Paul Langevin. *Licencié* in Science in 1900, he began his career at the Service Traction of the *Compagnie Française Thomson Houston* (C.F.T.H.), which allowed him to complete his training, and to broaden his expertise to a research on electric networks. He then contributed to the electrification of the railroad line *Paris-Invalides-Les Moulineaux-Versailles Rive gauche*, then of the line *Paris-Juvisy*. He thus perfected a graphical integration method using successive arcs of differential equations to operate trains. Parodi enlisted in the First World War and went back to his lieutenant uniform after refusing the rank of Major inherent to his position in the railway industry. He created shooting abaci, and at the beginning of 1917 he was appointed head of the *section Balistique et des tables de tir*. His calculations allowed him to solve Big Bertha's infamous bombings on Paris. After the war, performed a "world premiere": the electrification of a high-traffic network. At the same time, he established a project for the electrification of the French railroads. He became professor at the *École nationale des Télécommunications* (1927), the *Conservatoire National des Arts et Métiers* (1933), the *École Nationale de l'Aéronautique*, then the *École Supérieure d'Électricité*.[51] He published many articles on the electric traction in various periodic publications such as the *Lumière Électrique*, the *Revue Générale de l'Électricité*, the *Bulletin de la Société Française des Électriciens*, the *Revue Générale des Chemins de Fer*, ... he also published notes in the *C.R.A.S.* on the graphical integration and the equations for the operating of trains (Parodi 1931, 1942a,b,c; Parodi and Parodi 1942, 1943). During the Second World War, Hippolyte Parodi published several notes in the *C.R.A.S.*, one of which was co-written with his eldest son Maurice Parodi on the operating of trains and relaxation oscillations (Fig. 8.18).

[51] In 1944, he was called to sit at the *Commission de Mécanique du C.N.R.S.* After WWII, he became a member of the Academy of Science in 1949. He was honored with many distinctions: Ancel prize (1929), *prix Montyon de Mécanique de l'Institut de France* (1930) then Chevallier (1946) of the *Société des Ingénieurs Civils* which he later presided over.

Fig. 8.18 Hippolyte Parodi (Source: *Archives des Chemins de Fer*)

8.2.8.1 Phenomenological Analysis

Parodi's (1942) first written contribution to the *C.R.A.S.* was introduced on the 20th of July by Albert Caquot and aimed at linking the representation of the trains' motion with the phenomenon of relaxation oscillations. He began by writing the equation for the traction, which is actually the mathematical translation of the *Fundamental Principle of Dynamics or Newton's Second law*:

$$m\frac{d^2s}{dt^2} = \varphi_1(v) - \varphi_2(v) - \varphi_3(s) \tag{8.10}$$

where $\varphi_1(v)$ represents the forward force, $\varphi_2(v)$ resistance to forward motion, and $\varphi_3(s)$ positive or negative forces due to the gradients, curves, or slopes. He then recalls that "in preparation for its graphical resolution", he wrote, 40 years before Parodi (1909), the equation (8.10) with the form:

$$v\frac{dv}{ds} = \gamma_1(v, s) + \gamma_2(s) \tag{8.11}$$

Then, he explains that the equation for the relaxation oscillations is only one specific case of the equation (8.11).

> The equations for the relaxation are only one specific case of the equations for the electric traction in which $\gamma_1(v, s)$ would be written with the form $v\,f(s)$ and $\gamma_2(s)$ with the form ks. (Parodi 1942a, 170)

To demonstrate this, one must simply use Van der Pol's equation (3.8) (see *supra* Part I) and write (1926d, 982) $\dot{v} = z$. We thus obtain a first order differential equation that Van der Pol (1926d, 982) called "super Ricatti":

$$\frac{dz}{dv} - \varepsilon\left(1 - v^2\right) + \frac{v}{z} = 0 \tag{8.12}$$

By multiplying by z the equation (8.12) yields:

$$z\frac{dz}{dv} - \varepsilon\left(1 - v^2\right)z + v = 0 \tag{8.13}$$

By then substituting $z \to v$ and $v \to s$ it yields:

$$v\frac{dv}{ds} = \varepsilon\left(1 - s^2\right)v - s \tag{8.14}$$

By writing: $\gamma_1(v, s) = v\left[\varepsilon\left(1 - s^2\right)\right] = v f(s)$ and $\gamma_2(s) = -s$ the identity between (V_{18}) and (8.11) thus appears clearly. Parodi then explains that this "specific case" corresponds to the case of the automatic electric traction. He then demonstrates that, for a line with a parabolic profile, the previous change of variables is fully justified. We could nevertheless object that in the case of a train $v f(s)$ is not necessarily equal to $v\left[\varepsilon\left(1 - s^2\right)\right]$. He introduces in what he calls a "space-speed diagram", which corresponds to what is nowadays called a "phase plane", the solution to the equation (8.11).

He then adds:

> By giving positive values to the speed for one direction of the path, and negative ones for the return path, we find that the representation of the *motion*, conventionally used in railroad networks, blends into the representation of the solution for relaxation oscillations. The space-speed diagram is a cycle of the solution for the relaxation, and the graph for the running of trains is a simplified form, and not (depending on whether we take the signaling into account) of the space-time curves that are solutions to the original equation from which the equation for the relaxation derives. (Parodi 1942a, 171)

This excerpt shows that Parodi, as his predecessors did, used the analogy. Indeed, he identifies a "cycle of the solution for the relaxation". This allowed him to link the running of trains to this phenomenon. Nevertheless, there are no mentions of a "limit cycle" or "self-sustained oscillation" at that moment.

8.2.8.2 The Representation of the "Limit Cycle"

In this second note which extends the first, Parodi (1942b) replaces the nonlinear differential equation (8.11) with a piecewise linear differential equations system. Beforehand, he recalls the relaxation oscillation phenomenon, choosing terms that show his perfect integration of the concept by introducing (Fig. 8.19):

> We know that the phenomena represented by an equation for the relaxation are characterized, from a dynamic point of view, by energy absorption during part of the path, and by restitution during the rest of the path. (Parodi 1942b, 125)

This description of the relaxation phenomenon can be compared with the Tantalus cup example (see *supra*). The receptacle collects energy during the (slow) filling, and comes back during the (fast) emptying phase.

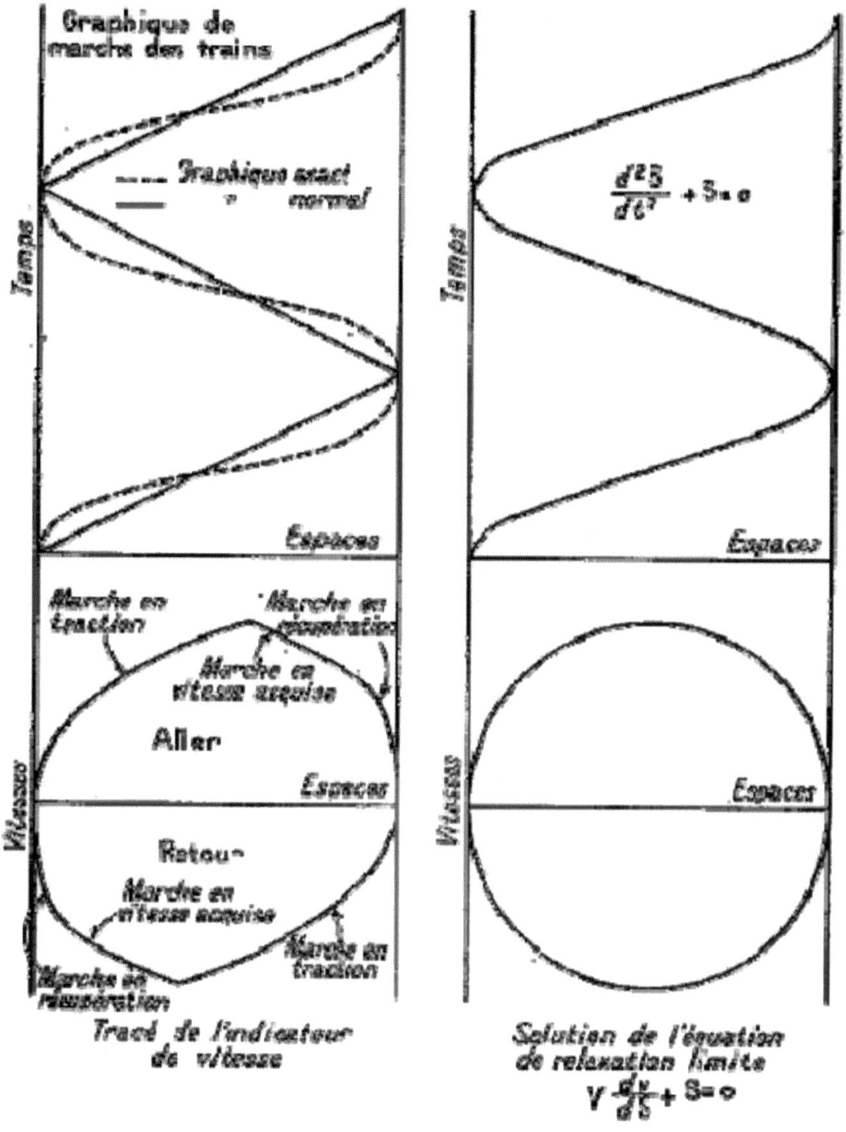

Fig. 8.19 Limit cycle representing the running of trains, by Parodi (1942a, 170)

By accounting for the forward force, constant during the travel from $-s_0$ to $+s_0$ and equal to γ, and the brake power, also constant and equal to f during the rest of the path on a parabolic-profile, low-downslope line, he obtains the following system of six integrable linear differential equations[52]:

$$(1)\ v\frac{dv}{ds} + f + s = 0 \text{ for } -\infty < s < -s_0$$

$$(2)\ v\frac{dv}{ds} - \gamma + s = 0 \text{ for } -s_0 < s < +s_0$$

$$(3)\ v\frac{dv}{ds} + f + s = 0 \text{ for } +s_0 < s < +\infty$$

$$(4)\ v\frac{dv}{ds} - f + s = 0 \text{ for } +\infty > s > +s_0$$

$$(5)\ v\frac{dv}{ds} + \gamma + s = 0 \text{ for } +s_0 < s < -s_0$$

$$(6)\ v\frac{dv}{ds} - f + s = 0 \text{ for } -s_0 > s > -\infty$$

By using a "piecewise linearization" in combination with the successive-arc graphical integration method which he perfected for ballistics and running of trains, Parodi was able to represent the "limit cycle" solution to the nonlinear differential equation (8.11). It must be noted that Rocard had already used this linearization process (1937a, 397) (see *supra*). He clarifies beforehand:

> We will show the analogy between this traction problem and relaxation oscillations. (Parodi 1942b, 126)

He then provides a representation reproduced on Fig. 8.20 and adds the following commentary:

> The figure shows that these successive curve arcs form a spiral coiling from the inside to the outside if the initial speed is low and from the outside to the inside if it is high. These two spirals strain towards a limit cycle. It therefore appears that after some time, for any starting speed, the motion strains towards a periodic regime and that, for a defined initial speed, this regime is instantly reached. (Parodi 1942b, 127)

By comparing this excerpt with the definition that Poincaré (1886b, 30) gave of the concept of limit cycle (see *supra* Part I) it is certain that Parodi used his terminology. It is however surprising that no reference to Poincaré, Liénard or Andronov can be found, especially as Parodi paraphrased almost all of Liénard's conclusions (1928, 906) (see *supra* Part I). He recalls the fundamental independence property, in relation to the initial conditions evidenced by Andronov (1928, 24, 1929a, 560) (see *supra*), yet does not mention self-sustained systems. It is very surprising that an engineer would not address the stability of these oscillations either, while he claims that there is a straining towards this limit cycle, whatever the initial condition chosen on the outside or on the inside.

[52]The numbers of the equations correspond to their validity domain. See Fig. 8.20.

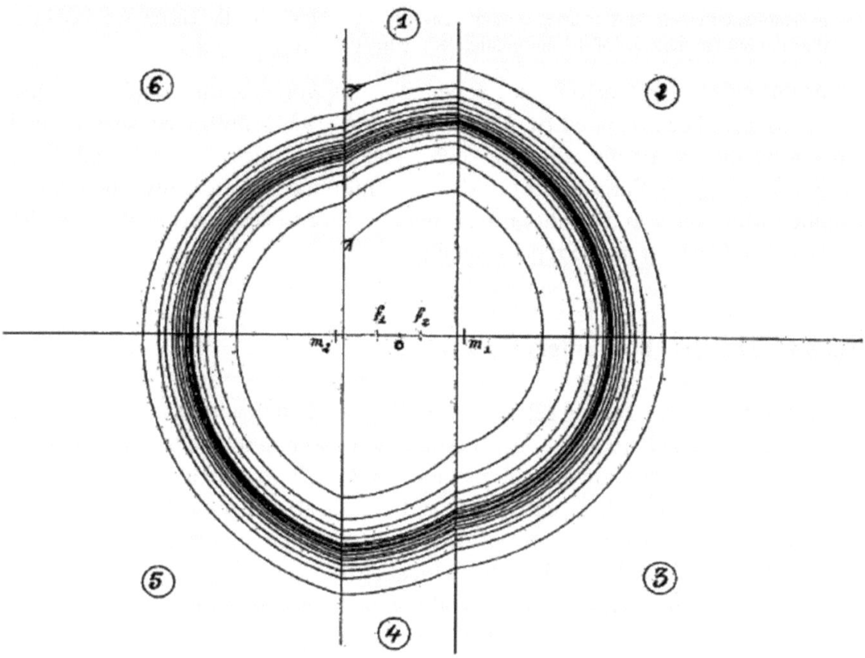

Fig. 8.20 Limit cycle representing the running of trains, by Parodi (1942b, 127)

8.2.8.3 The Graphical Integration I

This third note to the *C.R.A.S.* on the subject, entitled "Détermination graphique du cycle limite solution des équations de relaxation" ("Graphical determination of the limit cycle solution to the equations for the relaxation") (Parodi 1942c) aims at describing a graphical integration method based on the construction of "isogrype curves", rather than the "isocline curves"" construction created by Parodi several years earlier for the study on the running of trains and the calculation of ballistic trajectories. These methods were already presented in detail (see *supra* Part I). Parodi offers to obtain the solution to the equation for the relaxation (8.11), which he now writes by taking the changes of variables into account:

$$v\frac{dv}{ds} + v f(s) + s = 0 \tag{8.15}$$

and then explaining that the solution is easily found by using the graphical integration of the tangent and the center of curvature, and appears to be "in the form of spirals which have a turbulent appearance depending on the complexity of $f(s)$" (Parodi 1942c, 197). He then adds :

> A stable regime is nonetheless established for a specific value of the initial speed, *i.e.* the constant integration (...) This stable regime corresponding to Poincaré's limit cycle

is established when there is an exact compensation of the losses by the energy supplied by the network (in the case of the automatic electric traction). (Parodi 1942c, 197)

At this point, he refers to Poincaré, but not to his works, in this conclusion which seems to have been inspired by Andronov's note (1929a, 560) (see *supra*), this is proven by the use of the adjective "corresponding". Unless it is inspired by the reading of Le Corbeiller's article (1931b, 211), introduced at the Third Congress For Applied Mechanics in Stockholm (see *supra*). Nevertheless, he mentions neither, and did not demonstrate the regime stability.

8.2.8.4 The Graphical Integration II

In this fourth contribution on relaxation oscillations written with his eldest son, both Messrs (Parodi and Parodi 1942) provide the elements required in the understanding of the successive-arc integration method, presented in previous notes (Parodi 1942a,b,c). The following year Parodi and Parodi (1943) published an article in *La Revue Scientifique*, covering *in extenso* the contents of these four notes, but also containing more thorough mathematical developments. It is also possible that this work was fully written in 1942, and that it was cut into four parts because of the limited number of pages demanded by the formatting of notes in the *C.R.A.S.* Starting with the equation (8.15), they explain again the principle of piecewise linearization, considering that the function $f(s)$ is assumed constant in all the interval defining the areas (1)–(6) (see Fig. 8.20) and equal to its mean value ε_p. They add:

The solution to the equation (8.15), non-integrable, will therefore be reduced to the solution of numerous integrable differential equation

$$v\frac{dv}{ds} + \varepsilon_p v + s = 0 \tag{8.16}$$

ε_p mean value of $f(s)$ in the domain $s_p s_{p+1}$. (Parodi and Parodi 1942, 268)

It is then surprising to see that in order to solve the first order linear equation (8.16), they used the method developed by Gaston Darboux (1878) in his essay "Sur les équations différentielles algébriques du premier ordre et du premier degré" ("On algebraic differential equations of the first order and first degree"). Indeed, they write:

The equation (8.16) allows a solution to the form

$$(v + as)^{\lambda} (v + bs)^{\mu} = k$$

satisfying the relations

$$\lambda + \mu = 1, \quad ab(\lambda + \mu) = 1,$$
$$\mu a + b\lambda = 0, \quad \lambda a + \mu b = \varepsilon.$$

The integral (8.16) is therefore (^{2})

$$\left[v+s\left(\frac{\varepsilon}{2}-\sqrt{\frac{\varepsilon^2}{4}-1}\right)\right]^{\lambda-\frac{\varepsilon}{\sqrt{\varepsilon^2-4}}}\left[v+s\left(\frac{\varepsilon}{2}+\sqrt{\frac{\varepsilon^2}{4}-1}\right)\right]^{\lambda+\frac{\varepsilon}{\sqrt{\varepsilon^2-4}}}=K'$$

(^{2}) If $\varepsilon = \pm 2$ the solution is $L\frac{v-s}{v_0-s_0} = \frac{s}{v_0-s_0} - \frac{s_0}{v_0-s_0}$ (Parodi and Parodi 1942, 268)

In his memoir, Darboux (1878) studies differential equations with the form:

$$L\,(ydz - zdy) + M\,(zdx - xdz) + N\,(xdy - ydx) = 0 \tag{8.17}$$

where L, M and N are homogenous polynomials of degree m. Darboux (1878, 65) then shows that in the case of two variables, this equation is reduced to

$$(M - Ny)\,dx + (Nx - L)\,dy = 0 \tag{8.18}$$

Darboux's method (1878)[53] consists in searching for "particular algebraic integrants" i.e. polynomials, to determine a first integral[54] of this equation formed by the product of these polynomials. In his memoir, Darboux (1878, 79) indicates the number and degree of these polynomials.

- The number is yielded by the relation: $q = M_n + n - 1$ où $M_n = \binom{m+n-1}{n}$,
- The degree is: $m - 1$.

n represents the order of the differential equation before it is transformed (the number of variables) and m the degree of the polynomials.

By rewriting the equation (8.16) with Darboux's form, i.e. writing: $v = y$ and $s = x$ on a

$$\left(x + \varepsilon_p y\right) dx + ydy = 0 \tag{8.19}$$

It hence appears that: $M = x + \varepsilon_p y$, $N = 0$ and $L = -y$. Consequently, since $n = 2$ and $m = 1$ we obtain: $q = 2$. Darboux's method then indicates that the equation (8.19) possesses two "particular algebraic integrants" allowing the formation of a first integral, in other words, a general integral, which can be written in the same way as the one chosen by Messrs. Parodi:

$$\phi\,(x, y) = (x + ay)^{\lambda}\,(x + by)^{\mu}$$

[53]Summarized by Ginoux (2009, Ch. 5, 85).

[54]In the case of a first order equation, it is a general integral. See Ginoux (2009, 94).

According to Darboux (1878, 73), $\phi(x, y)$ is a first integrant if, and only if:

$$M\frac{\partial\phi}{\partial x} + L\frac{\partial\phi}{\partial y} = 0 \tag{8.20}$$

From this we deduce a degree of two homogenous polynomial.

$$[-a\lambda - b\mu + \varepsilon(\lambda+\mu)]\,x^2 + [(\lambda+\mu) - ab(\lambda+\mu) + (a\mu + b\lambda)\,\varepsilon]\,xy$$
$$+ (a\mu + b\lambda)\,y^2 = 0$$

By vanishing each of its coefficient, we obtain:

$$\begin{cases} -a\lambda - b\mu + \varepsilon(\lambda+\mu) = 0 \\ (\lambda+\mu) - ab(\lambda+\mu) + (a\mu + b\lambda)\,\varepsilon = 0 \\ (a\mu + b\lambda) = 0 \end{cases}$$

From this system, we easily deduce:

$$\begin{cases} a\lambda + b\mu = \varepsilon(\lambda+\mu) \\ ab(\lambda+\mu) = (\lambda+\mu) \\ (a\mu + b\lambda) = 0 \end{cases}$$

By simplifying $\lambda + \mu = 1$, we find the exact same system obtained by Parodi and Parodi (1942, 268), which leads to the determination of the constants a, b, λ and μ. By solving it, we indeed find:

$$a = \frac{\varepsilon}{2} - \sqrt{\frac{\varepsilon^2}{4} - 1}, b = \frac{\varepsilon}{2} + \sqrt{\frac{\varepsilon^2}{4} - 1},$$

$$\lambda = \frac{1}{2}\left[1 - \frac{\varepsilon}{\sqrt{\varepsilon^2 - 4}}\right] \text{ and } \mu = \frac{1}{2}\left[1 + \frac{\varepsilon}{\sqrt{\varepsilon^2 - 4}}\right]$$

It seems that this was indeed how both father and son (1942, 268) obtained the general integral (8.21) for the equation (8.16). However there are no references to Darboux (1878). It must be noted that Gaston Darboux's (1842–1917) research on the integration of differential equations have rarely been used in fields other than Mathematics. Moreover, they were only "rediscovered" at the beginning of the 1990s by Prof. Dana Schlomiuk (1993). Messrs. Parodi then clarify that depending on the values of ε^2 superior or inferior to 4, the solution is real or imaginary. Therefore, they made a change of variables, in order to transform the cartesian coordinate system into polar coordinates. The solution to the equation (8.16) is then written:

$$\rho^2\left(1+\frac{\varepsilon}{2}\sin 2\theta\right)e^{\frac{\varepsilon}{\sqrt{4-\varepsilon^2}}\omega} = K'' \text{ with } \omega = -2\arctan\frac{\tan\theta+\frac{\varepsilon}{2}}{\sqrt{1-\frac{\varepsilon^2}{4}}} \tag{8.21}$$

They then explain that the expression (8.21) represents the equation for a spiral, and that depending on the sign of ε its amplitude exponentially increases or decreases. Then, they added that in order to carry out the successive-arc integration method, tables of a function U comprising ω must be constructed beforehand. The article ends with this sentence:

> The limit cycle we are looking for can only be found by successive approximations. The method we used will be further explained in a later publication. (Parodi and Parodi 1942, 270)

Indeed, in an article published the following year, Parodi and Parodi (1943) went back over these four notes, and added a few more elements. The last paragraph especially, provides a method allowing to graphically determine the limit cycle, as it is not systematically reached from a randomly chosen initial condition.

Through Parodi's research, we see on the one side a relatively unexpected synthesis of Poincaré's, Van der Pol's and Andronov's work, and on the other side, an application of the concepts of limit cycles, relaxation oscillations and self-sustained oscillations. However, no reference, no interpretation in regards to characteristic time scales, and no stability condition of the limit cycle are suggested. Parodi only demonstrated that the running of trains is indeed a relaxation phenomenon.

The work of Misters Hamburger, Gause and Kostitzin, presented below, are part of the research conducted outside of France. Nevertheless, since they were published in France (and in French), it seems reasonable to assume they might have impacted (in some hardly quantifiable manner) the French scientific community.

8.2.9 *Ludwig Hamburger:* Relaxation Oscillations in the Economic Cycles

While one of Dr. Ludwig Hamburger's contributions (1930) was published in Dutch, a French version of this article was published in January 1931 in the *Supplément aux Indices du Mouvement des Affaires* of the Statistics Institute of the University of Paris (Hamburger 1931). In his article's introduction, Hamburger (1930, 1) pays homage to Van der Pol's work (1926a,b,c,d, 1927b,c, 1928a,b) whom he viewed as being the "discoverer" of relaxation oscillations. He then recalls that it was after

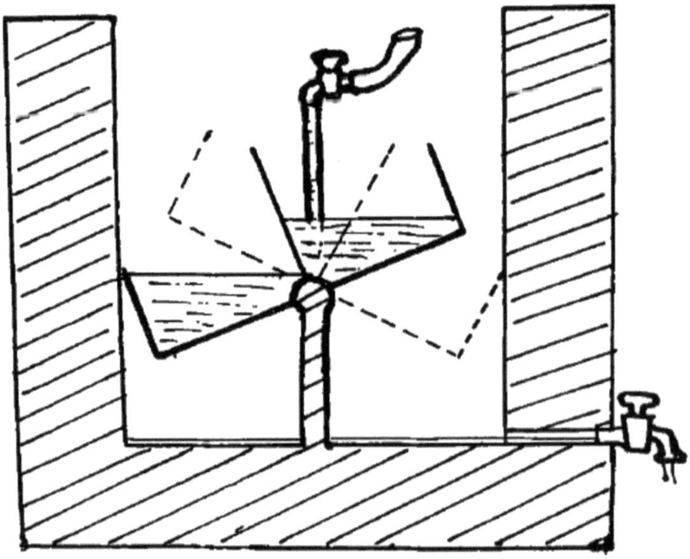

Fig. 8.21 Instance of relaxation oscillations, by Hamburger (1931, 15)

a lecture Van der Pol[55] gave on the 7th of May 1928 in front of the *société de philosophie*, that he "mentioned the possibility of the behaviour of chronic crises belonging to this category of oscillations" (Hamburger 1931, 4). He then explains that the chronic recurrence of economic recessions should be explained as essentially being a composite relaxation phenomenon. He then adds that this conception allows the understanding of the fundamental comparison between the amplitude of economic cycles at different times in a given country, with the function of the amplitude ratios persisting between the various technically developed countries. He also wrote a reminder of "the nature of relaxation oscillations" in which he follows the presentation of Van der Pol's article (1926d), adapting it to the problems of economic crises. In order to illustrate his words, he uses Van der Pol's example (1930, 307) (see *supra*), to which he adds a drawing (see Fig. 8.21).

He then explains that perturbations recurring in economics – sometimes "external" in nature, but most often "internal" – present the characteristics of relaxation oscillations, and play a crucial part in the genesis of economic cycles. He then notes that the possibility of influencing, through relatively limited means, the frequency of the economic relaxation oscillations matches the opinion stating that the effects of systematic credit policies are significant for the economic cycle. He suggests to the authorities, on the one hand, the adoption of a constructive social policy in which the

[55]Hamburger refers to the original dutch version of Van der Pol's article (1928a,b) on heart beats: De hartslag als relaxatietrillingen en een electrisch model van het hart, *Werk. Genoot. Nat.-, Genees- en Heelk*, **12** (1928), 614–618.

jerky course of the economic development would be partially smoothed out, and on the other hand, the beginning of a cooperation with a sufficiently developed statistics institute, and other various offices specializing in economics research, in order to provide forecasts on the economic situation rather than diagnostics (a posteriori). In this article, Hamburger (1931, 21) establishes an analogy between some curves representing the growth and decline of economic cycles, and the characteristic form of the relaxation oscillation (see *supra*), which allows him to state that they indeed belong to these relaxation phenomena. Hamburger appears to focus for one part on the "slow-fast" aspect of the concept introduced by Van der Pol allowing him to very accurately describe the evolution of these cycles, and for another part on Van der Pol's model's ability to predict their evolution through a generic equation (3.8).

8.2.10 Georgii F. Gause: Limit Cycles in Biological Associations

A Russian biologist at the Zoological Institute of the University of Moscow, Georgii Frantsevich Gause created the famous "competitive exclusion principle[56]", which is one of the founding principles of Ecology. In the middle of the 1930s, Gause (1934, 1935) carried out numerous in vitro experiments, in order to provide an experimental verification of the three laws formulated[57] by Vito Volterra (1931) in his "Lessons on the mathematical theory of the struggle for life". He then wrote several articles with Russian mathematician Aleksandr Witt, then with biologist Wladimir Wladimirovich Alpatov[58] who supervised his thesis in 1940.[59] In the middle of the 1930s, Gause (1934) published a research entitled "The struggle for existence", which became a cornerstone reference in this field. In 1935, Georges Teissier (1900–1972), then assistant director of the *Station Biologique* in Roscoff, and head of the collection "Actualités Scientifiques et Industrielles" for publisher Hermann, offered to present his research. In this sixty-page long booklet, Gause (1935) first recalls the equations established by Volterra (1926) and Lotka (1932) then considers the population growth of protozoans: *Paramecium aurelia* and *Glaucoma scintillans*. This study allowed him to firstly introduce the concept of

[56]This principle states that populations of two species exploiting a same resource cannot coexist indefinitely in a stable and homogenous environment, and the most competitive species will eliminate the other sooner or later.

[57]Periodic cycle law, law of conservation of means, law of perturbation of means (Volterra 1931, 19–27). From Israel (1996, 64), Gause, who seemed to be able to verify the first law, wrote to Volterra, who showed support. Unfortunately, what followed was disappointing.

[58]Alpatov (1898–1979) was the student of the statistician Raymond Pearl (1879–1940), and greatly influenced Gause.

[59]See Gause (1934, 1935) and Gause et al. (1936).

an "ecological niche". Secondly, Gause (1935, 7) used the "Poincaré method" to determine which species subsist in the niche. This resulted in a whole challenging of the predator-prey model used until then, which actually modifies functional responses of the prey and the predator in order to introduce a "stabilizing effect". He then evidenced the coexistence of two species, and wrote:

> (...) the reciprocal actions of the two species never stop (we already noted previously (GAUSE, 1934) that these fluctuations apparently possess "limit cycles". (Gause 1935, 47)

He thus experimentally demonstrated that the fluctuations of two species, one of which feeds on the other, could reach a stable periodic regime mathematically corresponding to a limit cycle. Although he did not refer to Poincaré (1882) or Andronov (1929a), this is one of the first times that this concept was used in mathematical biology.

By furthering his analysis of the nature of fluctuations, Gause writes:

> This way, a new type of fluctuation in biological associations is created, physicists call it *relaxation oscillation*. (Gause 1935, 59)

He then establishes a link with the concept introduced by Van der Pol (1926d) and gives it a relatively accurate definition:

> As noted by VAN DER POL (1934), what is characteristic for relaxations oscillations is the fact that during the longest part of the period, the phenomenon has an aperiodic or asymptotic behaviour (purely victims population) and, after this, the system suddenly becomes unstable (the aggressors or the bacterial infection might acclimatize), and the disorder goes in a discontinuous manner to another value, and then the same discharge phenomenon is again repeated, etc., we observe that a relaxation oscillation has the same characteristics of an ever-repeating discharging phenomenon. The time period is given by a relaxation time, hence the name of the phenomenon. (Gause 1935, 59)

He then reproduces the *Tantalus cup* example suggested by Le Corbeiller (1931a, 6) (see *supra*). He ends the paragraph with this sentence:

> Going back to our biological associations, it is noteworthy that the uninterrupted classic fluctuation and the discontinuous relaxation fluctuation are only two types which are expressed extremely distinctively, and that in the complex environment of genuine biological systems, elements from the two types can certainly participate. (Gause 1935, 60)

It is interesting to note that Gause uses the terminology introduced laterby Andronov (1901–1952) Andronov and Khaikin (1937) to describe relaxation oscillations (see *infra*). In this work, Gause (1935) establishes on the one hand, the correspondence between periodic fluctuations of biological associations and the limit cycles, and on the other hand, he characterizes the "kinetics" of these fluctuations, by showing that they belong to the class of relaxation phenomena (Fig. 8.22).

Fig. 8.22 Georgii Frantsevich Gause (Internet source)

8.2.11 *Vladimir Kostitzin:* Relaxation Oscillations in Biological Associations

A professor at the Faculty of Science of Moscow and Director of the Geophysics Institute of Moscow, Vladimir Aleksandrovich Kostitzin was a Russian mathematician who migrated to France. In 1929, he was invited at the newly created Henri Poincaré Institute, to present his mathematical work. His two lectures given on the 27th and 29th of May 1929 were then published in the Annals of the Henri Poincaré Institute (Kostitzin 1930). Half through the 1930s, after Lotka (1925) and Volterra (1926, 1928, 1931) published their research, he focused on mathematical biology, and published numerous contributions in this field (Kostitzin 1930, 1932, 1934). Kostitzin therefore published in 1936 a note in the *C.R.A.S.* entitled: "Sur les solutions asymptotiques d'équations différentielles biologiques" ("On the asymptotic solutions of biological differential equations") in which he covers a study carried out the same year by one of his compatriots: Andrei Nikolaievich Kolmogorov. In this study where we find the case of two species where one feeds on the other, Kolmogorov (1936) evidenced the existence of a limit cycle, although until then only Volterra (1926, 38, 1928, 1931, 14) had considered a (graphical) "center" type of solution for this differential equation representing a predator-prey type of interaction.

In this note, Kostizin demonstrates the existence of an asymptotic solution to these types of equations, and its existence "immediately is a result of H. Poincaré's and M. E. Picard's famous research" (Kostitzin 1936, 1124). He then considers the following system of differential equations:

$$\begin{cases} x' = \varepsilon x - \lambda y - x\left(\alpha^2 x^2 + \beta^2 y^2\right) \\ y' = \varepsilon y + \lambda x - y\left(\alpha^2 y^2 + \beta^2 x^2\right) \end{cases} \tag{8.22}$$

And clarifies that contrary to Volterra's "classic" model (1926, 38, 1928, 1931, 14), the functional response characterizing the predation is cubic rather than quadratic, which was incidentally legitimized by the recent research of one of his compatriots, Giorgii Fransevich Gause (1935). He made a change of variables in order to transform the cartesian coordinates system into polar coordinates, and found the following system:

$$\begin{cases} \dfrac{r'}{r} = \varepsilon - r^2 \left[\alpha^2 \cos^2 2\theta + \dfrac{1}{2} \left(\alpha^2 + \beta^2 \right) \sin^2 2\theta \right] \\ \theta' = \lambda - \dfrac{r^2}{2} \left(\beta^2 - \alpha^2 \right) \sin 2\theta \cos 2\theta \end{cases} \tag{8.23}$$

To demonstrate the existence of a closed curve, he assumed that the radius r is constant, and that the coefficient ε is positive, leading him to:

$$r' = 0\,, \quad r^2 = \frac{\varepsilon}{\alpha^2 \cos^2 2\theta + \dfrac{1}{2} \left(\alpha^2 + \beta^2 \right) \sin^2 2\theta} \tag{8.24}$$

He explains that the curve defined by $\theta' = 0$ cuts the plane in a kind of cross shape (see Fig. 8.23). He then adds:

> In these conditions, a state (x, y) inside the closed curve (Ko_3) passes, after a finite time, to the external area. Likewise, a state outside the closed curve passes after a finite time to the internal area of the curve (Ko_3). It is easily proven that, for t infinitely increasing, any integral curve of the system (Ko_1) approaches closer and closer to a limit closed curve, and the process strains towards a limit periodic state. (Kostitzin 1936, 1125)

It is striking to note that the reference to Poincaré (and Picard) is not accompanied by a quote and that he did not use the term "limit cycle", but rather chose to speak about "a certain limit closed curve". This is all the more surprising coming from a Russian mathematician who apparently was very aware of his compatriots' work.

In the above figure where all coefficient were chosen equal to the unity (see Fig. 8.23) we see that isocline curves indeed cut the phase plane in four areas, thus forming a cross, and that external curves (res internal) asymptotically approach the limit cycle (in red). Kostitzin then concludes:

> We therefore have, with the relaxation periodicities, a type of periodomorphism which must be quite frequent in nature, and contributes to the organization of chaos, and the local actualization of rare states. (Kostitzin 1936, 1126)

Here, he alludes to Van der Pol's concept (1926d) and qualifies this new type of oscillation as "periodomorphisms". He thus implies that this characteristic form of oscillations is quite common in nature, and that the predator-prey type of interactions can be linked to the relaxation phenomenon. Moreover, if the word nature is taken literally, he might have been referencing Van der Pol's various examples (see *supra*). Nevertheless, he did not establish a link with Andronov's

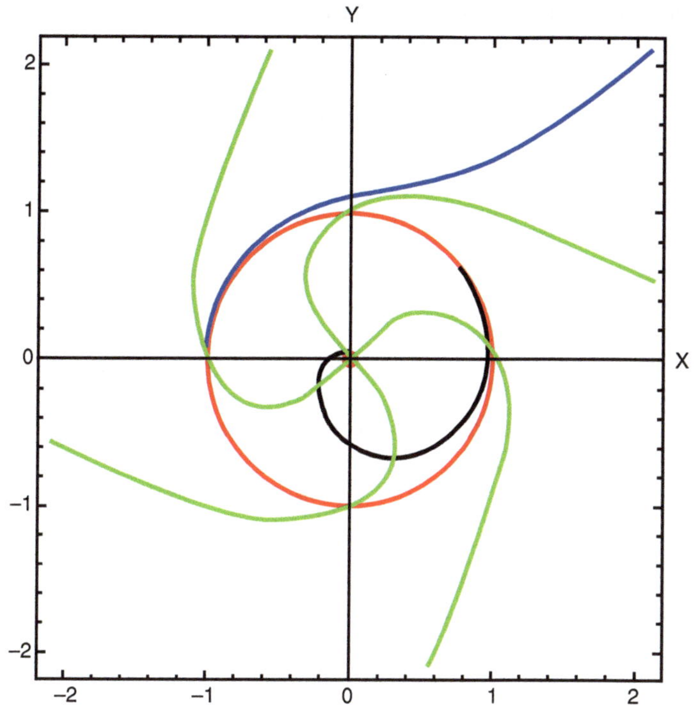

Fig. 8.23 Limit closed curve of the Kostitzin's system (1936)

concept of self-sustained system (1929a) and did not approach the question of stability. In his work called "Mathematical Biology" published the following year, Kostitzin (1937) focuses over several chapters on various mathematical models[60] likely to represent the a population's growth (Malthus law, Verhulst law or logistic distribution, for which he provided a generalization), the relations between species (competition, cooperation and predation) and Volterra's biological laws. While he goes over all the aspects of what is called nowadays the modelling of population dynamics: the influence of the preys' age, heterogeneity of the predator species, migration, seasonality, . . . he does not mention the situation in which the limitation of the functional response of the prey's and predator's growth[61] leads to a periodic solution which takes the form of a stable limit cycle of Poincaré's, even though Kolmogorov (1936) had already established it (see *infra*). It must be recalled that in 1937, Kostitzin produced, along with Jean Painlevé, a film entitled "Images mathématiques de la lutte pour la vie" ("Mathematical images of the Struggle for

[60]He also proposed models containing at least three species.

[61]Which is expressed by a saturation of the predation.

Table 8.1 Synoptic of relaxation oscillations examples

Device	Field	Authors
Wehnelt interrupter	Chemistry	Van der Pol (1926a,b,c,d)
Triode multivibrator	Electrotechnics	Van der Pol (1926a,b,c,d)
Gas triode		Eck (1936)
Series-dynamo		Van der Pol (1926b,c)
Singing arc		Van der Pol (1930)
Neon tube		Van der Pol (1926a), Morched-zadeh (1936), and Moussiegt (1949)
Piezoelectric quartz		Panc-Tcheng (1930)
Photoelectric cell		Hochard (1933)
Running of trains		Parodi (1942a,b,c) and Parodi and Parodi (1943)
Heartbeat	Biology Medicine	Van der Pol (1926b)
Nervous rythms		Fessard (1931)
Biological associations		Gause (1934, 1935) and Kostitzin (1936, 1937)
Garden hoses sound pipes	Hydrodynamics	Bourrières (1932, 1937a,b, 1939) and Auger (1938)
Vortex streets by Bénard-Karman		Foch (1935) and Castagnetto (1939)
Economic cycles	Economy	Hamburger (1930, 1931)

Life") kept in the archives of the *Palais de la Découverte*[62] which constitutes a priceless testimony of the profound knowledge held by "biomathematicians" at that time. It is both a picturization and a true synthesis of the work of Lotka (1925), Volterra (1926, 1928, 1931), Gause (1935) and Kostitzin (1936, 1937). The Table below (see Table 8.1) offers a synoptic representation of the various examples of relaxation oscillations covered in this part. In order to create a global visualization of the research carried out around the world in the field of nonlinear oscillations, the same type of study should be carried out, based on the work undertaken in England by Adrian (1930), Leyshon (1923, 1930, 1931a,b), Eccles et al. (1927), Richardson (1937), in Germany by Barkhausen (1920, 1929), Heegner (1924, 1927), ... in Italia by Tricomi (1933), Graffi (1940, 1942, 1951), Sansone (1949a,b), ... While such a study largely exceeds the frame of the present work, it is nevertheless possible to analyse the reception of Andronov's and Van der Pol's concepts in the United States, based on a publication that might be seen as being the most representative: the ones authored by Timoshenko (1928), Den Hartog (1934, 1936), Levinson and Smith (1942), Friedrichs and Stoker (1943), Flanders and Stoker (1946), Minorsky (1947),

[62]We were informed of the existence of this film by Prof. Giorgio Israel (1996, 63), and a copy was kindly sent to us by Prof. Pascal Acot and Drouin (1997, 47).

Lasalle (1949), and especially Andronov and Khaikin's book (1937), "translated[63]" by Solomon Lefschetz (Andronov and Khaikin 1949). In the U.S.S.R., the original version of this work allows to demonstrate that the concept of relaxation oscillations indeed appears in Andronov and Khaikin's book (1937) although with a completely different terminology from the one introduced by Van der Pol (1925, 793). In their French contributions, and in their book published in Russian[64] in 1937, Krylov and Bogolyubov (1932a,b,c, 1933, 1934a,b,c,d,e, 1935a,b, 1936a,b, 1937) show a complete adherence to this concept, as well as Van der Pol's research, especially regarding frequency demultiplication and synchronization (see *infra* Part III). In Europe, Mandel'shtam et al. (1935) presentation at the General Assembly of The International Union of Radio Science (U.R.S.I.) in London between the 12th and 18th of September 1934 helped showing his perception of relaxation oscillations.

8.3 Theses on *Nonlinear Oscillations* in France (1936–1949)

At the same time, several doctorate dissertations[65] were defended in France in the field of research on oscillations:

- "Étude des oscillations de relaxation et des différents modes d'oscillations d'un circuit comprenant une lampe néon" ("Study on relaxation oscillations and the various oscillation modes of a circuit comprising a neon lamp") defended by R. Morched-Zadeh in October 1936 at the Faculty of Science of the Toulouse University before the examination committee composed of Misters Charles Camichel (President), Marcel Lamotte and Pierre Dupin (Examiners).
- "Contribution à l'étude des tourbillons alternés de Bénard-Karman" ("Contribution to the study of Bernard-Kàrmàn's alternating vortices") defended by Luis Castagnetto in 1939 at the Faculty of Science of the Toulouse University before the examination committee composed of Misters Charles Camichel (President), Max Teissié-Solier and Pierre Dupin (Examiners).
- "Étude d'un système oscillant entretenu à amplitude autostabilisée et application à l'entretien d'un pendule élastique" ("Study of an self-stabilized-amplitude oscillating system and application to the sustaining of a pendulum") defended by Jean Abelé in 1943 at the Faculty of Science of the Paris University before the examination committee composed of Misters Henri Villat (President), François Croze and Yves Rocard (Examiners).

[63]Indeed, it was shown that Lefschetz's translation differed strongly from Andronov and Khaikin's original work (1937).

[64]It was also translated by Lefschetz and was more faithful to the original.

[65]These essays have been "found" based on bibliographical elements. The French dissertations found between these two wars have been digitalized and are yet accessible at: http://www.numdam.org.

- "Les oscillations de relaxation dans les tubes à décharge: application à l'étude de l'allumage" ("Relaxation oscillations in discharge tubes: application to the study on firing") defended by Jean Moussiegt in 1949 at the Faculty of Science of the Grenoble University.[66]

8.3.1 *Morched-Zadeh's Thesis*

We only have few biographical elements regarding R. Morched-Zadeh, an Iranian student, who is not however, related to Lotfi Zadeh.[67] His thesis dissertation is a spot-on synthesis of the work which played a fundamental part in the development of the theory of nonlinear oscillations. The detailed bibliography he provides thus allows the possibility to retrace the various steps of this process. The first chapter begins with a historical overview of "some natural phenomena considered as relaxation oscillations" (Morched-zadeh 1936, 1).

> Indeed, in the last few years, non-sinusoidal oscillations have caught the attention of many scientists, especially Mr. VAN DER POL's. The study of these oscillations has been very fertile and allowed to evidence the existence of a new type called "relaxation". (Morched-zadeh 1936, 1)

Morched-Zadeh therefore adopted Le Corbeiller's viewpoint (1931a, 22) (see *supra* Part I) and placed Van der Pol's results on the invention level, in other words considered that he discovered new type of oscillations. Then, he recalls the main devices in which relaxation oscillations occur:

> We can also cite, in the field of electricity: the oscillation of the electric arc studied by Mr. BLONDEL, M. JANET's experiment on the dynamo-series, and ABRAHAM and BLOCH's multivibrator; in hydraulics, BENARD-KARMAN's alternating vortices. (Morched-zadeh 1936, 1)

While the first three had already been presented by Van der Pol and Le Corbeiller (see *supra*), the fourth, Bénard-Kàrmàn's alternating vortices, was entirely new. However, it was not unknown to the members of his thesis committee, as Charles Camichel, Pierre Dupin and Max Teissié-Solier worked on the same problem[68] in the field of hydraulics. Consequently, they might have informed him of Adrien Foch's results (1935) evidencing the existence of relaxation oscillations in Bénard-Kàrmàn's alternating vortices (see *infra*). Moreover, this question was also part of the thesis submitted to Castagnetto that same year. The second paragraph, entitled "Un système de relaxation autoentretenu primitif" ("A primitive self-sustained relaxation system") (Morched-zadeh 1936, 2) corresponds to Le Corbeiller's presen-

[66]The examining committee was not specified in the document sent by the I.N.I.S.T.

[67]Personal communication with Mr. L. Zadeh.

[68]See Camichel (1927) and Camichel et al. (1927).

tation (1931a, 44) at the *C.N.A.M.*, in which he referred to Villard de Honnecourt's work.[69] Morched-Zadeh concludes with these words:

> Relaxation oscillations, as shown by BARKHAUSEN and VAN DER POL, have an irregular period. (Morched-zadeh 1936, 3)

Here, it seems that he confused Van der Pol's *relaxation oscillations* (1926d) with Barkhausen's *selbst schwingungen*, i.e. *self-oscillations* (1907), which may indicate that the distinction was not pronounced at the time. Then, he notes the fact that the period of the relaxation oscillations that Van der Pol (1926d, 987) had first deduced graphically, then approximated (Van der Pol 1927b, 114–115) (see *supra* Part I) is actually irregular.

In the following paragraphs, Morched-Zadeh recalls "M. Janet's experiment on the series-dynamo", nonetheless mentioning that Gérard-Lescuyer (1880a), related Witz's explanation (1889a), as well as Trutat and Bouttes's results (1925). Du Moncel's work (1880) was apparently forgotten. He continues his "state of the art" with a "theoretical study of self-sustained systems", for which the prototype is none other than Rayleigh's equation (8.5) (1883), then wrote, without any other explanation:

> From a theoretical point of view, H. POINCARÉ's limit cycles are crucially important when it comes to the theory of self-sustained oscillations, as A. ANDRONOW and A. WITT demonstrated. (Morched-zadeh 1936, 3)

While he recalls the correspondence established by Andronov (1929a), he unexpectedly cites Andronov and Witt's article (1930a,b) regarding systems with two degrees of freedom, and addresses the "Lyapunov stability" condition. Moreover, this is his only reference to Poincaré's and Andronov's work. He then "overviews" Blondel's work (1919b) on the oscillations of a triode, and reproduces his writing of the equation as well as the calculation of the value of the amplitude and period (see *supra*). The second chapter covers the main results presented in Van der Pol's various articles (1920, 1926a,b,c,d, 1927a,b,c, 1928a, 1930). In the second paragraph, entitled: "Differential equation of a self-sustained system" Morched-Zadeh establishes Van der Pol's equation (V_7) (1926d) (see *supra* Part I) he confuses once again a self-oscillator with a relaxation oscillator. However, it cannot be said with certainty that there was at this time, apart maybe for Haag (1943, 36) (see *supra*), an awareness of the fact that a relaxation oscillator is a self-oscillator. Morched-Zadeh then concludes by recalling the amplitude of the oscillations strains towards a precise and finite limit, and he clarifies that Strutt and Rayleigh (1883) and Blondel (1919b) had already obtained this result. Then, he addresses the oscillation characteristic of the triode again, and explains:

> We have seen how we came to choose the equation for the characteristic, with the form
>
> $$i = \alpha v - \gamma v^3$$

[69] See Fremont (1915).

and more generally, to

$$i = \alpha v - \gamma v^3 + \mu v^5$$

in order to have sustained oscillations. The presence of a term βv^2, which is well-known for expressing the detection and modulation, is not necessary. The terms γv^3 and μv^5 are necessary in order to get out of the elementary domain, and guide us in order to further the oscillatory phenomena which cannot be understood based on the linear theory of oscillations. (Morched-zadeh 1936, 27)

This excerpt calls for several important remarks. Firstly, the "choice" of an oscillation characteristic of a degree of five seems to have been influenced by the reading of Blondel's (1919b), Appleton and Van der Pol's (1922) as well as Rocard's articles (1932). Secondly, he provides an indication of the radio engineering signification of the term βv^2. Lastly, he shows a will to "get out of the elementary domain", in other words, the linear part of the characteristic, in order to study, not the starting, but rather the sustaining of the oscillations, by using a nonlinear theory, which was being constituted. The next paragraph deals with free and forced oscillations of Van der Pol's prototype equation, and recalls various results. In the third chapter, Morched-Zadeh presents the work of Messrs. Henri and Élie Cartan (1925), as well as the principle of Liénard's construction (1928). He thus goes over Liénard's demonstration (1928) leading to the existence and uniqueness of a stable periodic solution, i.e. of a *stable limit cycle* (see *supra* Part I), and writes:

We see that the external and internal curves traveled in the direction of the increasing times, asymptotically approach the closed curve D; hence, the corresponding periodic motion is stable. (Morched-zadeh 1936, 58)

This conclusion, which is almost a paraphrase, shows his perfect assimilation of Liénard's works but also his complete lack of awareness of the fundamental importance of Andronov's result. Indeed, Morched-Zadeh might as well have called the "closed curve D" a "limit cycle". He then goes over Liénard's calculations (1928) for the amplitude and period of the oscillations. The next paragraph, entitled: "Existence of several closed integral curves" is quite unexpected, as Morched-zadeh (1936, 67) implicitly refers to the second part of Hilbert's 16th problem (1900) regarding the maximal number and mutual position of limit cycles possessed by a planar polynomial differential equation with a given degree.[70] The chapter ends with him quoting Le Corbeiller (1931a, 21) (see *supra*) on the continuous passing from sinusoidal oscillations to relaxation oscillations, and a comparison of those two types of oscillations. The fourth chapter also comprises numerous reminders of the triode lamp, Abraham and Bloch's multivibrator (1919a,e), Van der Pol and Van der Mark's electric modeling of the heart (1928a,b), and the neon lamp, whose study was the central issue of Morched-Zadeh's thesis. The subject of the fifth and last chapter is the study of oscillations produced by a neon lamp Indeed, as Morched-Zadeh recalled (1936, 11) in the beginning of his essay, the neon lamp is

[70]Nowadays, this question is still open.

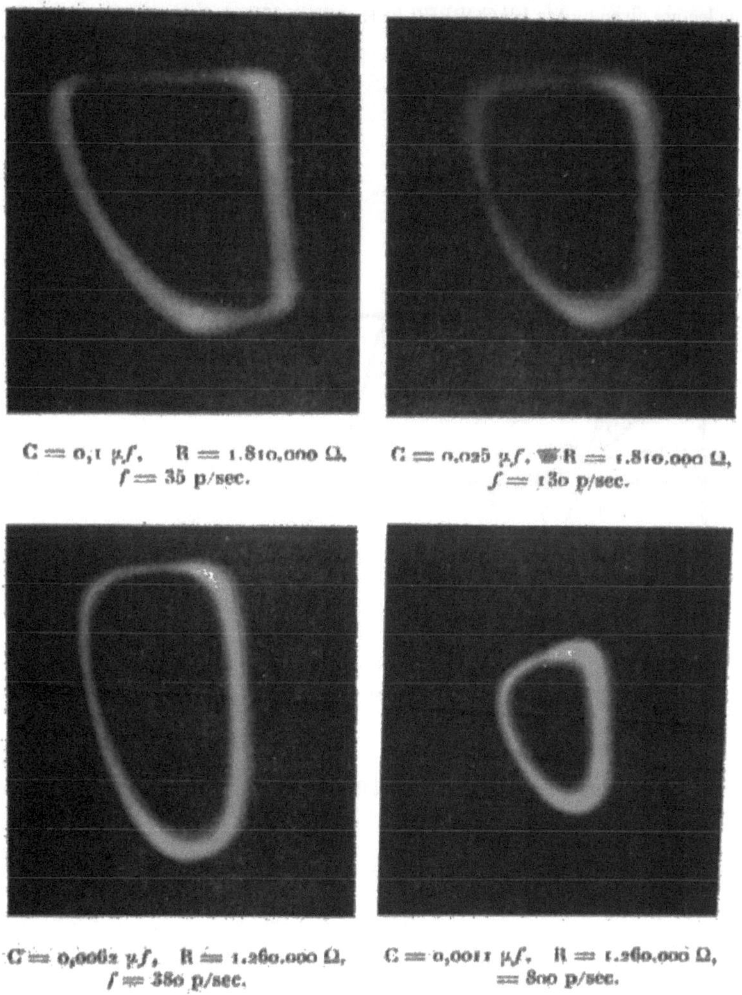

Fig. 8.24 Dynamic characteristics, by Morched-zadeh (1936, 127)

the seat of oscillations. The previous year, Le Corbeiller (1936, 366) clarified during his presentation before the Wireless Section (see *supra*), that they were relaxation oscillations, observed for the first time by Righi (1902). In order to successfully complete this experimental study of the various oscillatory regimes of the neon lamp, Morched-Zadeh uses a cathode-ray oscilloscope, and provides snapshots (see Fig. 8.24) of what he observes and calls "dynamic characteristics" (see *supra* Part I)

These snapshots (see Fig. 8.24) are the very first examples of experimental limit cycles. In his conclusion, Morched-zadeh (1936, 149) goes back over Le Corbeiller's work (1931a, 8–9) in order to establish the soundness of his experimental results.

In the lecture given by Mr. LE CORBEILLER on the self-sustained systems

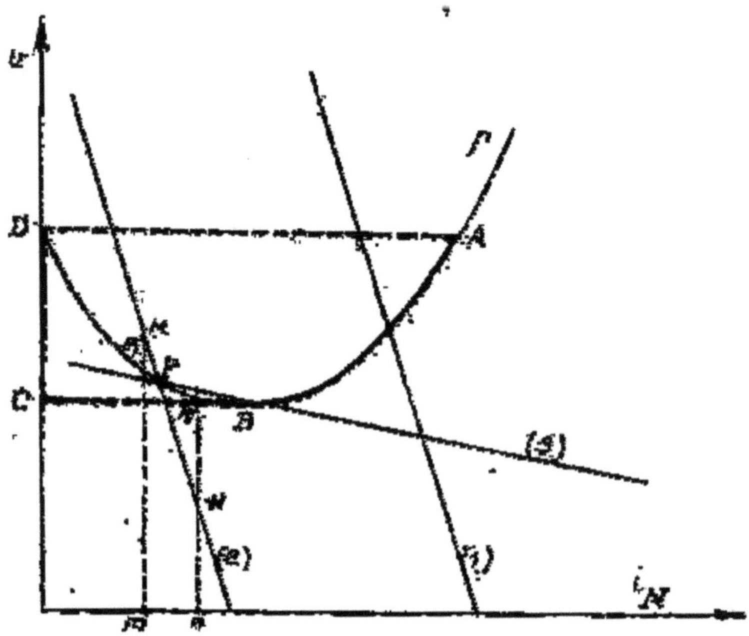

and the relaxation oscillations, at page 8 and page 9, we find the following equation:

$$(E - Ri_N) - v = CR\frac{dV}{dt} \qquad (1)$$

(i_N = current traversing the neon lamp: (see figure 41).
Let us use the characteristic of the neon lamp at a steady state (Γ), with an equation $V(i_N)$, i_N being the current in the lamp, and V the voltage at its terminals. The line $E - Ri_N = e$ meets the characteristic (Γ) at a given point. If this meeting point is on the rising part (line 1), we have a steady state in the lamp because the balance is stable, whereas if this point is on the falling part DB of the characteristic (line 2), we have an oscillating state, because this point is unstable. There is therefore a variable periodic state compatible with the equation (1), it is the one in which the representative point draws the cycle ABCDA indefinitely, by suddenly jumping from B to C, then from D to A. We note that the neon lamp periodically lets out a brief flash corresponding to the part AB of the cycle, during which the capacitor discharges. (Morched-zadeh 1936, 148)

Thanks to this analogy, Morched-Zadeh thus demonstrates that his study led to his experimental observation of relaxation oscillations in a neon lamp. Although this thesis mostly covers already research work such as Blondel's (1919b), the Cartans' (1925), Van der Pol's (1926d), Liénard's (1928), Andronov's (1929a), Le Corbeiller's (1931) ... it is one of the first scientific studies in the field of nonlinear oscillations, and thus provides precious insight on the way Andronov's and Van der Pol's were perceived.

Fig. 8.25 Bénard-Kàrmàn alternating vortices, by Van Dyke (1982, cover)

8.3.2 Castagnetto's Thesis

Regarding Louis Castagnetto's biography, very few elements can be found aside from the fact that he published several articles in the *C.R.A.S.*, during and after his thesis (Teissié-Solier et al. 1937; Teissié-Solier and Castagnetto 1938; Castagnetto 1958; Castagnetto and Matheau 1961a,b). His essay, entitled "Contribution à l'étude des tourbillons alternés de Bénard-Karman" ("A contribution to the study of Bénard-Karman's alternating vortices"), is comprised of three chapters. The first consists in a description of the equipment and various methods used to successfully complete his study. In the second, the "results obtained in the study of various phenomena accompanying the production of alternating vortices in an environment‘" (Castagnetto 1939, 71) are presented. The third and last chapter aims at studying specific characteristics of these vortices in order to connect them to relaxation oscillations.

In the first chapter, Castagnetto offers a historical reminder of the phenomenon of Bénard-Karman's alternating vortices[71]

> In 1908, Mr. Henri Bénard ([72]), in a note to the *Académie des Sciences*, displayed the results of his research started in 1906. Later, in 1912, Karman and Rubach ([73]) published an essay on the fluid and air resistance, in which they gave the theoretical stability condition for alternating vortices street. (Castagnetto 1939, 72)

[71]Bénard-Kàrmàn vortex street is a periodic pattern of vortices cause by the unstable separation of a flow around a body (generally a cylinder), and forming in its wake. See Fig. 8.25.

[72]Castagnetto refers to Bénard (1908).

[73]Castagnetto refers to Karman and Rubach (1912).

Here, we find the origin of the terminology *vortex street*. Then, Castagnetto recalls the main studies carried out in this field, especially by Misters Camichel, Dupin and Teissié-Solier (1927, 1928), and Mr. Villat (1929). After this "state of the art", Castagnetto addresses in the second chapter, the study of alternating vortices in the wake of a cylinder immersed in a vector field of unvarying speed. The third and last chapter, entitled "On alternating vortices and relaxation oscillation", aims at analytically demonstrating that during the establishing of a regime for which we observe alternating vortices, the oscillation's amplitude variation law is consistent with Van der Pol's results on the amplitude of oscillations in a triode lam. In the introduction, he already recalls that it was Adrien Foch who was one of the firsts to evidence the existence of relaxation oscillations in Bénard-Karman's alternating vortices:

> As highlighted by Mr. A. Foch ([74]), the alternating vortices belong to the category of relaxation phenomena. (Castagnetto 1939, 119)

Then, from the results previously established by Misters Camichel and Teissié-solier (1935), he explains

> a) That the law of amplitude variation y of oscillations in the wake of an obstacle in Poiseuille state following a perturbation is characterized by a defined value of the damping factor. They especially managed to verify the following law:
>
> $$y = Ea(t)\sin(\omega t + \varphi) \text{ with } Ea(t) = Ae^{-\varepsilon' t} \tag{8.25}$$
>
> b) That during the establishment of alternating vortices, the wake-oscillation's amplitude increase law on one point has the form:
>
> $$y = Ee(t)\sin(\omega t + \varphi) \text{ with } Ee(t) = \sqrt{\frac{a}{1 + be^{-\varepsilon t}}} \tag{8.26}$$
>
> These results are in accordance with Mr. Van der Pol's work ([75]) on the sinusoidal and relaxation oscillations. We will later see that the equation (8.25) is only one particular for of the equation (8.26) and that only one law can apply, either about the increase, or about the damping, of the oscillations. (Castagnetto 1939, 119)

It must be noted that the equation (8.26) exactly corresponds to the expression of the amplitude described by Van der Pol (1930, 255) (see *supra*), but that it is only valid for values of $\varepsilon \ll 1$.

In the first paragraph entitled "Relaxation oscillations" Castagnetto, denies trying to construct the history of this type of oscillation, but nevertheless recalls the names of the most prominent scientists such as Van der Pol (1926d, 1930), Van der Pol and Van der Mark (1927b), Liénard (1928), Henri and Élie Cartan (1925), Andronov (1929a), Andronov and Witt (1930a), Le Corbeiller (1931a, 1932), Krylov and Bogolyubov (1932a,b,c, 1933). This very thorough bibliography ends with a reference to Morched-Zadeh's thesis (1936), which he also defended at the

[74]Castagnetto refers to Foch (1935).

[75]Castagnetto refers to (1926c, 1927c).

Faculty of Science of the University of Toulouse, before the same committee. He then explains:

> Nonetheless, while the mathematical theory of relaxation oscillations has just started, several of the phenomena, which we will later discuss, being deeply linked to it, we must keep in mind some crucial results, owed for the most part to Mr. Van der Pol. (Castagnetto 1939, 120)

Contrary to what Castagnetto claims in this quote directly inspired by Le Corbeiller's (1931a, 45) (see *supra*), this is not the mathematical theory of relaxation oscillations which just started, but rather the assimilation of the theory of self-sustained oscillations by the French scientific community. Based on the differential equation characterizing the oscillations of a triode, he then obtains Van der Pol's equation (3.8), a prototype of relaxation oscillations:

$$y'' - \varepsilon\left(1 - y^2\right)y' + y = 0 \tag{8.27}$$

He then adds:

> For ε positive but very small, the oscillations are established slowly and have a somewhat sinusoidal form, Van der Pol gives for the approximate equation for the curve surrounding these oscillations
>
> $$y = \sqrt{\frac{a}{1 + be^{-\varepsilon t}}} \tag{8.28}$$
>
> For $\varepsilon = 1$, the appearance is slightly different, and the final amplitude is reached after a smaller number of oscillations than in the previous case. For ε is very superior to the unit, the form varies very strongly from the sine-wave, and its periodic form is reached almost after one period. For $\varepsilon = 0$, the oscillations are sinusoidal. (Castagnetto 1939, 121)

The equation (8.28) corresponds to the approximate value of the amplitude provided by Van der Pol (1926d, 980) in his famous article, then corrected during his presentation at the *École Supérieure d'Électricité* in March 1930 (Van der Pol 1930, 254) (see *supra*). He then recalls the proposal that Le Corbeiller credited (1931a, 21) (see *supra*) to Van der Pol, and used by Morched-zadeh (1936, 70) on the continuous passing from sinusoidal oscillations to relaxation oscillations, and reproduces Van der Pol's curves (1926d, 986) (see *supra* Part I) after a briefly mentioning Liénard's graphical method (1928). He then indicates characteristic properties of relaxation oscillations:

> The period is given by the product or the quotient of the two physical quantities characterizing the system, whereas in the case of sinusoidal oscillations, it is given by the square root of the product or the quotient of the two physical quantities characterizing the system. (Castagnetto 1939, 123)

Here, he wrote the basis of the demonstration which he planned on carrying out, as his work aimed at showing that the amplitude of the wake's oscillations at one point has the form (8.28), and that their period is the quotient of two values he defines afterwards. In order to ensure this, Castagnetto (1939, 123–126) first shows that the equation (8.25) is effectively a particular form of the equation (8.26), and substitutes these two expressions with the following ones:

- For the establishing:

$$y - \sqrt{\frac{a}{1 + be^{-\varepsilon t}}} \quad \text{with} \quad \varepsilon > 0 \tag{8.29}$$

- For the damping:

$$y = \sqrt{\frac{a}{1 + be^{\varepsilon t}}} \quad \text{with} \quad \varepsilon > 0 \tag{8.30}$$

Indeed, by using a limited Taylor expansion, on the order of one at ε, Castagnetto (1939, 127) shows that the equation (8.30) is reduced to the expression (8.25). He thus obtains one formula (to the sign) for the surrounding of the oscillations. However, he underlines:

> Lastly, let us note that for the ε with a value that is not very small in relation to the unit, the theoretical approximations carried out when establishing the formulae (8.29) and (8.30) are not valid anymore. (Castagnetto 1939, 131)

He then demonstrates that the curves defining the amplitude variation of the wake, experimentally measured, can be represented by the functions (8.29) and (8.30). He considers the curves $Ea\,(t)$ subscripted by the letter a for the damping, and the curves $Ee\,(t)$ subscripted by the letter *e* for the establishment (see Figs. 8.26 and 8.27).

He therefore determines, from these curves (Figs. 8.26 and 8.27), the three parameters a, b and ε of the functions (8.29) and (8.30). Moreover, in order for his calculations to stay valid, he also must verify that the parameter ε stays small before the unit. In addition, after writing $\varepsilon = 2\varepsilon'$, he constructs the variation curve of ε' depending of the speed V of the fluid (see Fig. 8.28).

We note that ε' stays small before the unit in a speed domain situated between 9 and 10 mm/sec (see Fig. 8.28 red circle). Which leads Castagnetto to writing the following statement:

> As indicated by Mr. C. Camichel, and as it results from the formulae (8.29) and (8.30), the theoretical definition of the criterion of the alternating vortices corresponds to $\varepsilon' = 0$, which, in physics, is expressed by the fact that an oscillation originating in the wake of the obstacle is likely to sustain itself indefinitely, without increasing nor decreasing. (Castagnetto 1939, 146)

Although his study is in the field of sinusoidal oscillations rather than relaxation oscillations, as the parameter ε' must stay small before the unit in order to keep the validity of the formulae (8.29) and (8.30), Castagnetto found one of the aspects of the relaxation oscillations, that is to say, the amplitude remains unchanged. He obtains, with a simple reading of Fig. 8.28, the value for the fluid speed V = 9.55 mm/sec for which $\varepsilon' = 0$.

Lastly, in order to demonstrate the wake oscillations, i.e. that the Bénard-Karman vortices are connected to the relaxation phenomenon, Castagnetto had to establish

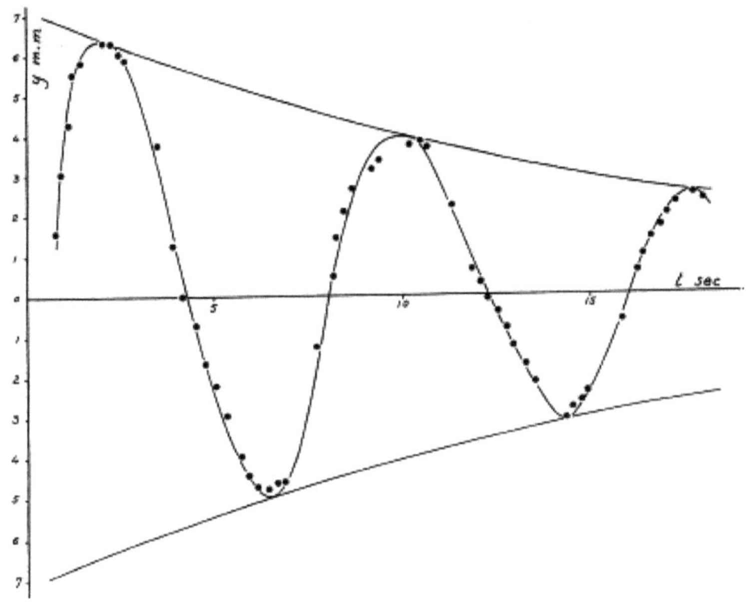

Fig. 8.26 Variation at one point of the wake's amplitude during a damping phase, by Castagnetto (1939, 132)

that their period is expressed as the product or the quotient of "hydraulics analogue" variables with one relaxation time (see *supra*). He finds:

> $T = \frac{5D}{V}$, relation defining the period of the alternating vortices with the quotient of the two quantities characterizing the system, which is indeed one of the characteristics of relaxation oscillations.[76] (Castagnetto 1939, 149)

8.3.3 *Abelé's Thesis*

Jean Abelé's biography was established thanks to an obituary notice written by Bouligand (1961) in the *Revue Générale des Sciences Pures et Appliquées*.[77] During

[76]D represents the diameter of the obstacle (cylinder) and V the fluid velocity.

[77]Jean Abelé (1886–1961) was a physicist, philosopher and writer. He first studied at the *Collège Saint-Joseph* in Reims before joining the *Compagnie de Jésus*. He then obtained a *Licence ès Lettres* and studied philosophy and science at Saint-Hélier (Jersey isle) then at the *École d'Antoing* (Belgium). He completed his specialization in Physics in Paris, under the supervision of Édouard Branly, and became *licencié ès Sciences*. He was mobilized during WWI along with General Gustave Ferrié in the *section de radiogogniométrie* (wireless telegraphy). He then studied theology at Enghien and was appointed Physics professor in Vals (1923–1961). He worked in a modest laboratory but nonetheless had several of his inventions patented, such as a device to receive

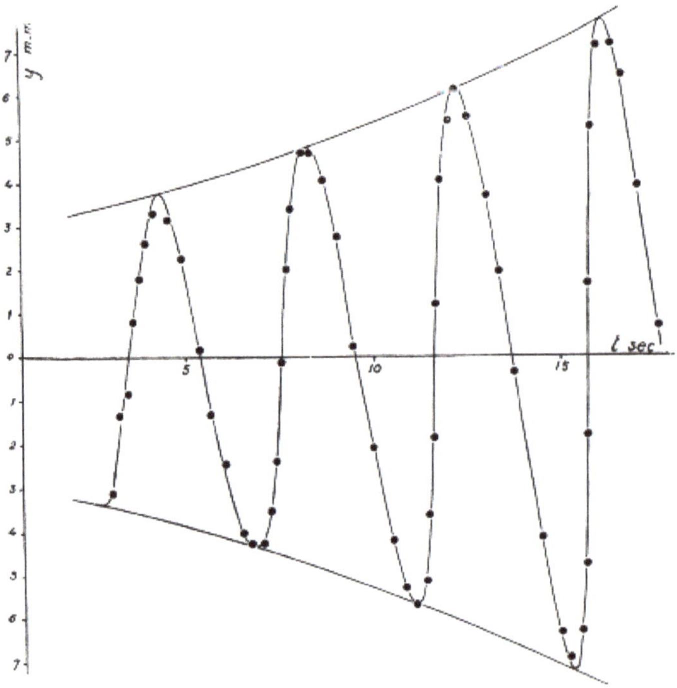

Fig. 8.27 Variation at one point of the wake's amplitude during a damping phase, by Castagnetto (1939, 132)

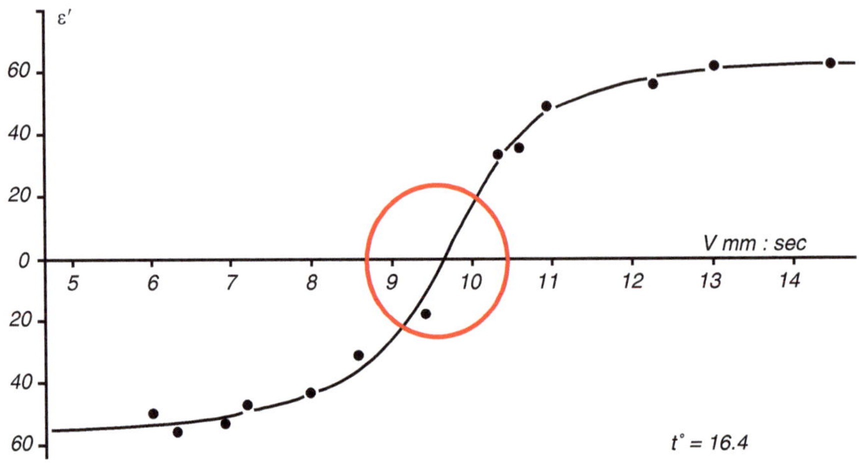

Fig. 8.28 Variations of ε' depending of the speed V of the fluid, by Castagnetto (1939, 145)

the 1940s, he published many scientific articles (Abelé 1942a,b, 1943, 1945a,b,c, 1946, 1947). Thus, aside from his thesis, his contribution to the theory of nonlinear oscillations is of great importance, especially in regards to the classification of oscillations (see *infra*).

His thesis dissertation is comprised of two parts: a theoretical part, and a technical part, which is of no interest to the present study. The theoretical part will therefore be the only one analysed. In his introduction, Abelé (193, 9–12) writes a firm assessment on the relation between the theory and engineering of self-sustained oscillations:

> However, the theory is outdated compared to engineering. (Abelé 1943, 12)

This sentence seems to quite accurately describe the situation of the French scientific community at the time, especially if we think for example about Abraham's work regarding the designing of the T.M. lamp, and the construction of the multivibrator with Bloch (see *supra* Part I). Abelé then recalls the development of the theory of self-sustained systems:

> We can reduce them to two main ones that we will call, based on Van der Pol,[78] *linear theory* and *nonlinear theory* of the self-sustained systems. The linear theory searches for the conditions which must be satisfied by a given oscillator for the self-sustaining to be *possible*. (...) The nonlinear theory intends to apply the resources of infinitesimal calculations to the more complex problem of the oscillations *stability*. (...) The equations representing the *stable* operating of a self-sustained oscillator can therefore only be *nonlinear* equations. (Abelé 1943, 13)

Here, Abelé gave a definition of the two main methods used hitherto in France to study sustained oscillations. The first one, which he called linear theory, simply aimed at determining, from the vicinity of the working point, a starting condition, also called sustaining condition of the oscillations. This condition was generally presented with the form of an inequality connecting the different variables of the system. The second one, which he called nonlinear theory, was at the next step. Once the system was "started", the concern was knowing if the amplitude of the oscillation would increase or decrease exponentially, or "stabilize" itself to a steady state corresponding to a condition which absolutely had to be determined in order to produce for example, reliable radio transmission systems. This is precisely what Poincaré (1908) and Andronov (1929a) (see *supra* Part I) did in their work, by providing an existence condition for a stable limit cycle, which is the mathematical expression of an oscillatory regime for which the amplitude is stabilized. The first

and amplify the high frequency of electric oscillations (1924), and a receiver for telegraphy and telephony (1926). He obtained a *doctorat ès Sciences* de la Faculté des Sciences at the University of Paris in 1943. His thesis dissertation won in 1948 the Prix du Général Ferrié. In 1948 he was appointed director of studies in Vals. In 1945, he became member of the *Centre national de la Recherche Scientifique*. In 1949, he presented his work at the *Congrès International de Philosophie des Sciences* held in Paris. Aside from the many studies he published in the periodic *Études*, he colaborated with the *Archives de Philosophie*, he was a member of its executive committee.

[78] Abelé refers to Van der Pol (1934).

chapter, entitled: "Founding principles of the nonlinear theory of self-sustained systems" begins with the presentation of the second order nonlinear differential equation for a mechanical oscillating system, with the following form:

$$m\frac{d^2z}{dt^2} + F\left(z, \frac{dz}{dt}\right) + sz = 0 \tag{8.31}$$

Although Abelé claimed to have based his work on Van der Pol's (1930, 1934), Liénard's (1928) and Le Corbeiller's (1931a, 1933a), this is not Van der Pol's equation (3.8), prototype of the relaxation oscillations, that he presents but a more general equation, which he considered as "fundamental for nonlinear theory" (Abelé 1943, 18). Moreover, Abelé (1943, 17) adds in a footnote that at the time his dissertation was finished, Rocard's book (1941), in which the second chapter dedicated to relaxation oscillations, had just been published. This is another mix up between self-sustained oscillation and relaxation oscillation (see *supra*). He then demonstrates that with a simple change of variables, i.e. by writing $v = dz\ dt$, the equation (8.31) is reduced to a first order differential equation:

$$mv\frac{dv}{dt} + F(z, v) + sz = 0 \tag{8.32}$$

He then suggests a "geometrical interpretation", consisting in placing in the plane (z, v), in other words, Poincaré's phase plane, and he recalls:

> A *stable* periodic motion corresponds to a closed integral curve, *asymptotically approached by the neighboring solutions, in spirals, from the outside and the inside, for t increasing.* One of the fundamental problems of nonlinear theory consists in each of these closed curves, called limit cycles.[79] (Abelé 1943, 18)

This excerpt paraphrases Liénard's (1928, 906), Andronov's (1929a, 560) and Haag's research (1934b, 906) (see *supra*), although the latter is not featured in the bibliography. Also surprising is the lack of any direct quote of Poincaré, or even any reference.

After having dedimensionalized (Ab_2) by writing $x = z\sqrt{s/2}$ and $y = v\sqrt{m/2}$: he considers an "energy interpretation" of the following equation:

$$xdx + ydy + E(x, y)\,dx = 0 \tag{8.33}$$

[79] Abelé refers to Andronov (1929a).

It is then interesting to compare this equation with the equation (8.34) written by Poincaré (1908, 390) (see *supra* Part I):

$$Ly\frac{dy}{dx} + \rho y + \theta(y) + Hx = 0 \tag{8.34}$$

By multiplying it by dx and neglecting the resistance ρ of the inductor, we obtain:

$$Hxdx + Lydy + \theta(y)\,dx = 0 \tag{8.35}$$

Abelé notes that the first two terms of the equation (8.33) represent the potential and kinetic energies of the oscillator during the interval dt. He then uses the *principle of conservation of mechanical energy* in order to explain that the third term represents the lost or gained energy during this interval, which corresponds to what Van der Pol (1928a, 367, 1930, 300) and Andronov (1928, 23) (see *supra*) called a "continuous, nonrenewable energy source". This was also based on this *principle* that Poincaré (1908) established the stability condition for the limit cycle (see *supra* Part I). Moreover, it appears that the equations (8.33) and (8.35) are completely identical. He then presents Liénard's equation and construction, as well as Van der Pol's equation, as special cases of his equation (8.35), prototype of self-sustained oscillations. Regarding Van der Pol's works, he recalls:

> This study, which constitutes a remarkable application of nonlinear theory, is interesting by the way it shows that by giving ε a continuous variation, we create a passage, also continuous, between two types of oscillations in appearance very different: the sinusoidal oscillations, and the ones which Mr. Van der Pol called: *relaxation oscillations*. In rigorously sinusoidal oscillations, a certain amount of energy is periodically transformed from potential energy to kinetic energy, and reciprocally, as the dissipated energy is null or exactly compensated by the supply from an energy source. In relaxation oscillations, the energy provided by the source is accumulated progressively as kinetic or potential energy, and, as soon as it reaches a determined value, is abruptly dissipated. While Van der Pol's equation allows the connection of these two opposed types of oscillatory states thanks to a vast intermediary domain, the sinusoidal oscillation and the relaxation oscillations are still *outer limits* to this domain. (Abelé 1943, 22)

While in turn, Abelé recalls that the proposal that Le Corbeiller (1931a, 21) (see *supra*) credited to Van der Pol, and which had been used in turn by Morched-zadeh (1936, 70) (see *supra*), presents the continuous passage from sinusoidal oscillations to relaxation oscillations being clearer and seems to constitute a kind of "hyphen" between the linear theory and nonlinear theory. This reversible *continuum* is quite unusual, as it allows the passage from linear oscillations to nonlinear oscillations, and vice versa. This is not the case, for example, in Mechanics. We can move, *via* Lorentz transform, from Relativistic Mechanics to Classic Mechanics, but not the other way round. Lastly, in this section Abelé appears to finally adopt Van der Pol and Andronov's "original" terminology in order to designate the third term of the equation (8.35).

In the second chapter, he proposes a "generalization of Liénard's equation" (Abelé 1943, 28) representing the operation of an self-stabilized-amplitude oscillator, and explains that:

> This oscillator indeed aims at reconciling the sinusoidality and stability of the generated equations. (Abelé 1943, 28)

By starting from symmetry considerations and using the *Duality Principle* (see *supra* Part I), he suggests the two following equations:

$$xdx + ydy + 2R(x, y)\, ydx = 0 \tag{8.36}$$

$$xdx + ydy + 2S(x, y)\, xdy = 0 \tag{8.37}$$

where the functions $R(x, y)$ and $S(x, y)$ depend on the circuit's layout, respectively serial and in parallel. Abelé also explains that the sinusoidality is expressed by the equation for a circle: $x^2 + y^2 = A^2$, and that the stability requires the circle to be the solution to both equations (8.36) and (8.37). This implies that the functions $R(x, y)$ and $S(x, y)$ contain, as a factor the expression $x^2 + y^2 - A^2$. It must be noted that the equation (8.36), which is, according to Abelé, a generalization of Liénard's equation (1928, 901), had already been obtained by Liénard (1931, 177) (see *supra*). In the next chapter, he establishes a link between the amplitude, or rather the "amplitude function" $A(x, y)$ and the functions $R(x, y)$ and $S(x, y)$ so that it can be "self-stabilized". He thus obtains the following equations for the serial circuit and the shunt circuit:

$$A(x, y) = +\sqrt{x^2 + 2Rxy + y^2} \tag{8.38}$$

$$A(x, y) = +\sqrt{x^2 + 2Sxy + y^2} \tag{8.39}$$

The fourth chapter is dedicated to the "setting up of the equation for an self-stabilized-amplitude oscillator", which he provides for the case of the serially assembled circuit.[80] To obtain it, he writes (8.36): $R = -a + bA(x, y)$, he hence obtains the following equation (Ab_8), by solving this equation in relation to R:

$$xdx + ydy + 2\left[-\left(a - b^2xy\right) + \sqrt{\left(a - b^2xy\right)^2 - a^2 + b^2\left(x^2 + y^2\right)}\right] ydx = 0 \tag{8.40}$$

By placing ourselves in Poincaré's phase plane, we can then transform the equation (8.40) into a system of two first order differential equations corresponding to Andronov's self-sustained system (1.10) (1929a, 560) (see Table 8.2).

In the last two chapters of the first part of his thesis, Abelé demonstrates that his oscillator possesses a *limit cycle* for which the equation is: $x^2 + y^2 = a^2/b^2$, which is

[80] A simple permutation of the variables x and y leads to the equation corresponding to the shunt circuit.

Table 8.2 Abelé's (1943) and Andronov's (1929a) differential equations systems.

Abelé (1943)	Andronov (1929a)
$\begin{cases} \dfrac{dx}{dt} = y \\ \dfrac{dy}{dt} = -x - 2R(x, y)\, y \end{cases}$	$\begin{cases} \dfrac{dx}{dt} = y + \mu f(x, y; \mu) \\ \dfrac{dy}{dt} = -x + \mu g(x, y; \mu) \end{cases}$

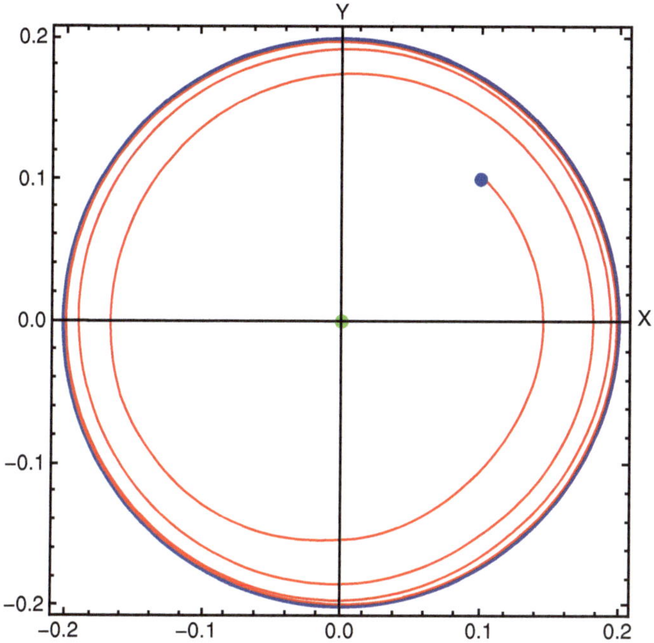

Fig. 8.29 Numerical integration (Mathematica, 7) of equation (Ab_8) for $a = 0.2$ and $b = 1$

the solution to the equation (Ab_8). He also provides the existence conditions for this limit cycle focusing on parameters a and b. He thus establishes that: $0 < a < 0.995$ and $b > 0$. By choosing for example $a = 0.2$ and $b = 1$, we verify by the means of a numerical integration (see Fig. 8.29) that the equation (Ab_8) representing an self-stabilized-amplitude oscillator indeed possesses a limit cycle.

The blue curve corresponds to the limit cycle for which the equation is $x^2 + y^2 = a^2/b^2$. The green dot represents the fixed point $O(0, 0)$, the blue one is the initial condition $I(0.1, 0.1)$. We note, as Abelé demonstrates (1943, 49), that the integral curves inside the circle approach it asymptotically.

We can nevertheless question Abelé's reasons for not using Liénard's method (1928) to demonstrate the existence and uniqueness of the limit cycle, and not Andronov's method (1929a) either to demonstrate its stability. This third thesis dissertation also provides precious indications on the assimilation level of the concept of relaxation oscillations and self-sustained oscillations by the French scientific community. He especially shows, as in the previous cases, that the importance of Poincaré's and Andronov's work was apparently not recognized at the time.

8.3.4 Moussiegt's Thesis

The few existing biographical elements on Jean Moussiegt were provided to me by his son (Denis). Jean Moussiegt defended his thesis, entitled: "Les oscillations de relaxation dans les tubes à décharges: application à l'étude de l'allumage" ("Relaxation oscillations in discharge tubes: application to ignition studies") in 1949 at the *Laboratoires d'Électronique et de Radioélectricité* of the Faculty of Science of Grenoble University.[81] His thesis dissertation consisted mostly in an experimental study of the imminent discharge of a current through a low-pressure gas contained in a neon tube. Nevertheless, the bibliographies, and some chapters in the second part, are extremely interesting as they provide a picture of how the concept of relaxation oscillations was received after the War.

In the introduction, Moussiegt (Moussiegt 1949, 593) first recalls that this intermittent discharge was usually seen as belonging to the relaxation-oscillation type, and then adds:

> The study of relaxation oscillations has advanced through the consideration of the *characteristic*, a curve representing the voltage variations depending on the current, when the system is in a steady state. A famous essay written by B. Van der Pol explains a general theory of oscillations, including relaxation oscillations, which appear as distorted sinusoidal oscillations. Later developments of this theory split farther and farther from the previous simple outline, and do not seem to hold concern for specifications, at least in regards to the oscillations produced by mean of a discharge tube. (Moussiegt 1949, 597)

Moussiegt credited Van der Pol with the merit of developing a "general theory of oscillations". This seems to be slightly excessive in light of the bibliographical references he provided, as he only refers to Van der Pol's work (1926c,d, 1927c, 1930), especially since there is no mention of the articles written by Andronov (1929a), and Andronov and Witt (1930a), or even Andronov and Khaikin's book (1937), whose translation by Lefschetz (Andronov and Khaikin 1949) had been published the same year. It therefore seems he focused his study on Van der Pol's concept of relaxation oscillations, in other words, the slow-fast aspect of the oscillatory phenomenon, rather than Andronov's concept of self-sustained oscillations, insofar as Poincaré's concept of *limit cycle* is entirely absent from this essay. Indeed, he replaced Poincaré's terminology with the expressions "limit relaxation capacity" (Moussiegt 1949, 601) and "relaxation cycle" (Moussiegt 1949, 606), which are rather remote from Poincaré's definition.

In the first paragraph of his general bibliography, entitled "relaxation oscillations and the general theory of oscillations" he mentions (amongst others): Abraham and Bloch (1919e), Abelé (1942a,b, 1946), Liénard (1928), Rocard (1937a, 1941), but also Richardson (1937) and Shohat (1943, 1944) (see *infra*). In chapter III

---

[81]In 1955, he was *Chef de travaux of the Laboratoire de Radioélectricité*. In 1965 he was appointed Director of studies of the *École d'ingénieurs de Grenoble*. He then became the very first director of the *Institut Universitaire Technologique* (I.U.T.) of Nice in 1970 and participated in the creation of the *Laboratoire d'Électronique de Nice* in 1973.

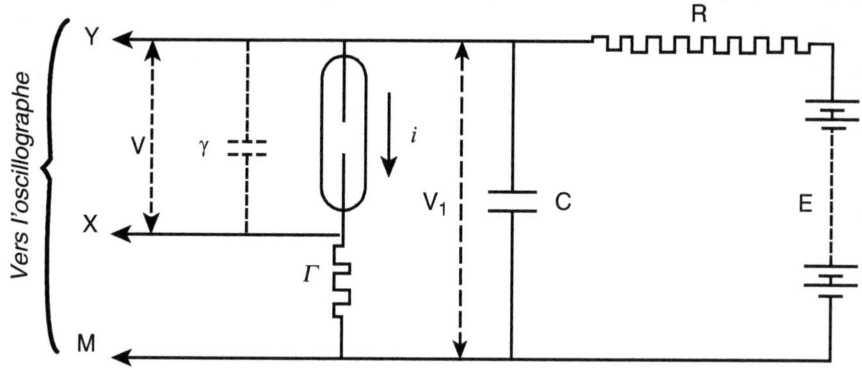

Fig. 8.30 Diagram of the assembly for the observation of the relaxation cycle with the cathode-ray oscillograph, from (Moussiegt 1949, 606)

of the first part, he suggests an "experimental study of the relaxation cycle", and provides the oscillographic readings of the dynamic characteristic (see Figs. 8.30 and 8.31), nonetheless explaining that such oscillograms had already been found by Richardson (1937) and Leyshon (1930) (see *infra*).

He then comments on these figures with the following sentence:

> The *dynamic characteristic* of the intermittent operation appears as a closed curve, which I will also call relaxation cycle. (Moussiegt 1949, 608)

It seems that he absolutely did not assimilate Andronov's correspondence (1929a) (see *supra*), probably due to ignorance. The first chapter of the second part, entitled "Theory of relaxation oscillations", begins with a reminder of the oscillations produced by an intermittent discharge. Moussiegt then reproduces a diagram which had been presented by Le Corbeiller (1931a, 9), and afterwards, Morched-Zadeh[82] (1936, 148) (see *supra*). He represents the static characteristic of the neon tube (see Fig. 8.32, red curve), as well as the equation lines $v = v_1 - ri$ and $v = E - (R + r)\,i$.

He then explains that the equation line $v = v_1 - ri$, which travels in parallel with itself in the direction of v increasing, becomes tangent at point M to the *static characteristic*. The starting occurs along the line MM' and the current then decreases until switch-off causing the passage of N to N'. He then notes that the equation line $v = E - (R + r)\,i$ cuts the phase plane (i, v) into two distinct domains I and II.

From this, he deduces the sign of v_1 depending on the time, which allows him to conclude:

> We therefore see how point P remains inaccessible and under which conditions the intermittent operating following the cycle of $MM'NN'$ is generated. (Moussiegt 1949, 623)

[82]It is strange that his work was not mentioned by Moussiegt.

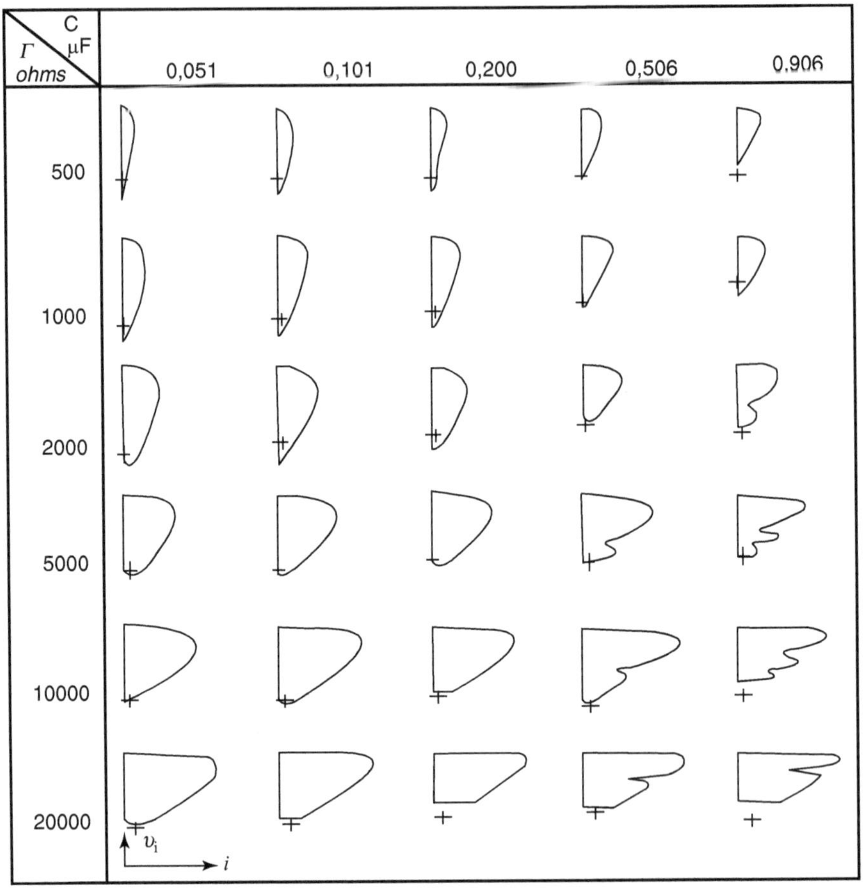

Fig. 8.31 Relaxation cycles, from (Moussiegt 1949, 606)

Again, we see that there is no mention of the expression limit cycle. He then refers to Le Corbeiller's work, expressing the following criticism:

> In a book written by Ph. Le Corbeiller (*Les systèmes auto-entretenus et les oscillations de relaxation*, Hermann, Paris, 1935), we find the mention of such a cycle on a figure, but the text does not provide any explanation. (Moussiegt 1949, 623)

He then analyses the "distortion of the small-capacity relaxation cycle", and continues his experimental study. This essay from the post-war years is more the work of an electronician than a mathematician, and shows a complete ignorance of Andronov's works (1929a) on self-sustained oscillations, which is expressed by a lack of awareness of the importance held by the concept of limit cycle in the oscillations theory. However, Van der Pol's concept of relaxation oscillation (1926d) appears to have been somewhat understood, although it seems to be more of a "relaxation effect".

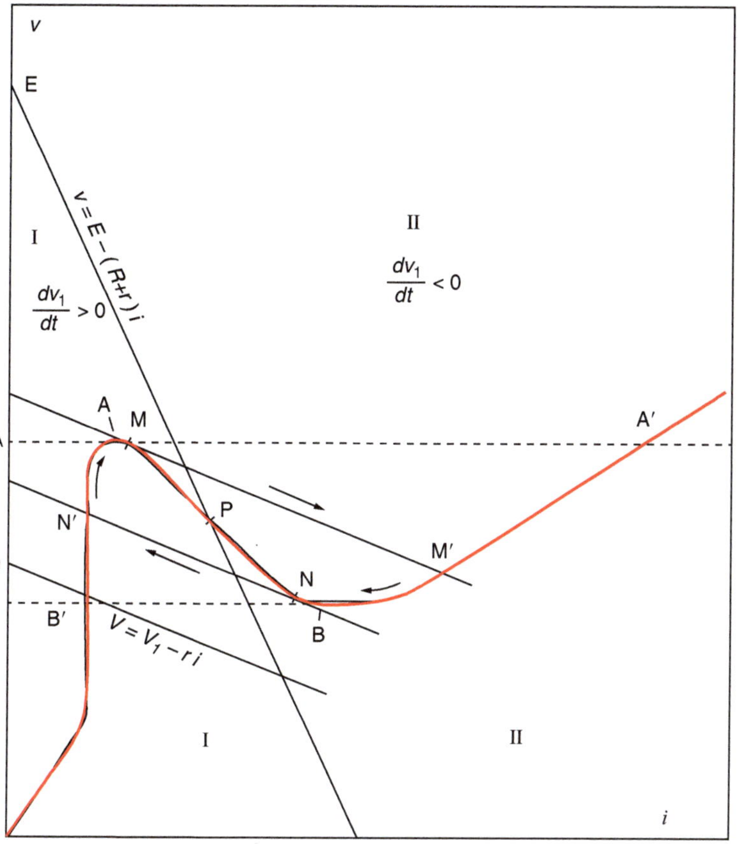

Fig. 8.32 Characteristic of the neon tube, by (Moussiegt 1949, 622)

Conclusion of Part II

In this second part, we followed the emergence of the concepts of relaxation oscillations and self-oscillations, their first demonstrations in France at the end of the 1920st and at the beginning of the 1930s, as well as their reception by the French scientific community.

The analysis of the lectures given in France by Van der Pol allows the distinguishing of the three steps which led him to the conceptualization of the relaxation oscillation phenomenon. By using the metaphorical example of the Tantalus cup, he first characterized this new type of oscillations that he had evidenced in his analysis of the prototype equation for the triode possessing two characteristic time scales: one slow, one fast. He established that their period is proportional to the product of two quantities analogous to a resistance and a capacitance, and gave them the name "relaxation oscillations". He multiplied the examples for apparently vastly different phenomena, such as the oscillations of a triode, the heartbeats, flowers' sleep, and finally suggested, using an analogy, that they were possibly represented by one same equation, for which he had established the prototype.

The extent of Andronov's work in France goes in a different path. It was first the note presented to the *Académie des sciences* in Paris on the 14th of October 1929, which in many ways was a reiteration of the previous note presented at the Congress of Russian physicists in 1928, which contributed to this spreading in France. We will see that the correspondence established by Andronov with Poincaré's work is not only the one that has been highlighted until now by historiography, regarding his essays "Sur les courbes définies par une équation différentielle" ("On curves defined by a differential equation"), and especially the chapter entitled "The theory of limit cycles", but also the one that used characteristic exponents introduced by Poincaré in the first volume of his *Les Méthodes Nouvelles de la Mécanique Céleste*, which allows the establishment of a stability condition for the limit cycle nowadays called "Poincaré criteria for orbital stability". This was precisely the condition that Poincaré had provided in 1908 during his lectures on wireless telegraphy.

The French scientific community responded in various different ways to these two conceptualizations of nonlinear oscillations. While Andronov's scientific work caused relative indifference, a great enthusiasm rose in favor of Van der Pol's ideas and caused a true "hunt for the relaxation effect". In the early 1930s, mathematicians Le Corbeiller and Liénard were among the rare French scientists to take interest in Andronov's work. While the first recognized – in accordance with Andronov's results – that the solution to Van der Pol's equation is indeed Poincaré's limit cycle, he did not address the question of its stability. It was Liénard who solved this problem, in a last study on "self-sustained oscillations", in which he demonstrated the stability of the solution to Van der Pol's equation, which he generalized (nowadays called a Liénard equation) by using, without naming it, the "Poincaré criteria", still not recognizing the limit cycle. Contrary to these two mathematicians, the French scientists, who after Van der Pol's lectures were focused on the study of oscillatory phenomena in various fields, were rather indifferent about the purely mathematical questions that arose from it.

The *conférence internationale sur les oscillations non linéaires* of 1933 was held in Paris at Van der Pol and Papalexi's initiative. Choosing the Henri Poincaré Institute was quite fitting, not just because of its tutelary figure's influence, but especially because the mathematicians, physicists and astronomers most likely to grasp the interest of pooling their skills to develop a true theory of nonlinear oscillations, generally gathered in Paris. Corroborating the idea of a French crossroads in the process of developing that theory of nonlinear oscillations, the failure of this conference highlights the inability of the French scientific community to federate its own results around a common thematic, prerequisite to the development of a theory.

Following this conference's failure, French mathematicians and engineers carried on in a sporadic and disordered fashion their research on nonlinear oscillations, focusing essentially on Van der Pol's concept of relaxation oscillations, which slowly turned into a sort of "paradigm". From a mathematical standpoint, Jules Haag and Yves Rocard highlighted the development of these methods (asymptotic expansion, piecewise linearization) leading to fundamental results. From a phenomenological point of view, the enthusiasm caused by Van der Pol's concept started what we called a "hunt for the relaxation effect", which consequently evidenced this new type of oscillations in a wide variety of fields of research, sometimes unexpected. Even if they did not bring many new elements to the theory of nonlinear oscillations, the four doctorate dissertations defended between 1936 and 1949 underlined, by the synthesis of older results they implemented, the paradigmatic aspect taken by the concept introduced by Van der Pol.

Part III
From Self-Oscillations to Quasi-periodic Oscillations

At the end of the First World War, the development of wireless telegraphy led the engineers and scientists to turn to the study of self-sustained oscillations in a three-electrode lamp subjected to a periodic "forcing", or a "coupling". It was a circuit comprised of a resistance and inductance linked, in the first case, to an alternating current generator,[1] or coupled, in the second case, by mutual inductance with an oscillating circuit comprised of another inductor and another capacitor. These two types of assembly were widespread at the time in radio engineering, and thus allowed to evidence on the one hand, the quasi-periodic oscillations, meaning the oscillations possessing at least two mutually independent frequencies, and on the other hand, the hysteresis, synchronization, drive and frequency demultiplication phenomena.

At the end of the nineteenth century, the research carried out in the field of Celestial Mechanics, and especially the studies on the "three -body problem", had led the astronomers and mathematicians to consider the effects of gravitational interaction applied by the most remote planets as a "perturbation". Poincaré and Lyapunov had then developed methods aiming on the one side, at expressing the periodic solution of the differential equations representing their motions by an asymptotic series following the potencies of a parameter corresponding to the mass of each planet, and on the other side, studying the stability. Nevertheless, these developments had two crippling flaws. Firstly, their convergence was only assured if this parameter was assumed sufficiently small. Then, they caused the appearance of terms called "secular" in the solution, in other words, of the form $t^n sin\,(\omega t)$, which completely prevented the construction of a periodic solution. In order to rectify this, Poincaré and Lyapunov used a technique introduced by the Swedish mathematician Anders Lindstedt (1854–1929), which led to the development of the "Poincaré-Lindstetdt method",[2] also called the "méthode du petit paramètre"

[1] Which means, for which the *e.m.f.* is represented by a sinusoidal function.

[2] Krylov and Bogolyubov called it the "Poincaré-Lyapunov method".

("small-parameter method[3]") in French. Therefore, at the beginning of the twentieth century, the physicists and engineers had two methods at their disposal in order to study the "nonlinear problems" that arose in the field of radio engineering. The harmonics analysis developed by Joseph Fourier (1822) and implemented by Strutt and Rayleigh (1877), and the "Poincaré-Lindstedt method". While the first allowed to obtain some fundamental results, such as the value of the period and amplitude of the triode's oscillations as a first approximation (see *supra* Part I), the second one was invalidated by the nature of the observed oscillatory phenomena itself. Indeed, the parameter involved in the developments sometimes took values much larger than the unity, for example in the case of the relaxation oscillations evidenced by Van der Pol (1926d) for $\varepsilon \gg 1$ (see *supra* Part I), and the solution therefore quickly differed. The "Poincaré-Lindstedt method", and the obstacles it generated, will be briefly described in the first paragraph.

Van der Pol (1920, 1922, 1927a) studied these problems as early as the beginning of the 1920, and bypassed the two obstacles of the Poincaré-Lindstedt method, by developing another one, called "method of the slowly varying amplitudes[4]", likely to provide the "zeroth order approximation" for the amplitude and period of a radio device with either free or forced oscillations. His research in this field, presented in the second paragraph, represent the most important part of his contribution to the development of the *Theory of Nonlinear Oscillations*. It will also be established that Van der Pol's research cannot be reduced to the sole discovery of the relaxation oscillations at their representation by way of the prototype equation nowadays named after him.

A few years later, the representatives of the School of Kiev in France: Nikolai Krylov and his student Nikolai Bogolyubov, questioned in a first phase, the "Poincaré-Linsdtedt method", which they called "Poincaré-Lyapunov method", in order to highlight the importance of the latter's contribution in the development of stability conditions for the periodic solutions.

In a note presented to the C.R.A.S. by the mathematician Jacques Hadamard on 22 February 1932, they therefore wrote:

> Consequently, the application of the famous methods ([2]) of H. Poincaré, based on the expansions of the functions representing the oscillations as per the potencies of parameters, is not sufficient anymore in this case. Indeed, the expansions of the quasi-periodic functions as per said parameters, on which also depend the frequencies and amplitudes, *do not give the adequate representation*, and may even misrepresent the character of the sought functions, not to mention the fact that the expansions in question do not converge uniformly over the whole real axis, as they contain polynomials in *t*.
>
> ---
>
> ([2]) Recently used for the study of the periodic solution in radio engineering. (Krylov and Bogolyubov 1932a, p. 957–958)

---

[3]This expresses the condition ensuring the convergence of these developments. See *infra*.

[4]Also called "Van der Pol method", Mandel'shtam et al. (1935, 89).

In a second phase, the developed a new theoretical framework that they named "Nonlinear Mechanics[5]", in which they inscribed Van der Pol's results, after substantially developing them. On 15 January 1933, exactly two weeks before the first *Conférence Internationale de Non linéaire* was held Paris at the initiative of Balthazar Van der Pol and Nikolai Papalexi, and in which they were supposed to participate (see *supra* Part II), and article titled "Fundamental problems of Nonlinear Mechanics" was published in the *R.G.S.P.A.*, in which they reminded:

> In his own works in the field of nonlinear oscillations in radio engineering, Mr. Balth. van der Pol, who did not use the mentioned results of Poincaré-Lyapunov,[6] obtained his results by using processes that were ingenious, albeit lacking in the required mathematical rigor. Yet, it is fair to note that the inaccurate processes of the eminent Dutch scientist, which he applied *ad hoc*, nonetheless provide indications on the nature of quasi-periodic oscillations, for which the study of the Poincaré-Lyapunov methods in their present state hardly seem practical. In his many works, Mr. Balth. van der Pol developed the theory of drive, as well as the theory of synchronization (of the natural frequency with the external force), and his research led him to study the frequency demultiplication phenomenon in relaxation systems (from M. van der Pol's terminology). (Krylov and Bogolyubov 1933, 10)

At the same time, the French scientific community did not seem to react to Krylov and Bogolyubov's repeated and vehement criticisms (1932a, 1933, 1934e, 1935a, 1936b) against the Poincaré-Lindstedt method. However, this seemingly passive attitude is not surprising, as a "long line" of French astronomers and mathematicians such as Ernest Esclangon, Jean Favard or Hervé Fabre had been studying the quasi-periodic functions for a long time, while others such as Pierre Fatou or Jean Chazy worked on furthering Poincaré's methods for the case where the parameter is largely above the unity.

Indeed, while Piers Bohl (1893, 1900, 1910) may be seen as the "founder of the theory of quasi-periodic functions" (Krylov and Bogolyubov 1935b, 107), it was actually Ernest Esclangon (1904) who introduced the terminology "quasi-periodic function", in his thesis dissertation, titled "Les fonctions quasi-periodiques' ("Quasi-periodic functions") presented before the *Faculty of Science* of Paris, and defended on 29 June 1904 before Misters. Paul Appell (President), Paul Painlevé and Henri Poincaré (Examiners). These fundamental researches were then continued in Copenhagen, around the Danish mathematician Bohr (1923a,b, 1924, 1925a,b, 1926) by Bochner (1925, 1927, 1929, 1930, 1931, 1933), in the Soviet Union by Besicovic (1925, 1931, 1932a,b), Kovanko (1928a,b, 1929a,b), Markov (1929), Pontrjagin (1933), Stepanov (1925), or Stepanov and Tychonov (1933), in Germany by Weyl (1926), and in France by Fatou (1927a,b, 1928), Favard (1927a,b, 1933) then Fabre (1937a,b). Aside from the "forcing", quasi-periodic solutions had also been observed in radio engineering, when two oscillating circuits were coupled, represented by an autonomous differential equation system with two degrees of freedom. In this case, this led to the problem of the "characteristics

[5]Krylov and Bogolyubov's works will be discussed in the fourth paragraph.

[6]Poincaré-Lindstedt method, to which Lyapunov brought his contribution regarding the stability of the periodic solutions. See *infra*.

on the surface of a torus[7]" previously studied by Henri Poincaré (1885a, p. 220), and completed by Arnaud Denjoy (1932a,b). This is probably the reason why he was tasked with the analysis, on behalf of the *R.G.S.P.A.*, the works of Krylov and Bogolyubov, about which he wrote:

> The case that is the most remarkable, most perfect, but also most accidental, and yet least important in regards to physics or celestial mechanics applications, is the one of the periodic solutions. H. Poincaré and Lyapunov showed the fundamental role of these integral when they exist, and from this, founded the theory in essays that have been classics for a long time. A more general case is the one of the quasi-periodic solutions, considered for the first time by P. Bohl, founder of the theory of functions of the same nature. But before N. Krylov's works, the existence of quasi-periodic solutions had been established with a rigor sufficient to only a very narrow class of differential equations. However, by studying in depth the remarkably simple case of the characteristics on the surface of a torus, H. Poincaré had evidenced specific orders in fact likely to express themselves with an analogous form in much broader cases. In 1932, I myself completed the results of H. Poincaré on this matter, by establishing the quasi-periodicity in all the cases where there were no periodic solutions. (Denjoy 1935, 390)

While Krylov and Bogolyubov challenged the validity of the Poincaré-Lyapunov methods, which they were modifying in order to expand their range of application to the quasi-periodic oscillations, the Mandel'shtam-Papalexi School showed a more moderate attitude towards Poincaré's method, of which it claims to be the inheritor and kept on applying it to radio engineering for the case of the coupling, in other words, the case of systems with two degrees of freedom.[8] Regarding the case of the forcing, Mandel'shtam and his disciples seemed to prefer Van der Pol's method[9] to the one developed by Krylov and Bogolyubov in their "Nonlinear Mechanics".

> (...) engineering readily agrees to a quantitative theory lacking in rigor and being only approximated, as long as it sufficiently represents practically important cases. Hence the necessity to develop approximate methods for the study of nonlinear systems, which must, of course, account for the specificities of these systems. A quantitative approximate method, with slow variation, or, as we will call it, the van der Pol method. Although in essence this method has been used for a long time in celestial mechanics, it was van der Pol who was the first to systematically apply it to problems of radio engineering, and found a series of fundamental results on forced synchronization, "streaking", etc.[10] (Mandel'shtam et al. 1935, 89)

The first paragraph thus aims to briefly remind of the principles of the Poincaré-Linstedt method, and to explain the reasons why its application to problems of radio engineering is not always feasible. The two next paragraphs focus on the presentation of Van der Pol's and Krylov-Bogolyubov's methods. The various research carried out or published in France in the field of quasi-periodic oscillations

---

[7]The simplest example is the one of the coupled harmonic oscillator for which the solution coils over the surface of a torus with a motion resulting from the two rotation frequencies. See Bergé et al. (1988, 312).

[8]See Andronov and Witt (1930a, 1935).

[9]See Andronov and Khaikin (1937, 435).

[10]Mandel'shtam referred to Van der Po's publications (1922, 1927a).

will be analyzed in detail in paragraph four, in which it will be established that Van der Pol's method, which had already been applied to problems of Celestial Mechanics, holds its founding principles in Pierre Fatou's works. The last paragraph will then allow to show that the "Méthodes Nouvelles de la Mécanique Céleste" (Poincaré 1892, 1893, 1899), as well as the essays "Sur les courbes définies par une equation différentielle" (Poincaré 1881, 1882, 1885a,b, 1886a,b), were the subject of many works in France during this time. It will be thus demonstrated that, contrary to what Dahan Dalmedico (1996) claimed, Henri Poincaré's difficult legacy regarding dynamical systems had indeed been inherited by a whole line of astronomers and mathematicians.

Chapter 9
The Poincaré-Lindstedt Method: *The Incompatibility with Radio Engineering*

9.1 The Poincaré-Lindstedt Method

These methods introduced by Henri Poincaré (1892) in his "Méthodes Nouvelles de la Mécanique Céleste" ("New Methods of Celestial Mechanics") and by Aleksandr Lyapunov (1892, 1907) in his "General Problem of Motion Stability" are part of *Asymptotic Theories*[1] (see *supra* Part II). Faced with the three-body problem, (Poincaré 1892, 51) considered the approximation of the solution to the equations for the motion by using series expansions according to the increasing potency of a parameter μ assumed sufficiently small. The main drawback of these series, aside from the problem of the value of μ, was the introduction of *secular terms*, well known to astronomers. These terms, expressed with the form $t^n sin(\omega t)$, shattered any hope of developing a periodic solution. Therefore, during the late nineteenth century, Lindstedt developed a method (1883) in order to eliminate the *secular terms* in expansion.

Poincaré's method is based on a theorem securing the fact that "the expansion converges for any value of t, as long as $|\mu|$ is sufficiently small" (Poincaré 1892, 60). This resulted in it being called "méthode du petit paramètre" ("small parameter method"). The obstacles of its application to radio engineering noted by Krylov and Bogolyubov (see *supra*) were evidenced with a simple example, which will be studied in later paragraphs. These difficulties can be highlighted from a simple example which will be resumed in the following sections. This is Van der Pol's equation (3.8), which corresponds neither to the case of the forcing nor to the coupling, but whose periodic solution, represented by a limit cycle,[2] can be approximated using this method. Considering Van der Pol's equation (3.8) which

[1] Nowadays called Perturbations Theory.

[2] Construction of a series expansion of $v(t)$ allowing to obtain an approximation of the value and period of the oscillations.

J.-M. Ginoux, *History of Nonlinear Oscillations Theory in France (1880–1940)*,
Archimedes 49, DOI 10.1007/978-3-319-55239-2_9

can be written in the following manner:

$$\ddot{x} - \varepsilon\left(1 - x^2\right)\dot{x} + x = 0 \tag{9.1}$$

The application of Poincaré's method consists in searching for a solution with the form:

$$x(t, \varepsilon) = x_0(t) + \varepsilon x_1(t) + \varepsilon^2 x_2(t) + \ldots \tag{9.2}$$

The convergence of this series expansion is assured by Poincaré's theorem as long as $\varepsilon \ll 1$. This is not the case anymore for relaxation oscillations, which occur when $\varepsilon \gg 1$. By substituting (9.2) in (9.1), we obtain:

$$\ddot{x}_0 + x_0 + \varepsilon\left[\ddot{x}_1 + x_1 - \left(1 - x_0^2\right)\dot{x}_0\right] + \varepsilon^2\left[\ddot{x}_2 + x_2 - \left(1 - x_0^2\right)\dot{x}_1 + 2x_0\dot{x}_0x_1\right] + \ldots = 0 \tag{9.3}$$

This equation is identically verified for $\varepsilon \ll 1$. Consequently, all the power coefficients of ε must cancel term to term. This leads to the following differential equation system:

$$\begin{cases} \ddot{x}_0 + x_0 = 0 \\ \ddot{x}_1 + x_1 = \left(1 - x_0^2\right)\dot{x}_0 \\ \ddot{x}_2 + x_2 = \left(1 - x_0^2\right)\dot{x}_1 - 2x_0\dot{x}_0x_1 \end{cases} \tag{9.4}$$

The solution to the first equation (harmonic oscillator) is written:

$$x_0(t) = a_0 sin(t)$$

By substituting in the second member of the system's second equation (9.4) we obtain

$$\ddot{x}_1 + x_1 = \left(1 - a_0^2 sin^2(t)\right)\left(a_0 cos(t)\right) \tag{9.5}$$

By taking the trigonometric relations into account:

$$cos^2(t) + sin^2(t) = 1 \text{ and } cos^3(t) = \frac{1}{4}cos(3t) + \frac{3}{4}cos(t)$$

This equation yields the following form:

$$\ddot{x}_1 + x_1 = a_0\left(1 - \frac{a_0^2}{4}\right)cos(t) + \frac{a_0^3}{4}cos(3t) \tag{9.6}$$

Unless we cancel the first term $cos(t)$ of the right hand side of the equation (9.6), it will cause in the solution $x_1(t)$, a *secular term* with the form $tsin(t)$. This implies we must write $a_0 = 2$. The solution is then written:

$$x_1(t) = \frac{1}{4}cos(t) - \frac{1}{4}cos(3t) + a_1 sin(t)$$

By substituting the second member of the system's third equation (9.4) we find

$$\ddot{x}_2 + x_2 = \frac{1}{4}sin\,(t) - 2a_1cos\,(t) - \frac{3}{2}sin\,(3t) + 3a_1cos\,(3t) + \frac{5}{4}sin\,(5t) \tag{9.7}$$

It is clear that regardless of a_1, the term $\frac{1}{4}sin\,(t)$ cannot be canceled. Consequently, the integration of (9.7) will generate a *secular term* which makes the construction of a periodic solution infeasible, therefore invalidating the method. This deficiency is linked to the fact that the method was developed in order to provide a periodic solution with the unknown period $T = \frac{2\pi}{\omega}$. However, the period is not rigorously equal to T. In the case of Van der Pol's equation (9.1) the period is only equal to T when $\varepsilon = 0$, and differs for any other value of ε (see *supra* Part II). We must therefore use Lindstedt's modification on the method, which consists in a change in time scale allowing the step-by-step elimination of secular terms. Indeed, Lindstedt considered that the period, or rather the frequency, is developed in series depending on the increasing potencies of ε.

$$\omega\,(t,\varepsilon) = \omega_0\,(t) + \varepsilon\omega_1\,(t) + \varepsilon^2\omega_2\,(t) + \ldots\ textwith\ \omega_0\,(0) = 1 \tag{9.8}$$

By writing: $\omega t = \tau$, Van der Pol's equation yields:

$$\omega^2 x'' - \varepsilon\left(1 - x^2\right)\omega x' + x = 0 \tag{9.9}$$

where the primes represent the derivation in relation to τ.

By substituting (9.8) in the equation (9.9) we obtain by canceling term to term all the potency coefficients of ε, the following differential equation system:

$$\begin{cases} x_0'' + x_0 = 0 \\ x_1'' + x_1 = -2\omega_1 x_0'' + \left(1 - x_0^2\right)x_0' \\ x_2'' + x_2 = -2\omega_1 x_1'' - \left(\omega_1^2 + 2\omega_2\right)x_0'' + \left(1 - x_0^2\right)x_1' - 2x_0x_0'x_1 + \omega_1\left(1 - x_0^2\right)x_0' \end{cases} \tag{9.10}$$

The solution to the first equation is again written: $x_0\,(\theta) = a_0\,sin\,(\theta)$ By replacing it in the second member of the system's second equation (9.10) we find:

$$x_1'' + x_1 = 2\omega_1 a_0 sin\,(\theta) + a_0\left(1 - \frac{a_0^2}{4}\right)cos\,(\theta) + \frac{a_0^3}{4}cos\,(3\theta) \tag{9.11}$$

To eliminate the *secular terms* of the equation's second member, i.e. in order to obtain a periodic solution, the following must be chosen:

$$\omega_1 = 0 \quad \text{and} \quad a_0 = 2$$

Once again, the solution is written:

$$x_1\,(\theta) = \frac{1}{4}cos\,(\theta) - \frac{1}{4}cos\,(3\theta) + a_1 sin\,(\theta)$$

By substituting in the second member of the system's third equation (9.10) we have:

$$x_2'' + x_2 = \left(4\omega_2 + \frac{1}{4}\right) sin\,(\theta) - 2a_1 cos\,(\theta) + 3a_1 cos\,(3\theta) - \frac{3}{2} sin\,(3\theta) + \frac{5}{4} sin\,(5\theta) \tag{9.12}$$

The application of Lindstedt's method now allows the choosing of ω_2 and a_1, so that the secular terms, i.e. $sin\,(\theta)$, $cos\,(\theta)$ and $cos\,(3\theta)$ are eliminated. The solution $x_2\,(\theta)$ is therefore periodic, provided that we write:

$$\omega_2 = -\frac{1}{16} \quad \text{and} \quad a_1 = 0$$

We can deduce an approximation of the angular frequency, and therefore, of the periodic solution to Van der Pol's equation (9.1).

$$\omega = 1 - \frac{\varepsilon^2}{16} + o\left(\varepsilon^4\right) \quad \Leftrightarrow \quad T = 2\pi\left(1 + \frac{\varepsilon^2}{16}\right) + o\left(\varepsilon^4\right) \tag{9.13}$$

The approximation of the periodic solution to Van der Pol's equation (9.1) is then written:

$$x\,(t) = 2sin\,(\omega t) + \frac{\varepsilon}{4}\left[cos\,(\omega t) - cos\,(3\omega t)\right] + o\left(\varepsilon^2\right) \tag{9.14}$$

By using the trigonometric form: $cos\,(3t) = 4cos^3\,(t) - 3cos\,(t)$, the equation (9.14) is written:

$$x\,(t) = 2sin\,(\omega t) + \varepsilon sin^2\,(\omega t)\,cos\,(\omega t) + o\left(\varepsilon^2\right) \tag{9.15}$$

It is then interesting to compare this expression with the zeroth order and first order for approximations of the solution to Van der Pol's equation (3.8), i.e. (9.1) which had been provided by Van der Pol (1926d, 980, 1930, 254) and by Liénard (1928, 948) (see *supra* Part II), for the case where $\varepsilon \ll 1$. In order to establish such a comparison, it must first be considered that the period of the oscillations is $T \approx 2\pi$ as Van der Pol supposed (1926d, 987), and that consequently, the angular frequency is $\omega \approx 1$. Then, by noting on the one hand that Van der Pol's solution (1926d, 980) can be written, when $\varepsilon \ll 1$ assuming that the phase at the origin $\varphi = 0$:

$$x\,(t) = \frac{2sin\,(t+\varphi)}{\sqrt{1 + Ce^{-\varepsilon(t+C)}}} = 2sin\,(t) + o\,(\varepsilon)$$

By writing on the other hand: $\omega = 1$ and $k = \varepsilon$ in Liénard's periodic solution (1928, 948) (see *supra* Part II) in order to compare it with the previous one, we obtain, when $\varepsilon \ll 1$:

$$x\,(t) = \frac{2sin\,(t)}{\sqrt{1 + Ke^{-\varepsilon t}}}\left[1 + \frac{\varepsilon}{2}\frac{sin\,(t)\,cos\,(t)}{1 + Ke^{-\varepsilon t}}\right] = 2sin\,(t) + \varepsilon sin^2\,(t)\,cos\,(t) + o\left(\varepsilon^2\right)$$

It therefore appears that Liénard (1928) used the Poincaré-Lindstedt method in order to obtain the first order approximation in ε of the solution to Van der Pol's equation (3.8), i.e. (9.1) for the case where $\varepsilon \ll 1$. Moreover, it has been established (see *supra* Part II) that Van der Pol (1930, 254) provided a similar approximation two years later. Prior to this, Blondel (1919b) was the first to write the equation for the oscillations of a triode, and to use the harmonic analysis (AH) in order to calculate an approximation for the angular frequency (period) and the amplitude (see *supra* Part I). This method consists in searching for the solution with the form of a Fourier series expansion. As for Van der Pol's equation (3.8), i.e. (9.1), it consists in taking the symmetries into account and calculating for the case where $\varepsilon \ll 1$, this development is written:

$$x(t) = a_0 sin(\omega t) + b_0 cos(\omega t) + a_1 sin(2\omega t) + b_1 cos(2\omega t) + a_2 cos(3\omega t) + \dots \tag{9.16}$$

where the coefficients a_i and b_i are assumed constant. By replacing (9.16) in Van der Pol's equation (3.8), i.e. (9.1), then canceling the various polynomials in factors before $sin(\omega t)$, $sin(2\omega t)$, $sin(2\omega t)$ we deduce the values of the coefficients a_i and b_i, and therefore obtain the first, second and third harmonics.

The results are presented in Table 9.1 below.

The following year, while Van der Pol (1920) was doing the same thing, Blondel (1920) suggested a higher order approximation for the amplitude. A few years later Rocard (1932, 1941, 40; 943, 217) provided the third order approximation for the amplitude of the oscillations of a triode, represented by Van der Pol's equation (3.8), i.e. (9.1) for the case where $\varepsilon \ll 1$. A comparison of different values of the amplitude obtained by using harmonic analysis and of the Poincaré-Lindstedt method is presented in Table 9.2.

In order to allow such a comparison, the internal resistances r_1 and r_2 should be neglected in Blondel's equation (1919b, 947), which should be integrated depending on time. We therefore obtain the simplified equation (9.17) presented in Part I:

$$C\frac{d^2u}{dt^2} - \left(b_1 h - 3b_3 h^3 u^2 - \dots\right)\frac{du}{dt} + \frac{u}{L} = 0 \tag{9.17}$$

Table 9.1 Zeroth order and first order approximations in ε of the Van der Pol's equations (3.8) and (9.1) (1926d, 979)

Zeroth order and first order approximations in ε of the solution to Van der Pol's equation: $\ddot{x} - \varepsilon\left(1 - x^2\right)\dot{x} + x = 0$ Case $\varepsilon \ll 1$	
$x(t) = 2sin(t) + o(\varepsilon)$	Van der Pol (1926d, 980)
$x(t) = 2sin(t) + \varepsilon sin^2(t)\, cos(t) + o\left(\varepsilon^2\right)$	Liénard (1928, 948)
$x(t) = 2sin(t) + \varepsilon sin^2(t)\, cos(t) + o\left(\varepsilon^2\right)$	Poincaré-Lindstedt method

Table 9.2 Zero, first and second order approximations of the amplitude of the oscillations of a triode represented by Van der Pol's equation (3.8) and (9.1)

Zero, first and second order approximations in ε of the amplitude of the oscillations of a triode represented by Van der Pol's equation: $\ddot{x} - \varepsilon\left(1 - x^2\right)\dot{x} + x = 0$ Case $\varepsilon \ll 1$	
$a_0 = \frac{2}{h}\sqrt{\frac{b_1}{3b_3}} = 2$	Blondel (1919b, 947)
$a_0 = \sqrt{\frac{4}{3}\frac{\alpha}{\gamma}} = 2$	Van der Pol (1920, 704)
$\begin{cases} a_0 = \frac{2}{\sqrt{\beta}} = 2 \\ a_1 = \frac{2}{3}\varepsilon\frac{\alpha}{\beta} = 0 \\ a_2 = \frac{\varepsilon}{4\sqrt{\beta}} = \frac{\varepsilon}{4} \end{cases}$	Rocard (1932, 1941, 40, 1943, 217)
$\begin{cases} a_0 = 2 \\ a_1 = 0 \\ a_2 = \frac{\varepsilon}{4} \end{cases}$	Poincaré-Lindstedt method

By writing: $b_1 h = 3b_3 h^3 = \varepsilon$ and $L = C = 1$, (9.17) is reduced to the equation (3.8), i.e. (9.1).

Disregarding the resistance R, Van der Pol's equation (1920, 704) presented in Part I, leads to the simplified equation (9.18):

$$C\frac{d^2v}{dt^2} - \left(\alpha - 3\gamma v^2\right)\frac{dv}{dt} + \frac{1}{L}v = 0 \tag{9.18}$$

By writing: $4\alpha = 3\gamma = \varepsilon$, $\beta = 0$ and $L = C = 1$ (9.18) is reduced to the equation (3.8), i.e. (9.1).

Lastly, Rocard's equation (1932, 1326) presented in Part II

$$\omega^2 v - \varepsilon\left[1 - \alpha v - \beta v^2 - \gamma v^3 - \delta v^4 - \ldots\right]v' + v'' = 0 \tag{9.19}$$

is reduced, by writing $\alpha = \gamma = \delta = 0$ and $\omega = \beta = 1$ to the equation (3.8), i.e. (9.1).

Studying Tables 9.1 and 9.2 allows to demonstrate that the results obtained by using the Fourier harmonic analysis and the Poincaré-Lindstedt methods concord perfectly. From there, we can question the reasons why Krylov and Bogolyubov challenged them, and their inadequacy for this type of problem. As for the harmonic analysis, it seems that it was linked to the complexity of the calculations, which increases with the approximation order, as already noted by Blondel who writes about this method:

> The mathematical solution is therefore theoretically possible, but is too complicated in practice from a physicist's point of view. I will therefore settle for considering a simplified first approximation, from physical considerations, the characteristic form indeed leading to the certitude that the series will be convergent (...) (Blondel 1919b, 947)

The following year, after providing the third harmonic, Blondel adds: "This is an extremely long and tedious calculation (...)" (Blondel 1920, 72). Ten years later, Jean Mercier recalls in the same context: "Unfortunately, the calculations are inextricable." (Mercier 1929, 35)

Following Van der Pol's observation (1926d) on the relaxation oscillation phenomenon, which only occurs for values of the parameter $\varepsilon \gg 1$ Fourier's method was faced with another problem described by Le Corbeiller:

> *Case of ε being very large.* – Here the oscillation curve visibly has a great number of harmonics. In mathematical language, the corresponding Fourier series converges very slowly. It is therefore absolutely unrealistic in this case to laboriously calculate the first, second or third terms of the series. (Le Corbeiller 1931a, 22)

Regarding the Poincaré-Lindstedt method, the convergence of the series expansion likely to represent an approximation of the solution is therefore not assured anymore, contrary to what Blondel assumed above. Moreover, during the 1930s, most of the research conducted in the field of nonlinear oscillations was carried out by using radio engineering devices comprised of essentially "forced" or "coupled" systems. Although the Fourier and Poincaré-Lindstedt methods are still valid for the study of periodic oscillations, the quasi-periodic case is impervious to them, as recalled by Krylov and Bogolyubov:

> The critical statements apply mainly to theories based on the use of Poincaré and Lyapunov's rigorous methods. These famous methods are very important when it comes to mathematical analysis, but are however hardly manageable for a practitioner, as they make the study of oscillators, which have a somewhat complicated diagram, almost inaccessible from a practical point of view. Moreover, these methods, at least in their present state, hardly allow for more than the study of simply-periodic states, which means that the quasi-periodic states, of which we have established the existence, are apparently excluded from their application field. (Krylov and Bogolyubov 1936b, 509)

In the case of the forcing, represented by a non-autonomous differential equation, or the coupling, the principle of Poincaré-Lindstedt method, i.e., the process described above, remains the same, but the calculations are however much longer and complex.[3] This is also one of the probable reasons why other methods were developed, such as Van der Pol's or Krylov-Bogolyubov's methods.

[3]See Andronov and Khaikin (1937, 455) or Minorsky (1947, 138).

9.2 Forcing or Coupling: *Towards Quasi-periodic Oscillations*

In the middle of the 1920s, the existence of quasi-periodic oscillations in radio engineering devices (wireless telegraphy) was evidenced by using two different methods:

- either by periodic forcing,
- or by coupling.

9.2.1 *Forced Oscillators*

The study of three analogous devices: series-dynamo machine, singing arc, and triode, by Blondel, Janet, Cartan, Van der Pol and Liénard led to the development of a second order nonlinear differential equation which was *autonomous*, i.e. in which time did not explicitly appear (see *supra*, Parts I & II). This equation, a prototype of a self-oscillator, allowed Van der Pol to discover the relaxation oscillation phenomenon.

However, as early as the beginning of the 1920s, Van der Pol (1920) also worked on the oscillations of a circuit (see Fig. 9.1) comprised of an inductance coil L, a capacitor of capacitance C, and a triode, powered by a voltage generator with an *f.e.m.* of type $v(t) = E_s in(\omega_1 t)$.

Van der Pol's equation (3.8) is written in this case:

$$\ddot{v} - \alpha\left(1 - v^2\right)\dot{v} + \omega_0^2 v = \omega_1^2 E_s in\left(n\omega_1 t\right) \tag{9.20}$$

Later, this *non-autonomous* second order nonlinear differential equation took the following form:

$$\ddot{x} + \mu f(x, \dot{x}) + \omega_0^2 x = F(t) \tag{9.21}$$

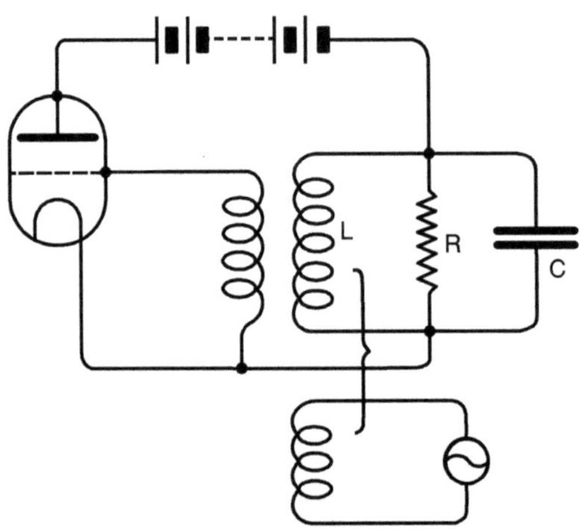

Fig. 9.1 Forced oscillations of a triode, from Van der Pol (1920, 759)

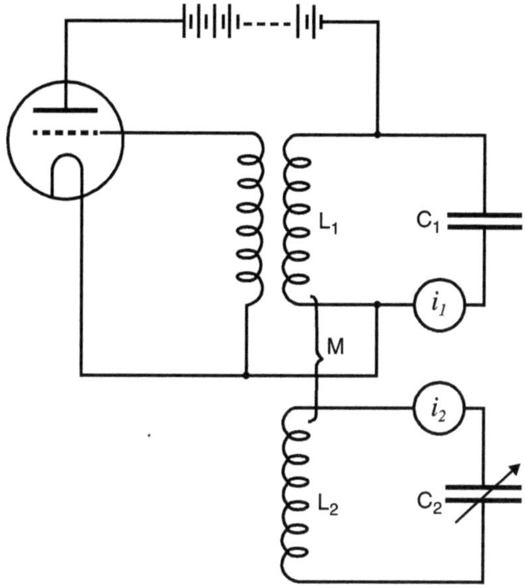

Fig. 9.2 Oscillations of a triode with two degrees of freedom, by Van der Pol (1922, 701)

9.2.2 *Coupled Oscillators*

Two years later, Van der Pol (1922) studied the coupling by mutual induction of two oscillating circuits comprising, for the first one an inductance coil L_1, a capacitor of capacitance C_1 and a triode, powered by a D.C. generator, and for the second an inductance coil L_2, and a capacitor of variable capacitance C_2 (see Fig. 9.2).

Van der Pol then obtains the equations for an oscillator with two degrees of freedom:

$$\begin{cases} \ddot{v}_1 - \alpha_1\left(1 - v_1^2\right)\dot{v}_1 + \omega_1^2 v_1 + k_1\omega_1^2 v_2 = 0 \\ \ddot{v}_2 + \alpha_2\dot{v}_2 + \omega_2^2 v_2 + k_2\omega_2^2 v_1 = 0 \end{cases} \quad \text{with} \quad \begin{cases} \varepsilon_1 = \dfrac{\alpha_1}{\omega_1} \ll 1 \\ \varepsilon_2 = \dfrac{\alpha_2}{\omega_2} \ll 1 \end{cases} \tag{9.22}$$

Later, this *autonomous* second order nonlinear differential equations system is written:

$$\begin{cases} \ddot{x} + \omega_1^2 x = \mu f\left(x, \dot{x}, y, \dot{y}, \mu\right) \\ \ddot{y} + \omega_2^2 y = \mu g\left(x, \dot{x}, y, \dot{y}, \mu\right) \end{cases} \tag{9.23}$$

***Note*:**

By applying Alembert's (1748) theorem, the *Coupled* system (9.23) may be replaced by a system of four autonomous first order differential equations. As for the *Non Autonomous* equation (9.21), it can also be transformed into a system of four autonomous first order differential equations, by noting that the sinusoidal forcing is the solution to a harmonic oscillator. The two problems are therefore perfectly equivalent and lead to periodic or quasi-periodic solutions.

Chapter 10
Van der Pol's Method: *A Simple and Classic Solution*

As early as the year 1920, Van der Pol (1920) worked on the free and forced oscillations of a triode. In the first case, he tackled the determination of an approximate value of the amplitude and period problem, by using the Poincaré-Lindstedt method and harmonic analysis. In the paragraph entitled "First Method for finding the Amplitude of the Fundamental", Van der Pol (1920, 704) recalls that he followed a solving method suggested by Professor Hendrik Antoon Lorentz (1853–1928). It was actually a "variation" of the Poincaré-Lindstedt method that Rocard (1932) incidentally used a few years later, but which cannot pass the obstacle linked to the presence of secular terms, which "disturb the periodic character of the solution", as noted by Van der Pol (1920, 706) in a footnote. He thus obtains the value a_0 of the amplitude of the fundamental, presented in Table 9.2 above. In the next paragraph, he uses the harmonic analysis and finds the same result. He then suggests a geometrical method based on Witz's construction (1889b), which again leads him to the same value a_0 for the amplitude (see *supra* Part I). In the second case, he writes a non-autonomous second order differential equation, which is written in a simplified form:

$$\ddot{v} - \alpha \left(1 - v^2\right) \dot{v} + \omega_0^2 v = \omega_1^2 E_s in \left(n\omega_1 t\right) \text{ with } \varepsilon = \frac{\alpha}{\omega_0} \ll 1 \tag{10.1}$$

By using the harmonic analysis, he managed to express the amplitude of the forced oscillations depending on the circuit's constants and the forcing amplitude.

10.1 The Slowly Varying Amplitudes Method and the Hysteresis Phenomenon (I)

In January 1922, Edward Appleton and Balthazar Van der Pol (1922) published an article entitled: "On a type of Oscillations-Hysteresis in a simple triode generator", in which they firstly write the equation for the oscillations of a triode with a generic

J.-M. Ginoux, *History of Nonlinear Oscillations Theory in France (1880–1940)*,
Archimedes 49, DOI 10.1007/978-3-319-55239-2_10

form, which was soon called "Van der Pol equation". Secondly, they presented a method consisting in slowly varying amplitudes,[1] Mandel'shtam et al. (1935) soon called it the "Van der Pol method", which helped evidence the hysteresis phenomenon. The method of slowly varying amplitudes is based on the principle of harmonic analysis, i.e. on the search for a solution with the form of a Fourier series expansion, but for which the coefficients, which represent the amplitude of the oscillations, are considered as functions varying with the time.[2] Appleton and Van der Pol (1922, 180) wrote:

$$x(t) = a(t)\, sin(\omega t) \tag{10.2}$$

where the amplitude $a(t)$ varies slowly with time. This means that during a period, the variations of the amplitude are slow, i.e. that:

$$\frac{da}{dt} \ll \omega_a \quad \text{and} \quad \frac{d^2a}{dt^2} \ll 1 \tag{10.3}$$

By replacing (10.2) in the equation (3.8), i.e. (9.1) and accounting for (10.3) they obtain:

$$\left[2\frac{da}{dt} - \varepsilon\left(a - \frac{a^3}{4}\right)\right] cos(\omega t) + a\left(1 - \omega^2\right) sin(\omega t) = 0$$

Then, by multiplying this equation by a, they write:

$$\left[\frac{da^2}{dt} - \varepsilon\left(a^2 - \frac{a^4}{4}\right)\right] cos(\omega t) + a^2\left(1 - \omega^2\right) sin(\omega t) = 0 \tag{10.4}$$

By then noting that the frequency of the fundamental ω which has previously been calculated is approximately neighboring the unity, it yields:

$$\frac{da^2}{dt} - \varepsilon\left(a^2 - \frac{a^4}{4}\right) = 0 \tag{10.5}$$

This first order differential equation is easily solved and leads to:

$$a^2 = \frac{4}{1 + Ce^{-\varepsilon(t+C)}} \tag{10.6}$$

By restoring the variables' dimensions, the value of the zeroth order approximation of the amplitude is then written:

[1]The acronym (AV) refers to "Amplitude Variable" or Appleton and Van der Pol whereas the acronym (AH) referred to "Analyse Harmonique" ("Harmonic Analysis").

[2]This is reminiscent of Lagrange's method for the "variation of the constant".

$$a_0 = \frac{\sqrt{\dfrac{4}{3}\dfrac{\alpha}{\gamma}}}{\sqrt{1 + Ce^{-\varepsilon(t+C)}}} = \sqrt{\frac{4}{3}\frac{\alpha}{\gamma}} + o(\varepsilon) \tag{10.7}$$

The solution is written as first approximation for $\varepsilon \ll 1$:

$$x(t) = \frac{2sin(t)}{\sqrt{1 + Ce^{-\varepsilon(t+C)}}} = 2sin(t) + o(\varepsilon) \tag{10.8}$$

Which allows to somewhat quickly find the values obtained with the Poincaré-Lindstedt method or harmonic analysis. The higher order approximations can then be obtained by searching for a solution with the form

$$\begin{aligned} x(t) &= \sum_{n=1}^{+\infty} a_{n-1}(t)\, sin(n\omega t) \\ &= a_0(t)\, sin(\omega t) + a_1(t)\, sin(2\omega t) + a_2(t)\, sin(3\omega t) + \ldots \end{aligned} \tag{10.9}$$

It must however be noted that during the two years following the publication of this article, Appleton and Greaves (1923), then Greaves (1924) obtained higher orders by using the Poincaré-Lindstedt method, as Van der Pol later recalls (1934, 1063).

By using the method of slowly varying amplitudes that Appleton (1922) established, based on Routh's (1877) and Hurwitz's criteria (1895), the stability conditions of a triode for which the oscillation characteristic is not limited to a cubic anymore but extends to a quantic. They also evidenced the hysteresis phenomenon, which is expressed by the fact that there is a relation between the amplitude of oscillations and a parameter of the circuit (resistance or capacitance, or more generally the frequency). This allowed them to evidence that the evolution of the amplitude can be subjected to a "sudden" transition, and can "jump" from a finite value to another from a specific value of the parameter (see *infra*).

10.2 The Mode Competition and Hysteresis Phenomena (II)

In April 1922, Van der Pol (1922) studied, by himself this time, the hysteresis phenomenon generated by the oscillations of a triode with two degrees of freedom, which can be described by the following differential equation system:

$$\begin{cases} \ddot{v}_1 - \alpha_1\left(1 - v_1^2\right)\dot{v}_1 + \omega_1^2 v_1 + k_1\omega_1^2 v_2 = 0 \\ \ddot{v}_2 + \alpha_2 \dot{v}_2 + \omega_2^2 v_2 + k_2\omega_2^2 v_1 = 0 \end{cases} \quad \text{with} \quad \begin{cases} \varepsilon_1 = \dfrac{\alpha_1}{\omega_1} \ll 1 \\ \varepsilon_2 = \dfrac{\alpha_2}{\omega_2} \ll 1 \end{cases} \tag{10.10}$$

From the outset of his article, he gives a precise definition of the hysteresis phenomenon:

> Now it is found experimentally that, when the system oscillates in one of the two modes of vibration and the natural frequency of the secondary circuit is varied gradually, the system suddenly jumps at a certain point from the first mode of vibration to the other. If afterwards the natural frequency of the secondary is varied in the reverse direction it is found that the system jumps from the second to the first mode of vibration, but at a point which is *not* identical with the first one mentioned above, and thus a kind of oscillation hysteresis is obtained, which, apart from its importance in technical applications, is of interest from a physical point of view. (Van der Pol 1922, 700–701)

He then describes the relation between intensities i_1^2 and i_2^2 occurring in each circuit (see *infra* Fig. 9.2) and the angular frequencies ω_1^2 and ω_2^2, and represented them as diagrams (see Figs. 10.1 and 10.2). He then explains that the condition for which the system starts oscillating with the largest of the two angular frequency is represented by EFB, whereas the one corresponding to the oscillations with the smaller angular frequency is represented by DCA. He then recalls:

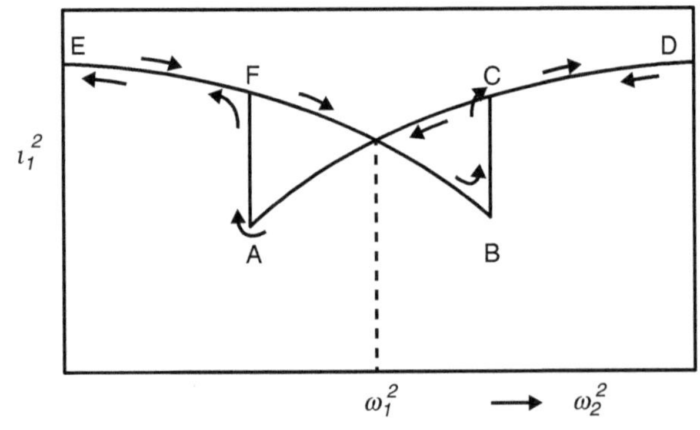

Fig. 10.1 Relation between the square module of the amplitude of the intensity in circuit 1 and the angular frequency of circuit 2, by Van der Pol (1922, 702)

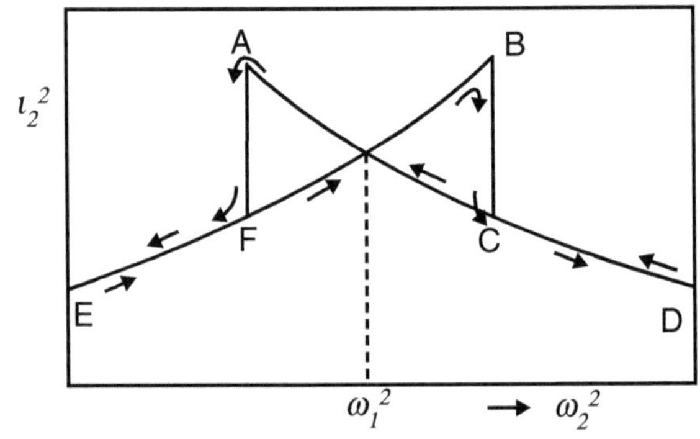

Fig. 10.2 Relation between the square module of the amplitude of the intensity in circuit 2 and the angular frequency of circuit 1, by Van der Pol (1922, 702)

> These phenomena were noticed by the author in February 1920, but it was felt that no satisfactory explanation could be given unless progress was made in the development of a non-linear theory of sustained oscillations. For it is obvious that, when the problem is treated with linear differential equations, the principle of superposition is valid, and in this case oscillations in the one mode are uninfluenced by oscillations in the other. (Van der Pol 1922, 702)

When a dozen years later, in an article entitled "Nonlinear Theory of Electric Oscillations", Van der Pol (1934, 1067) went back over his first researches, especially this article, he writes:

> The obvious consequence is, therefore, that the mode of oscillation which will actually occur depends on the way in which the particular set of parameters is reached, and a very pronounced oscillation hysteresis may occur, as is also well known in practice. The explanation of this hysteresis can, however, only be given on the basis of a nonlinear theory such as explained above, because this typically involves the interaction of one oscillation by another, as the principle of superposition is no more valid in nonlinear systems. (Van der Pol 1934, 1067)

The Figs. 10.1 and 10.2 perfectly illustrate the hysteresis phenomenon, which is closely linked to the concept of bifurcation[3] introduced by Poincaré (1885b, 270). Indeed, on these figures, Van der Pol (1922) qualitatively represents the relations between the square module of the intensity in each circuit, and the angular frequency in circuit 2. In the first circuit (Fig. 10.1), when the square module of the angular frequency ω_2^2 is increases, the amplitude decreases from the value E to B then "jumps" suddenly to the value C and evolves towards D. By decreasing the square module of the angular frequency ω_2^2, the amplitude decreases from the value D to A then "jumps" suddenly to the value F and evolves towards E. The same phenomenon is observed in the second circuit (Fig. 10.2).

In order to study this system (10.10), Van der Pol (1922, 706) transforms it into a fourth order[4] in $v\left(t\right)$ differential equation. Then, he applies the slowly varying amplitudes method and searches for a solution with the form:

$$v\left(t\right) = a\left(t\right) sin\left(\omega_I t\right) + b\left(t\right) sin\left(\omega_{II} t + \lambda\right) + a_2\left(t\right) sin\left(3\omega t\right) + \dots \tag{10.11}$$

This leads him to express the evolution of the amplitude in each of the circuits with the form of a system[5]:

$$\begin{cases} \dfrac{da^2}{dt} = E_I a^2\left(a_0^2 - a^2 - 2b^2\right) \\ \dfrac{db^2}{dt} = E_{II} b^2\left(b_0^2 - b^2 - 2a^2\right) \end{cases} \quad \text{where} \quad \begin{cases} E_I = \dfrac{3}{4}\gamma \dfrac{\omega_I^2}{\omega_1^2} \dfrac{\omega_I^2 - \omega_2^2}{\omega_I^2 - \omega_{II}^2} \\ E_{II} = \dfrac{3}{4}\gamma \dfrac{\omega_{II}^2}{\omega_1^2} \dfrac{\omega_2^2 - \omega_{II}^2}{\omega_I^2 - \omega_{II}^2} \end{cases} \tag{10.12}$$

[3] See Bergé et al. (1984, 40–42).

[4] In accordance with Alembert's theorem 1748.

[5] The system (10.12) is perfectly analogous to predator-prey types of models which were only developed three or four years later by Lotka (1925) and Volterra (1926). Moreover the effect Van der Pol was since then called "mode competition". For instance, see Abarbanel et al. (1993, 51).

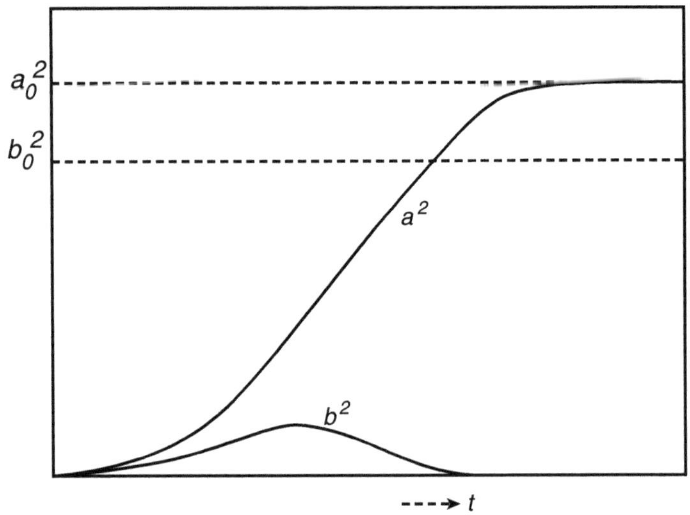

Fig. 10.3 Graphical integration of the system (10.12), by Van der Pol (1922, 702)

From this he deduces a value for the amplitudes a and b as a first approximation:

$$\begin{cases} a_0^2 = \dfrac{\alpha_1}{\dfrac{3}{4}\gamma} - \dfrac{\alpha_2}{\dfrac{3}{4}\gamma} \dfrac{\left(\omega_I^2 - \omega_2^2\right)^2}{k^2\omega_2^4} \\ b_0^2 = \dfrac{\alpha_1}{\dfrac{3}{4}\gamma} - \dfrac{\alpha_2}{\dfrac{3}{4}\gamma} \dfrac{\left(\omega_{II}^2 - \omega_1^2\right)^2}{k^2\omega_2^4} \end{cases} \tag{10.13}$$

He then graphically integrates the system (10.12). He notes, based on Fig. 10.3, which represents the evolution of the square module of the oscillations' amplitudes in each circuit, that the variation rate of b^2 is smaller than the variation rate of a^2. He deduces that when the square module of the amplitude a^2 of the oscillations in the circuit I increases, the square module of the amplitude b^2 of the oscillations in the circuit II disappears. This leads to the establishment of the stability for each oscillation mode.

Thus, the slowly varying amplitudes method seems more adapted to the study of coupled systems, such as for example the oscillations of a triode with two degrees of freedom, and allows, firstly, to calculate the zeroth order approximations of the amplitude of the oscillations in each circuit, and secondly, to establish their stability.

10.3 The Automatic Synchronization and Drive Phenomenon

Two years later, in October 1924, Van der Pol (1924)[6] began studying the forced oscillations of a triode (see *supra*) for which the equation takes the following form:

$$\ddot{v} - \alpha\left(1 - v^2\right)\dot{v} + \omega_0^2 v = \omega_1^2 E_s in\left(n\omega_1 t\right) \text{ with } \varepsilon = \frac{\alpha}{\omega_0} \ll 1 \tag{10.14}$$

He especially studied the case where the frequency of the receiving circuit ω_0 is very close to the frequency ω_1 of the sine-wave voltage generator, and searched, by applying his "slowly-varying amplitude method", a solution with the form:

$$v\left(t\right) = b_1\left(t\right) sin\left(\omega_1 t\right) + b_2\left(t\right) cos\left(\omega_2 t\right) \tag{10.15}$$

with

$$\frac{db_i}{dt} \ll \omega_1 b_i \text{ and } \frac{d^2 b_i}{dt^2} \ll \omega_1 \frac{db_i}{dt} \text{ for } i = 1, 2 \tag{10.16}$$

so $\ddot{b}_i$ can be neglected in the calculation. He then replaces (10.15) in (10.14) and only retains the terms of the fundamental frequency. Therefore, the nonlinear term $v^2\dot{v} = \frac{d}{dt}\left(v^3\right)$, which intervenes in many phenomena, is modified, taking into account the fact that v^3 is written:

$$v^3 = \frac{3}{4}\left(b_1^2 + b_2^2\right)\left(b_1\left(t\right) sin\left(\omega_1 t\right) + b_2\left(t\right) cos\left(\omega_2 t\right)\right)$$

He again obtains a system of two first order differential equations:

$$\begin{cases} 2\dot{b}_1 + zb_2 - \alpha b_1\left(1 - \dfrac{b^2}{a_0^2}\right) = 0 \\ 2\dot{b}_2 - zb_1 - \alpha b_2\left(1 - \dfrac{b^2}{a_0^2}\right) = \omega_1 E \end{cases} \quad \text{where} \quad \begin{cases} z = \dfrac{\omega_0^2 - \omega_1^2}{\omega_1} \approx 2\left(\omega_0 - \omega_1\right) \\ a_0^2 = \dfrac{\alpha}{\frac{3}{4}\gamma} \end{cases} \tag{10.17}$$

By writing $\gamma = 0$ in (10.17) the system becomes linear, and can therefore be integrated. This allows him to deduce b_1 and b_2.

$$\begin{cases} b_1\left(t\right) = e^{(\alpha/2)t}\left(C_1 sin\frac{z}{2}t + C_2 cos\frac{z}{2}t\right) - \dfrac{z\omega_1 E}{z^2 + \alpha^2} \\ b_2\left(t\right) = e^{(\alpha/2)t}\left(-C_1 cos\frac{z}{2}t + C_2 sin\frac{z}{2}t\right) - \dfrac{\alpha\omega_1 E}{z^2 + \alpha^2} \end{cases} \tag{10.18}$$

[6] The English version of this article was published three years later. See Van der Pol (1927a).

By then replacing $b_1(t)$ and $b_2(t)$ in the expression of the solution (10.15) he obtains:

$$v(t) = C_3 e^{(\alpha/2)t} \sin(\omega_0 t + \phi) + \frac{\omega_1 E}{\sqrt{(\omega_0 - \omega_1)^2 + \alpha^2}} \sin\left(\omega_1 t + \tan^{-1}\frac{2\alpha}{\omega_0 - \omega_1}\right) \tag{10.19}$$

which corresponds exactly to the solution to the equation (10.1), linear when $\gamma = 0$. Van der Pol thus demonstrated the efficacy of his methods, which allows on the one hand, to obtain more directly the various approximations of the amplitude of forced systems, and on the other hand, to construct a solution to the equation more easily than by using the Poincaré-Lindstedt or Fourier methods. Considering the general problem ($\gamma \neq 0$) he searches for the stationary solutions to the system (10.17), i.e. $\dot{b}_1 = \dot{b}_2 = 0$. The solution is then written:

$$z^2 + \alpha^2 \left(1 - \frac{b^2}{a_0^2}\right)^2 = \frac{\omega_1^2 E^2}{b^2} \tag{10.20}$$

Van der Pol deduces that the equality (10.20) can be interpreted as corresponding to the case where the forced oscillations are the only ones present since this equation shows that the functions b are from now independent from the time. The free oscillations represented by the periodic parts of b are entirely absent. This leads to the establishment of *a* stability condition for the forced oscillations:

$$\begin{cases} b^2 > \frac{1}{2} a_0^2 \\ \alpha^2 \left(1 - \frac{3b^2}{a_0^2}\right)\left(1 - \frac{b^2}{a_0^2}\right) + z^2 > 0 \end{cases}$$

By replacing: $z \to \alpha x$, $y \to \frac{b^2}{a_0^2}$ and $\frac{\omega_1^2 E^2}{a_0^2} \to \alpha^2 E$, the equation (10.20) is written:

$$x^2 + (1 - y)^2 = \frac{E}{y} \tag{10.21}$$

The solution to the equation (10.21) then takes the form of "resonance curves" drawn for different values of E (see Fig. 10.4). The stable and unstable parts are respectively represented as full lines and dotted lines.

The stable and unstable areas are limited by the horizontal line represented by the equation: $y = 1\ 2$ and by the ellipsis represented by the equation: $x^2 + (1 - -3y)(1 - -y) = 0$ (represented in red on Fig. 10.4).

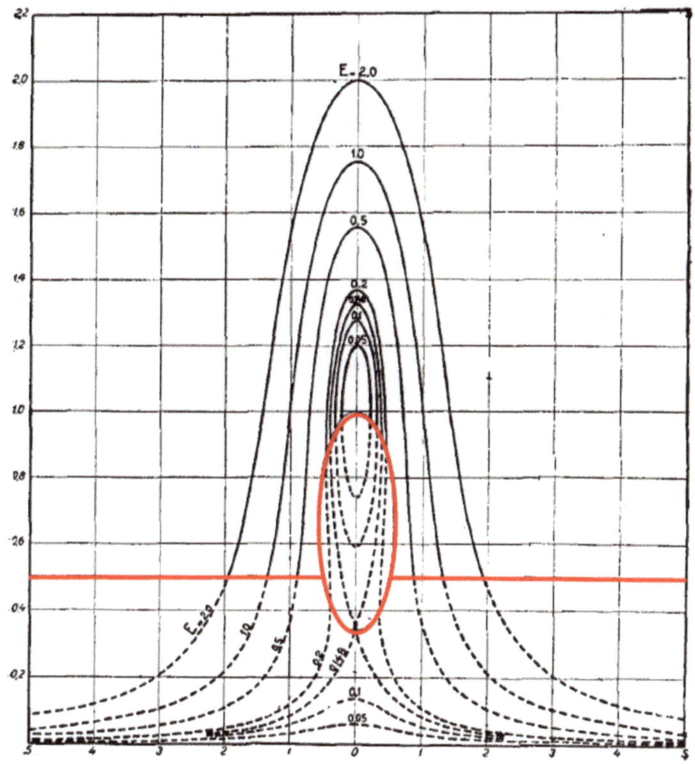

Fig. 10.4 Resonance curves, by Van der Pol (1927a, 73)

Van der Pol (1927a, 75) then defines the field included between this ellipsis and the resonance curve of equation (10.21) as a "silent zone"[7] (see Fig. 10.5 where the "silent area" corresponding to the value $E = 2$ is represented).

The "silent area" corresponds to the field in which the free oscillation is canceled and where we cannot hear a frequency beat $(\omega - \omega_0)/2\pi$ between the signal imposed by the voltage generator of sinusoidal *e.m.f.* and the free oscillation. Van der Pol (1927a, 74) then demonstrated that when the signals each have a strong amplitude, the "silent area" is determined by the relation (10.21) and the line $y = 1/2$. By substituting this value in (10.21) and restoring the dimensions of the variables, he obtains:

$$(\omega_0 - \omega_1)^2 + \frac{1}{16}\alpha_2 = \frac{\omega_1^2 E^2}{2a_0^2} \tag{10.22}$$

[7]This terminology was most likely introduced by Appleton (1922, 232).

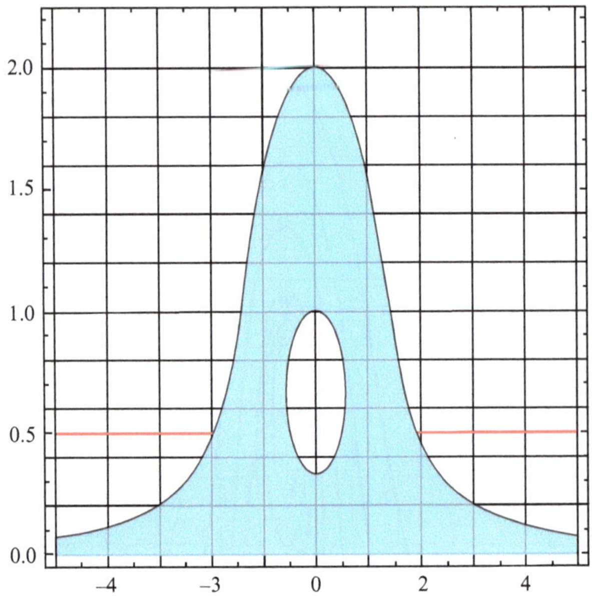

Fig. 10.5 Silent area for the value $E = 2$ (drawn with Mathematica 7)

Now, in the case of signals with a strong amplitude $\frac{\alpha_2}{16} \ll (\omega_0 - \omega_1)^2$, which yields:

$$\frac{\omega_0 - \omega_1}{\omega_1} = \pm \frac{E}{\sqrt{2}a_0} \tag{10.23}$$

Van der Pol (1927a, 75) deduces[8] that when the difference in frequency of the two signals is inferior to this value, an *automatic synchronization* phenomenon occurs and the two circuits oscillate with the same frequency.

By then going to the edge of the silent area, Van der Pol (1927a, 76), determines the precise point where the free oscillation originates, and demonstrates that its appearance frequency is not equal to, but rather superior to the value: $\omega_1 - \sqrt{(\omega_0 - \omega_1)^2 - \alpha^2/16}$. This led him to evidence the frequency drive or tailing phenomenon, which he defines thus:

> Hence the free frequency undergoes a correction in the direction of the forced frequency, giving the impression as if the free frequency were being attracted by the forced frequency. (Van der Pol 1927a, 76)

Van der Pol then completes his study by searching for solutions to the equation (10.1), outside the resonance area, with the form of a linear combination of forced and free oscillations, with the following form:

[8]He nevertheless recalls that this result had been confirmed by Appleton's experiments (1922).

$$v(t) = a(t)\, sin\,(\omega_0 t + s) + b(t)\, sin\,(\omega_1 t + \lambda) \tag{10.24}$$

By substituting (10.23) in (10.1) and equating with zero the coefficients of the terms containing $sin\,(\omega_0 t)$, $cos\,(\omega_0 t)$, $sin\,(\omega_1 t)$ and $cos\,(\omega_0 t)$, he obtains a new set of conditions:

$$\begin{cases} \omega_0 - \omega_0 = 0 \\ a\left(1 - \dfrac{a^2 + 2b^2}{a_0^2}\right) = 0 \\ z^2 + \alpha^2\left(1 - \dfrac{b^2 + 2a^2}{a_0^2}\right)^2 = \dfrac{\omega^2 E^2}{a_0^2} \end{cases} \tag{10.25}$$

The second equation of (10.25) is easily solved and leads to the equations:

$$a = 0 \tag{10.26}$$

$$1 - \frac{a^2 + 2b^2}{a_0^2} = 0 \tag{10.27}$$

The first represents the canceling of the free oscillation by the previously considered forced oscillation, and in this case the third equation (10.25) is reduced to (10.20). The second, which is written: $a^2 = a_0^2 - 2b^2$ corresponds from (10.20) to the solution to the equation (10.1) containing free oscillations. Consequently, the amplitude of the forced oscillations is given by:

$$z^2 + \alpha^2\left(1 - \frac{3b^2}{a_0^2}\right)^2 = \frac{\omega_1^2 E^2}{a_0^2} \tag{10.28}$$

This expression, which differs from the previous (10.20), of factor three, leads Van der Pol to conclude that the apparent resistance of the system in forced oscillation is increased due to the presence of the free oscillations. Moreover, he demonstrated from (10.27) that when the amplitude of the forced oscillation reaches the value $b^2 = a_0^2/2$, the free oscillation is canceled by the forced oscillation.

By replacing: $z \rightarrow \alpha x$, $y \rightarrow \dfrac{a^2 + b^2}{a_0^2}$ and $\dfrac{\omega_1^2 E^2}{a_0^2} \rightarrow \alpha^2 E$, the equation (10.28) is written:

$$x^2 + (-2 + 3y)^2 = \frac{E}{1 - y} \tag{10.29}$$

He then draws the solution to this equation for various values of E. The graphical representation is more complicated, since in order to obtain it, we must superimpose the solutions to (10.21) represented in the field defined by $-E \leq x \leq E$ and $y \geq 1/2$ and the solutions to (10.29) represented in the complementary. On Fig. 10.6, the

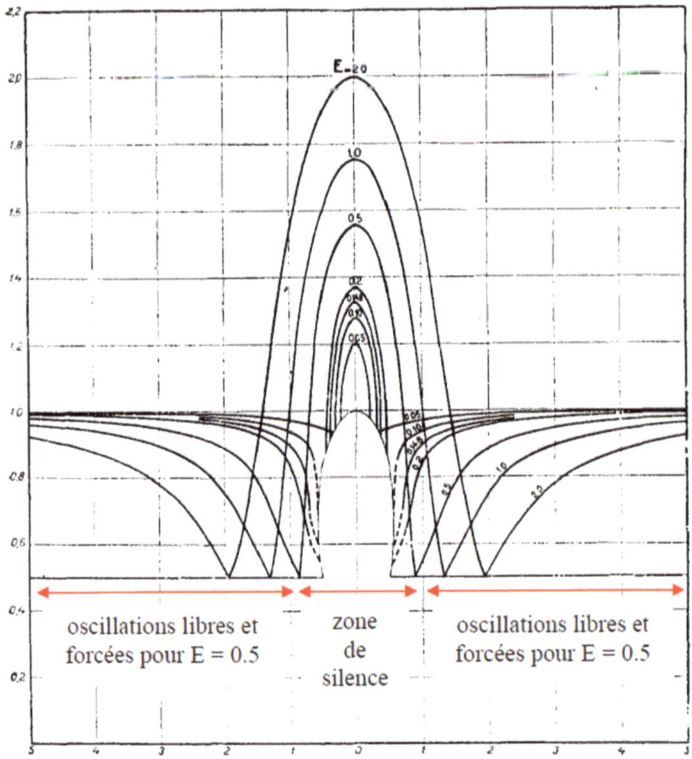

Fig. 10.6 Resonance curves, by Van der Pol (1927a, 79)

curve solution to (10.21) for the value $E = 2$, in the interval $[-2, 2]$ and the solution to (10.28) were drawn in the complementary (Fig. 10.7).[9]

In 1926, Van der Pol then evidenced, in many publications,[10] the oscillation relaxation phenomenon occurring for values of the parameter $\varepsilon \gg 1$.

10.4 The Frequency Demultiplication Phenomenon

The next year, Van der Pol and Van der Mark (1927a) published an article entitled "Frequency Demultiplication", in which he again studied the forced oscillations of a triode, but in the field of relaxation oscillations, i.e. for $\varepsilon \gg 1$. The equation representing the oscillations keeps the same form as previously:

$$\ddot{v} - \alpha\left(1 - v^2\right)\dot{v} + \omega_0^2 v = \omega_1^2 E_s in\left(n\omega_1 t\right) \text{ with } \varepsilon = \frac{\alpha}{\omega_0} \gg 1 \tag{10.30}$$

[9]For more details, see Lawden (1954, 350–354).

[10]See Van der Pol (1926a,b,c,d). See *supra* Parts I and II.

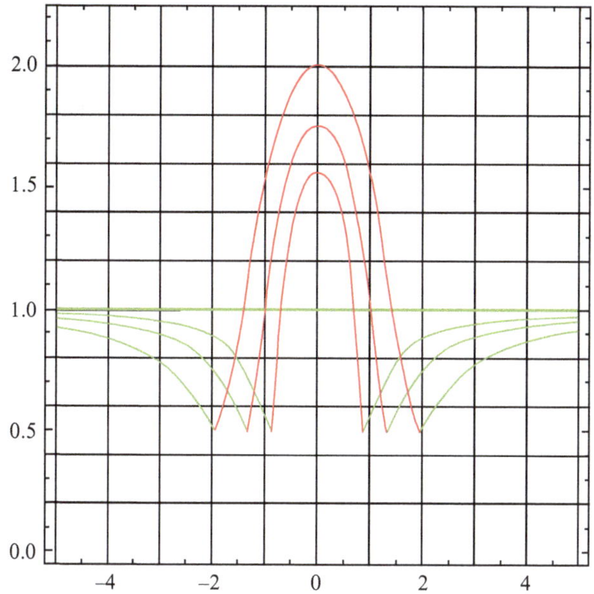

Fig. 10.7 Resonance curves (10.21) and (10.28) drawn with Mathematica 7

Van der Pol and Van der Mark (1927) then explain that the *automatic synchronization* phenomenon, observed in the case of the forced oscillations of a triode, can also occur for a range of the parameter corresponding to the relaxation oscillations ($\varepsilon \gg 1$), but in a much wider frequency field. They also explain that the resonance phenomenon is almost inexistent in forced relaxation oscillations, and that consequently, the sinusoidal *e.m.f.* inducing the forcing influences the period (or frequency) of the oscillations more than it does their amplitude, and add:

> (...) it is found that the system is only capable of oscillating with *discrete frequencies*, these being determined by *whole submultiples of the applied frequency*. (Van der Pol and Van der Mark 1927, 363)

Indeed, they show through experiments that when the frequency ω_1 of the generator of the sinusoidal *e.m.f.* increases inside the automatic synchronization area, the frequency of the relaxation oscillations suddenly "jumps" end synchronizes with a sub harmonic value of the frequency equal to $\frac{1}{2}\omega_1$. If the frequency of the forcing increases again, the one for the relaxation oscillations synchronizes on the following subharmonic: $\frac{1}{3}\omega_1$, and so on. They named this phenomenon *frequency demultiplication*.[11] A few years later, Van der Pol (1934, 1081) gave it an analytical illustration, by explaining that if, in the equation (10.30), we write $n = 3$ and $\omega_0 = \omega_1$:

[11]In literature it also appears with the terminology *subharmonic oscillations*.

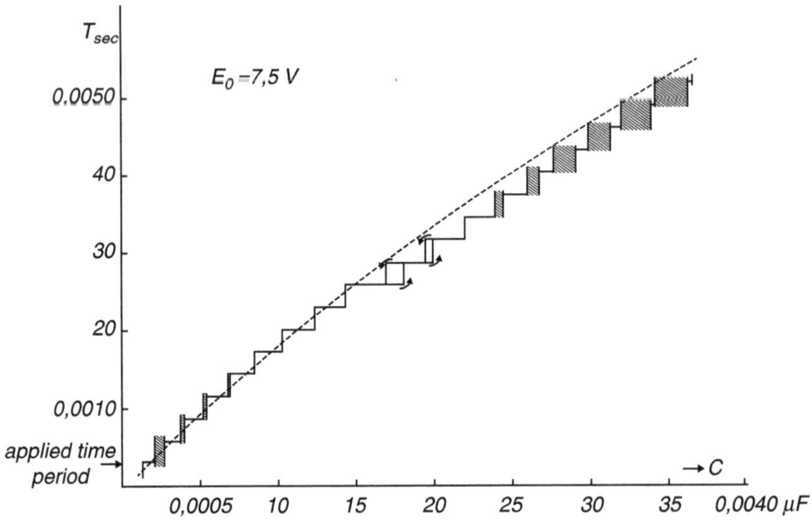

Fig. 10.8 Representation of the frequency demultiplication frequency, by Van der Pol and Van der Mark (1927a, 364)

$$\ddot{v} - \alpha\left(1 - v^2\right)\dot{v} + \omega_0^2 v = \omega_0^2 E_s in\left(3\omega_0 t\right) \text{ with } \varepsilon = \frac{\alpha}{\omega_0} \gg 1 \tag{10.31}$$

We can then demonstrate that it admits $v\,(t) = 2cos\,(\omega_0 t)$ as an exact solution for $E = 2\varepsilon$. It therefore appears that the frequency of this solution is three times the forcing's.

In their article, Van der Pol and Van der Mark (1927a) had proposed, in order to evidence the frequency demultiplication phenomenon, the following construction (see Fig. 10.8) on which we can see a "jump" of the period for each increase in the value of the capacitor's capacitance.

In order to evidence this frequency demultiplication phenomenon, Van der Pol and Van der Mark used a phone, not because, as Gleick (1989, 72) naively claimed, they did not have an oscilloscope,[12] but in order to allow a direct and quick detection of a phenomenon, as reminded by Van der Pol in 1920:

[12]The use of the phone was indeed not to make up for a lack of galvanometric oscillometers or cathode-ray oscilloscopes (see *supra* Part I) the use of which was already widespread at the start of the twentieth century. As early as 1920 Van der Pol (vol. 1, 173 1960) wrote in his PhD thesis:

(…) hebben wij bij onze verdere proeven een Duddell-thermogalvanometer aangewend.
(…) we also used a Duddel thermo-galvanometer.

The next year Appleton and Van der Pol (1921, 201) clarified:

(…) the current-time records being obtained with a Duddell Oscillograph.

> (...) we determined with the aid of a telephone connected to the induction solenoid through a three-stage amplifier, the point on the hysteresis curve where the discontinuities for the first time occur and the spot where they disappear again. (Van der Pol 1960, vol. 1, 248)

They then described the phenomenon they heard in the receiver:

> Often an irregular noise is heard in the telephone receivers before the frequency jumps to the next lower value. However, this is a subsidiary phenomenon, the main effect being the regular frequency multiplication. (Van der Pol and Van der Mark 1927a, 364)

The irregular noise was actually the sound manifestation of the transition which was taking place. Indeed, as the frequency varied, the solution to the differential equation (10.1), which had been until now represented by a limit cycle, i.e. by a periodic attractor, would draw a "strange attractor" transcribing the chaotic behavior of the solution. Van der Pol seemed to have reached the limits of deterministic physics with how far he went in the exploration of nonlinear and non-autonomous systems. He "flirted", as Mary Lucy Cartwright and John Edensor Littlewood did twenty years later (1945, 1947, 1948), with the first signs of chaos, when they found "strange" the behavior of the solution to the differential equation (10.1) for specific values of the parameters.

Thus, the analysis of this part of Van der Pol's work allowed the demonstration of the slowly varying amplitudes method that he developed in the early 1920s, and appears more suited to the study of forced or coupled nonlinear systems in the case of sinusoidal oscillations ($\varepsilon \ll 1$), as in the case of relaxation oscillations ($\varepsilon \gg 1$) than the Fourier and Poincaré-Lindstedt methods. Moreover, this method's application to the study of these systems also led to the evidencing and characterizing of hysteresis, mode competition, automatic synchronization and frequency demultiplication phenomena. However, it will be established in the following that Van der Pol's method holds its mathematical founding principles in the astronomer Pierre Fatou's works.

Chapter 11
The Krylov-Bogolyubov Method: *Towards a Nonlinear Mechanics*

During the 1930s, Russian mathematicians Nikolai Krylov (1879–1955) and Nikolai Bogolyubov (1909–1992), inscribed the slowly varying amplitudes method unveiled a few years earlier by Van der Pol in the framework of a theory they named "Nonlinear Mechanics". They also studied in detail the various phenomena evidenced by Van der Pol.

Thus, while Van der Pol (1920, 1922, 1927a) withheld this part of his research (except for the frequency demultiplication phenomenon which he exemplified with the model of the heart[1]) during his many lectures in France (see *supra* Part II), it was through the notes that Krylov and Bogolyubov (1932a,b,c, 1933, 1934e, 1935a,b, 1936a,b) presented to the *C.R.A.S.*, that it will be spread in the French scientific community. During this time, Krylov and Bogolyubov (1934a,b,c,d) formalized their theory in a series of reports published in Russian, but which always included at least a summary in French (the second report is entirely in French), there they demonstrated what they called "New methods of nonlinear mechanics" (Krylov and Bogolyubov 1936a,b, 509). In 1937, the same year the "Theory of Oscillators" written by Andronov and Khaikin (1937) was published, Krylov and Bogolyubov's "Introduction to Nonlinear Mechanics" (1937) was also published in Russian. This monograph, which was the synthesis of their work in this field, was partially translated into English in 1943 by Solomon Lefschetz.[2] Nevertheless, neither the language barrier, nor the distance, refrained the spreading of these researches. Indeed, aside from the fact that Krylov and Bogolyubov published in *C.R.A.S.*, as they did in the *R.G.S.A.*, the *Annales de la Faculté des Sciences de Toulouse*, the *Mémorial des Sciences Mathématiques*, and the *Onde Électrique*,[3] they were invited to present their works at the first *Conférence Internationale de Nonlinéaire in Paris*

[1]See Van der Pol and Van der Mark (1928a) and Van der Pol (1930, 1938a).

[2]See Krylov and Bogolyubov (1937).

[3]See Krylov (1925, 1927, 1931) and Krylov and Bogolyubov (1936b).

J.-M. Ginoux, *History of Nonlinear Oscillations Theory in France (1880–1940)*,
Archimedes 49, DOI 10.1007/978-3-319-55239-2_11

(see *supra* Part II) in 1933, they also presented their "New Methods" at the Henri Poincaré Institute over four lectures in 1935.

Their main interlocutor at the time was Arnaud Denjoy, who invited them and summarized their work to the *R.G.S.A.* (Denjoy 1935). The research carried out at the Henri Poincaré Institute[4] allowed the finding of part of the correspondence between Misters Krylov and Denjoy. Moreover, the report n° 8 written by Krylov and Bogolyubov (1934d) held Krylov's handwritten mention:

> À Monsieur le Professeur A. Denjoy, cordial et respectueux souvenir.
> ("To Professor A. Denjoy, we convey the cordial best wishes.")

and the subtitle: *pour le résumé en français V. p 100–110*. Therefore, these various publications will provide precious information on the way these "New methods of Nonlinear mechanics" spread in France in the 1930s.

11.1 Slowly Varying Amplitudes and Phase Method

In Eastern Europe, the followers of the Kiev School of thought as well as those of the Mandelstam-Papalexi School were studying systems which they called "quasi-linear", and which are represented by Van der Pol's equation (3.8) with the following form:

$$\ddot{x} + \varepsilon f(x, \dot{x}) + \omega^2 x = 0 \tag{11.1}$$

This autonomous nonlinear first order differential equation is a first generalization[5] of Van der Pol's prototype equation (1926d, 979), and is named after the fact that is becomes linear when $\mu \ll 1$, i.e. $\varepsilon \ll 1$. Indeed, in this case the solution is: $x(t) = a_sin(\omega t + \varphi)$. Therefore, as Van der Pol did (1922, 180) and Krylov and Bogolyubov (1933, 18, 1937, 8, 44) searched for a solution with the form:

$$x(t) = a(t)\, sin(\omega t + \varphi) \tag{11.2}$$

where the amplitude $a(t)$ slowly varies with time. But Krylov and Bogolyubov (1933, 18, 1937, 10, 45) furthermore that the phase $\varphi(t)$ also slowly varies with the time. Thus, by deriving (11.2) depending on the time, it yields:

$$\dot{x}(t) = a(t)\, \omega cos(\omega t + \varphi) + \dot{a}(t)\, sin(\omega t + \varphi) + a(t)\, \dot{\varphi}(t)\, cos(\omega t + \varphi) \tag{11.3}$$

So, since $\dot{a}(t) \ll \omega_a(t)$, $\ddot{a}(t) \ll 1$ and $\dot{\varphi}(t) \ll \omega_\varphi(t)$, $\ddot{\varphi}(t) \ll 1$, (11.3) implies on the one hand that:

[4] With the kind assistance of Madame Brigitte Yvon-Deyme and her collaborators.

[5] This is the equation suggested by Alfred Liénard (1931, 177).

$$\dot{a}(t)\,sin\,(\omega t+\varphi)+a(t)\,\dot{\varphi}(t)\,cos\,(\omega t+\varphi)=0 \tag{11.4}$$

and on the other that:

$$\dot{x}(t)=a(t)\,\omega cos\,(\omega t+\varphi) \tag{11.5}$$

By deriving (11.5) depending on time, the equation (11.1) is written, by taking (11.4) into account:

$$\begin{aligned}&\dot{a}(t)\,\omega cos\,(\omega t+\varphi)-a(t)\,\omega\dot{\varphi}(t)\,cos\,(\omega t+\varphi)\\&+\mu f\,[a(t)\,sin\,(\omega t+\varphi),a(t)\,\omega cos\,(\omega t+\varphi)]=0\end{aligned} \tag{11.6}$$

By combining with (11.5) Krylov and Bogolyubov (1937, 11, 45–46) obtain:

$$\begin{cases}\dot{a}(t)=-\dfrac{\mu}{\omega}f\,[a(t)\,sin\,(\omega t+\varphi),a(t)\,\omega cos\,(\omega t+\varphi)]\,cos\,(\omega t+\varphi)\\ \dot{\varphi}(t)=\dfrac{\mu}{a(t)\,\omega}f\,[a(t)\,sin\,(\omega t+\varphi),a(t)\,\omega cos\,(\omega t+\varphi)]\,sin\,(\omega t+\varphi)\end{cases} \tag{11.7}$$

It therefore appears that the seconds members of the equations (11.7) contain periodic functions of the period $T = 2\pi\ \omega$. Krylov and Bogolyubov (1937, 11, 46–47) expand them with a Fourier series. This yields, by writing $\psi = \omega t + \varphi$:

$$\begin{cases}f\,[asin\,(\psi),a\omega cos\,(\psi)]\,cos\,(\psi)=K_0(a)+\sum\limits_{n=1}^{\infty}[K_n(a)\,cos\,(n\psi)+L_n(a)\,sin\,(n\psi)]\\ f\,[asin\,(\psi),a\omega cos\,(\psi)]\,sin\,(\psi)=P_0(a)+\sum\limits_{n=1}^{\infty}[P_n(a)\,cos\,(n\psi)+Q_n(a)\,sin\,(n\psi)]\end{cases} \tag{11.8}$$

where the coefficients K_n, L_n, P_n and Q_n are Fourier's coefficients. With, for K_0 and P_0:

$$\begin{cases}K_0(a)=\dfrac{1}{2\pi}\int_0^{2\pi}f\,[asin\,(\psi),a\omega cos\,(\psi)]\,cos\,(\psi)\,d\psi\\ P_0(a)=\dfrac{1}{2\pi}\int_0^{2\pi}f\,[asin\,(\psi),a\omega cos\,(\psi)]\,sin\,(\psi)\,d\psi\end{cases} \tag{11.9}$$

The system (11.7) is written by taking (11.8) and (11.9) into account:

$$\begin{cases}\dot{a}(t)=-\dfrac{\mu}{\omega}K_0(a)-\dfrac{\mu}{\omega}\sum\limits_{n=1}^{\infty}[K_n(a)\,cos\,(n\psi)+L_n(a)\,sin\,(n\psi)]\\ \dot{\varphi}(t)=\dfrac{\mu}{a\omega}P_0(a)+\dfrac{\mu}{a\omega}\sum\limits_{n=1}^{\infty}[P_n(a)\,cos\,(n\psi)+Q_n(a)\,sin\,(n\psi)]\end{cases} \tag{11.10}$$

Krylov and Bogolyubov (1937, 12, 48) carried out the integration of the system (11.10) over a period T by supposing the amplitude $a(t)$ and the phase $\varphi(t)$

are constant. This operation consists in an "averaging" of the second member of the system (11.10) which is then written:

$$\begin{cases} \dfrac{a(t+T)-a(t)}{T} = -\dfrac{\mu}{\omega}K_0[a(t)] \\ \dfrac{\varphi(t+T)-\varphi(t)}{T} = \dfrac{\mu}{a\omega}P_0[a(t)] \end{cases} \tag{11.11}$$

The slowly varying amplitude and phase led Krylov and Bogolyubov (1937, 12, 48) to consider that the duration of the phenomenon is large in comparison to T, and to consider T as an infinitesimal duration ΔT. By taking the limit of the left members of the system (11.11) they obtain "the equations of the first approximation" (Krylov and Bogolyubov 1937, 12, 48):

$$\begin{cases} \dot{a}(t) = -\dfrac{\mu}{\omega}K_0[a(t)] \\ \dot{\varphi}(t) = \dfrac{\mu}{a\omega}P_0[a(t)] \end{cases} \tag{11.12}$$

By writing again $\psi = \omega t + \varphi$ and taking into account the system (11.12), they write the derivative with respect to the time ψ:

$$\frac{d\psi}{dt} = \omega + \frac{\mu}{a\omega}P_0[a(t)] \tag{11.13}$$

By replacing K_0 and P_0 by the expressions given by (11.9), they obtain the following system:

$$\begin{cases} \dot{a}(t) = -\dfrac{\mu}{\omega}K_0[a(t)] = -\dfrac{\mu}{\omega}\dfrac{1}{2\pi}\int_0^{2\pi} f[a\sin(\psi), a\omega\cos(\psi)]\cos(\psi)\,d\psi \\ \dot{\psi}(t) = \omega + \dfrac{\mu}{a\omega}P_0[a(t)] = \omega + \dfrac{\mu}{a\omega}\dfrac{1}{2\pi}\int_0^{2\pi} f[a\sin(\psi), a\omega\cos(\psi)]\sin(\psi)\,d\psi \end{cases} \tag{11.14}$$

Consequently, the first approximation of the solution to (11.1) is written:

$$x(t) = a(t)\sin(\psi) \tag{11.15}$$

It must be noted that if it is necessary to have $\mu \ll 1$ in the Krylov-Bogolyubov method, the reasons are dissimilar to Poincaré-Lindstedt's. In the Poincaré-Lindstedt method, the smallness μ ensured the convergence of the series. In Krylov-Bogolyubov's method, this smallness allows the treating of the trigonometric series intervening in the expressions (11.10) "approximately" as Fourier series. They replaced the second member of these equations by their averages, and thus transform them into simpler expressions that Mandel'shtam et al. (1935, 90) called "Van der Pol equations or "truncated equations"".

In order to illustrate the effectiveness of their method, Krylov and Bogolyubov (1937, 14, 52) considered the equation of the simple pendulum in the case of small oscillations, and Van der Pol's equations for the oscillations of a triode. By comparing Van der Pol equation (3.8) to the general form (11.1), they deduced that $\omega = 1, f(x, \dot{x}) = -\left(1 - x^2\right)\dot{x}$. They obtain from (11.9) in the case $\mu \ll 1$ i.e. $\varepsilon \ll 1$:

$$\begin{cases} K_0\left[a(t)\right] = -\dfrac{a}{2}\left(1 - \dfrac{a^2}{4}\right) \\ P_0\left[a(t)\right] = 0 \end{cases} \tag{11.16}$$

They then deduct from (11.14)

$$\begin{cases} \dot{a}(t) = \mu\dfrac{a}{2}\left(1 - \dfrac{a^2}{4}\right) \\ \dot{\psi}(t) = 1 \end{cases} \tag{11.17}$$

The integration of this system leads to the solution (10.8) obtained by Appleton and Van der Pol (1922) (see *supra*). Later on, Krylov and Bogolyubov (1937) extended their method to higher-order approximations and to cases of forced or coupled systems. The complexity of the calculations then increases as it was seen previously with the approximation order and nature of the system. Therefore, although the Krylov-Bogolyubov method seems much more suited than the Poincaré-Lindstedt method to the study of radio engineering devices, it is still based on the same hypothesis of the existence of a small parameter ($\mu \ll 1$ i.e. $\varepsilon \ll 1$) and therefore cannot claim that it challenges it, as it was the case many times, and especially in the periodic publication *Onde Électrique* at the end of the 1930s:

> The critical statements apply mainly to theories based on the use of rigorous Poincaré-Lyapunov methods. These famous methods hold a very specific importance in mathematical analysis, but are however hardly manageable for a practitioner, as they turn the study of oscillators, which have a somewhat complicated diagram, almost inaccessible from a practical point of view. Moreover, these methods, at least in their present state, hardly allow for more than the study of simply-periodic states, which means that the quasi-periodic states, of which we have established the existence, are apparently excluded from their application field. (Krylov and Bogolyubov 1936b, 509)

While it indeed appears that Poincaré-Lindstedt expansions are, in the case of radio engineering, "hardly manageable for a practitioner", they nevertheless stay entirely valid, and applicable to systems with one degree of freedom possessing a small parameter, as is the case for Van der Pol's equation (3.8). However, in the case of systems with several degrees of freedom, such as coupled or forced oscillators, the Poincaré-Lindstedt method must step aside for Krylov-Bogolyubov's. Nevertheless, it will be established (see *infra*) that the followers of the Mandelstam-Papalexi School continued to use the Poincaré-Lindstedt method for this case.

11.2 The First Note in the *C.R.A.S.* of 1932: *The Problem of Nonlinear Mechanics*

In a note in the *C.R.A.S.* entitled "Quelques exemples d'oscillations linéaires" ("Instances of nonlinear oscillations"), Krylov and Bogolyubov (1932a) present their method's principle. In their introduction, they recall the importance of *Nonlinear mechanics*:

> Nowadays, the problem with linear oscillations, in linear circuits, is that they hold little mathematical difficulties. We believe that this is not the case with nonlinear oscillations (which occur in nonlinear circuits), in spite of recent research. Nonlinear oscillations intervene significantly, even sometimes sovereign, in various applications, amongst others, in biology (Mr. V. Volterra's and his continuators' work), chemistry, physics, astronomy (for instance the Cepheids), and evidence indicates that the problem of Nonlinear Mechanics is on the agenda. (Krylov and Bogolyubov 1932a, 957)

It is surprising to see that Krylov and Bogolyubov suggests, in order to illustrate nonlinear oscillations, the same examples as the ones used by Andronov two years earlier (1929a) without referencing them. Amongst others, we can again find the examples of, the problem of Cepheids in astronomy, Volterra's work in biology, with the same ambiguity as seen previously (see *supra* Parts I & II). Concerning "continuators", it seems likely that Krylov and Bogolyubov were hinting at Kostitzin's (1930) and Gause and Alpatov's research (1931). They then studied the oscillations of a triode subjected to a forcing, which can be represented by the Van der Pol[6] equation (10.1) (see *supra*) or by a simplified form (9.21):

$$\ddot{x} - \varepsilon f(x, \dot{x}) + \omega_0^2 x = E sin(\alpha t) \tag{11.18}$$

in which we write: $\mu = \varepsilon$ and $F(t) = E sin(\alpha t)$. By applying their method, Krylov and Bogolyubov demonstrate that:

> Stationary oscillations are composed of natural oscillations, forced oscillations and, contrary to the case of linear oscillations, combined oscillations as well (...) (Krylov and Bogolyubov 1932a, 958)

Then, they deduce a second order approximation of the solution to the equation[7] ε^2 (11.18):

$$x(t) = \frac{E}{\omega_0^2 - \alpha^2} sin(\alpha t) + B sin(\omega_0 t + \varphi) + O(\varepsilon) \tag{11.19}$$

It must be noted that the Poincaré-Lindstedt method would have led to the same expression, but with much longer developments (see *infra*). Krylov and Bogolyubov also provided a second order approximation of the angular frequency:

---

[6]He did not refer to his works.

[7]This expression is only presented at the first order in order to simplify its reading and understanding.

$$\omega^2 = \omega_0^2 - \frac{\varepsilon^2}{2\omega_0^2 B^2}\chi \tag{11.20}$$

where χ is a coefficient depending on the function $f(x, \dot{x}) = -\omega_0 \dot{F}(x)$ which represents the differential in relation to the time of the oscillation characteristic of the triode $F(x)$. They then explained that:

> The frequency, called natural, ω_0 depends not only on the constant, the circuit, but also the amplitude of the external force and the latter dependence is proportional, for small values of the parameter ε, to ε^2, such that for this natural frequency, the smaller the value of the parameter, the more stable it is. (Krylov and Bogolyubov 1932a, 959)

Lastly, they recall that for some values of the angular frequency, phenomena varying in nature occur:

> Aside from the resonance $\omega_0 \approx \alpha$, there exist other resonances, the demultiplication resonances for ω_0, neighboring n/α, for instance $\alpha/2, \alpha/3, 2\alpha/2, \ldots$ (Krylov and Bogolyubov 1932a, 959)

11.3 The Second Note in the *C.R.A.S.* of 1932: *On the Drive Phenomenon*

This second note's aim is, on the one hand, to evidence an existence condition for quasi-periodic oscillations in Van der Pol's forced equation (11.18) and on the other, to define the value for which the drive phenomenon occurs. Krylov and Bogolyubov (1932b, 1064) then assume that the oscillation characteristic has a cubic form:

$$F(x) = x + \frac{1}{2}Ax^2 - \frac{\gamma^2 x^3}{3} \quad \text{hence} \quad f(x, \dot{x}) = -\dot{F}(x) = -\left(1 + Ax - \gamma^2 x^2\right)\dot{x}$$

The equation (11.18) is written for the case where $A = 0$ and $\gamma = 1$

$$\ddot{x} - \varepsilon\left(1 - x^2\right)\dot{x} + \omega_0^2 x = E_s in\left(\alpha t\right) \tag{11.21}$$

They deduce from this that the equation (11.21) possesses quasi-periodic solutions when the following condition is met:

$$\left(\frac{E\alpha\gamma}{\omega_0^2 - \alpha^2}\right)^2 < 2 \tag{11.22}$$

A quasi-periodic solution of (K_{22}) with two frequencies is represented on Fig. 11.1.

Krylov and Bogolyubov (1932b, 1064) then explain that if the inequality is reversed, the solution is periodic with the frequency of the external force.

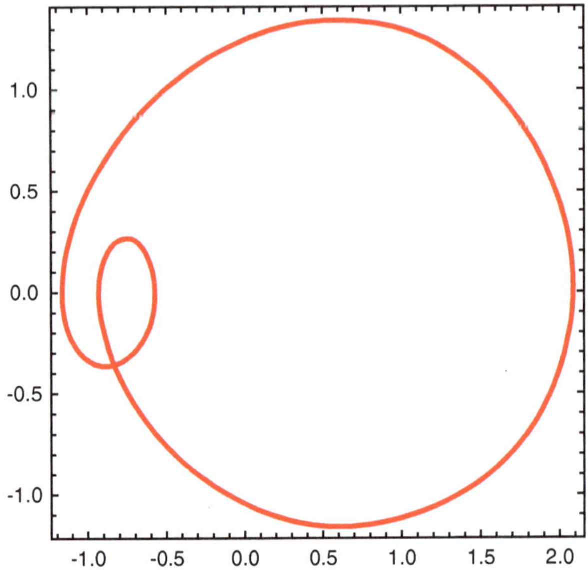

Fig. 11.1 Quasi-periodic solution of (11.21), for $E = 1$, $\omega_0 = 1, \alpha = 0.5$ and $\varepsilon = 0.05$

> (...) this inequality is also found with Mr. Andronow (*Journal of theoretical and Applied Physics* 7, IV), who obtained it by applying H. Poincaré's method (...) (Krylov and Bogolyubov 1932b, 1064)

Here, Krylov and Bogolyubov are referring to Andronov and Witt's article (1930c) published in Russian in the *Journal of Technical Physics of the USSR* (and not in the *Journal of theoretical and Applied Physics*), in which they used the Poincaré-Lindstedt methods to determine the condition (11.22). Therefore they're used in order to study the systems with several degrees of freedom. However Krylov and Bogolyubov otherwise:

> In conclusion, the results of this Note, related to the properties of the main resonance, can also be obtained by applying the research methods of periodic solutions. Yet this only occurs because here, we chose the oscillating system with only one degree of freedom and the external force with only one frequency. If these conditions are not met, the above methods cease to be applicable while ours will adapt without difficulty. (Krylov and Bogolyubov 1932b, 1066)

Indeed, they thus uncovered the limits of the application field of the Poincaré-Lindstedt methods.

11.4 The Third Note in the *C.R.A.S.* of 1932: *On the Demultiplication Phenomenon*

The aim of this third note is to study the frequency demultiplication phenomenon also evidenced by Van der Pol (1927a) (see *supra*). Before moving on to its content, it is significant that Krylov and Bogolyubov could enjoy such a place in the *C.R.A.S.* in order to spread their work. We ought to recall that it was mathematician Jacques Hadamard, who was at the time correspondent with the Russian Academy of Science, and foreign member of the Academy of Science of the U.S.S.R.[8] who presented Krylov and Bogolyubov's note, as well as Andronov and Witt's two years before (see *supra* Part II). Moreover, still according to Maz'ya and Shaposhnikova (2005, 181–183), Jacques Hadamard was invited to the Soviet Union to participate in the first Congress of U.S.S.R. Mathematicians in Kharkiv in 1930. Amongst others, he went along with Arnaud Denjoy and Élie Cartan, and then visited Kiev, where he was Nikolai Krylov's guest (1879–1955). The latter attended Hadamard's lectures in Paris from 1907 to 1908, and also participated in Hadamard's seminar in 1926 and 1927. This privileged relation between the two of them, from the same generation,[9] could explain the somewhat large proportion[10] of notes published by Krylov in the *C.R.A.S.* In this note, Krylov and Bogolyubov (1932c) study the resonance of the demultiplication by choosing $\omega_0 = \alpha\ 2$. They then demonstrate, by using Van der Pol's cubic as the oscillation characteristic, that:

> (...) for sufficiently strong external forces, the demultiplication phenomenon does not occur. (Krylov and Bogolyubov 1932c, 1120)

They also analyzed the stability of the demultiplication state, and established the conditions under which the beats disappear. Lastly, barely a year after Le Corbeiller (1933b, 328) called for the creation of a *Theory of Oscillations* (see *supra* Part II), they did likewise:

> The methods we used to reach the results included in these Notes are applicable and efficient, as we insured during our research, to many other applications (for instance the oscillations of synchronous machines the longitudinal stability of planes, etc.) and Chapters of modern mathematical (Quantum mechanics), and can open the way, in our opinion, to the creation of general Nonlinear mechanics. (Krylov and Bogolyubov 1932c, 1122)

This drive to have a new discipline emerge was unfortunately not followed by concrete action.

---

[8]See Maz'ya and Shaposhnikova (1998, 181).

[9]Aleksandr' Andronov (1901–1952), Aleksandr' Witt (1902–1938) and Nikolaï Bogolyubov (1909–1992) were part of the "young generation" of Russian mathematicians.

[10]In 1932 only, there were no less than three notes of his in the *C.R.A.S.*

11.5 The Article in the *R.G.S.A.* of 1933: *Towards a Nonlinear Mechanics*

This article was published exactly two weeks before the first *Conférence Internationale de Non linéaire* held in Paris, in which Krylov and Bogolyubov should have participated (see *supra* Part II), and shows several peculiarities. Firstly, it is both a homage and challenge of the Poincaré-Lindstedt methods. Secondly it recalls the importance of the phenomena evidenced by Van der Pol, but it also laments the lack of mathematical rigor of his approach. Thirdly it is a general presentation of the founding principles of *Nonlinear mechanics*, and especially of the method developed by Krylov-Bogolyubov, as well as the main results connected to its application to radio engineering. After recalling the principles of the Poincaré-Lindstedt methods, and the problems linked to the existence of *secular terms*, they state:

> This elaboration's greatest merit is its rigorous methodical search for periodic solutions to nonlinear problems which belong to H. Poincaré and A. Lyapunov, who must consequently be considered as being true creators of this new chapter of Mechanics, - the "Nonlinear mechanics", which aims at creating a general theory of nonlinear oscillations. (Krylov and Bogolyubov 1933, 10)

They immediately add:

> Until recent days, H. Poincaré and A. Lyapunov's fundamental results were not used by physicists and engineers who continued, even for essentially nonlinear problems, endorsing the use of classic systems of the linear theory as a sovereign medicine, valid for all cases. (Krylov and Bogolyubov 1933, 10)

Indeed, in regards to what was explained in Parts I & II of this work, almost no engineer or physicist[11] seems to have used the Poincaré-Lindstedt method in France until 1933, in the field of radio engineering, Krylov and Bogolyubov then clarify:

> In spite of their importance, the famous Poincaré-Lyapunov methods, at least in their present form, are only valid for the periodic oscillations whereas in radio engineering for example, as well as in other branches of science of course, we find ourselves faced, due to the essence of things itself, with quasi-periodic oscillations, in other words, oscillations possessing at least two mutually independent frequencies. Quasi-periodic oscillations are due to the existence of quasi-periodic solutions to the nonlinear differential equations, which represent an extreme difficulty, to which Bohl has dedicated memorable research ("On some general differentials applicable in Mechanics", *Thesis*, Dorpat, 1900). Yet, in the work of the father of the theory of quasi-periodic functions, quasi-periodic solutions answering to types of nonlinear differential equations which are not found in radio engineering or in general in systems opening to the generalization of natural oscillations have been searched for. The problem therefore stayed posed despite its oft-reported importance.[1]
>
> ---
>
> 1. These oscillations were also studied, in a more specific case, by M. Balth. Van der Pol thanks to his methods. (Krylov and Bogolyubov 1933, 11)

---

[11] Excepted for possibly Liénard (1928, 948) (see *supra*).

Thus, they continue to question the Poincaré-Lindstedt methods, which they call 'Poincaré-Lyapunov methods', by implicitly referring to Lyapunov's contribution regarding the stability of periodic solutions. This approach has two aims: to present the Krylov-Bogolyubov method as the only mathematically rigorous approach likely to provide an adequate representation of the solution and to show that there exists an alternative to the methods suggested by the Mandelstam-Papalexi School. A competition of sorts took place, and reached its climax in 1937 with the simultaneous publication of Andronov and Khaikin's *Theory of Oscillations* (1937) and Krylov and Bogolyubov's *Introduction to Nonlinear Mechanics* (1937). Krylov and Bogolyubov also refer to Piers Bohl's research (1900), which is the source of the theory of quasi-periodic functions, and which will be discussed in the next paragraph. They then summarize Van der Pol's work, by following the same process consisting in emphasizing the lack of mathematical rigor in comparison with their own method, and reminding the fundamental importance of the phenomena he discovered: synchronization, frequency demultiplication, relaxation. Lastly, they explain the founding principles of *Nonlinear Mechanics* as well as the method they developed, by using "symbolic calculus", which makes it even harder to comprehend. However, they shed light on the motives of radio engineers and the problems to be solved:

> In the previous parts, we only detailed sinusoidal approximations, which are in many ways useful for practical calculations in radio engineering, but are hardly enough for the present state of radio engineering, as the frequency of the oscillator depends as we know on the operating state, and the aforementioned sinusoidal approximation does not show this dependence. Moreover, the current multiplication of wireless telegraph transmitters urgently requires the consistency of the oscillators' frequencies, hence the necessity of calculating the frequencies by using formulae providing the highest possible approximation in order to realize the conditions ensuring the aforementioned consistency in the most perfect manner. (Krylov and Bogolyubov 1933, 14)

They end their article by explaining that:

> For lack of space, all these results have not been listed in this article; its main aim was to call researchers' attention to the founding principles of "Nonlinear Mechanics" in their application to radio engineering. We believe that the results obtained with the methods used in this article may sometimes, get ahead of the experimental laboratory results, and in this might also contribute to the direction of the research, and at times guide the experiment themselves. (Krylov and Bogolyubov 1933, 14)

Again, Krylov and Bogolyubov tried to raise awareness among members of the French scientific community concerning the importance and necessity of continuing research in the field of nonlinear oscillations.

11.6 The Note in the *C.R.A.S.* of 1934: *The Second "Topological Excursion"*

It has been established in Part II that Andronov's note (1929a), in which he demonstrates the stability of the limit cycle representing the periodic solution for a self-oscillator, was a first "topological excursion[12]" in a field that was not yet called Dynamical systems. By solving the problem of the existence and stability of quasi-periodic solutions for forced or coupled oscillators, based on 's Poincaré (1885a, 220), Birkhoff (1927) and Denjoy's work (1932a,b), Krylov and Bogolyubov (1934e) carries out a "second topological excursion".

> For instance, interrogations concerning the existence and stability of quasi-periodic solutions, which play a very particular role in many problems of oscillations dealt with in nonlinear mechanics, are almost completely unexplored, especially for the differential equations corresponding to systems able to produce natural oscillations (self-periodic systems). In our recent researches pertaining to the subject, we attempted to contribute to the filling of this gap by developing verifiable criteria for the existence and stability of the quasi-periodic solutions. We succeeded in this endeavor by first using a special method which allowed us to establish the very distinct correspondence between the properties of the quasi-periodicity of the exact solutions and their first approximations formed from our nonlinear mechanics processes. By continuing our studies in this direction, we largely used the deep researches of H. Poincaré and M. A. Denjoy ([13]) on the characteristics on the surface of a torus, as well as essential ideas of M. G. Birkhoff related to the correspondence between the existence of the invariant curve of an isolated transformation and the existence of the quasi-periodic solution. (Krylov and Bogolyubov 1934e, 1592)

In this last sentence, Krylov and Bogolyubov implicitly refers to Birkhoff's work (1920, 79), as it is later confirmed in their note. Indeed, considering a differential equation of the type:

$$\ddot{x} + \omega_0^2 x = \sum_{n=1}^{\infty} \varepsilon^n f_n \left(t, x, \frac{dx}{dt} \right) \tag{11.23}$$

where $f_n\,(t, x, dx/dt)$ are integer polynomials of $sin\,(t)$, $cos\,(t)$, x and dx/dt, they demonstrate that it possesses a family of quasi-periodic solutions with the form

$$x = z\,(t, \nu t) \tag{11.24}$$

where ν indicates the "rotation number" as defined by Birkhoff (1920, 87). They add:

> It must be noted that the solutions (11.24) degenerate into periodic solutions each time the rotation number becomes rational. Considering the continuity of $\nu = \nu\,(\varepsilon)$, from this stems the fact that the aggregation of values of ε, for which the quasi-periodic solutions (11.24)

[12]See Chenciner (1985).

[13]Krylov and Bogolyubov referred to Poincaré (1885a, 220) and Denjoy (1932b).

> become simply periodic, is dense everywhere, and we also showed that this aggregation generally consists in intervals (synchronization areas); in each of which ν remains constant. (Krylov and Bogolyubov 1934e, 1593)

The link with the emerging *Theory of Dynamical Systems* is thus established. They later reinforced it in various articles published in French in the *C.R.A.S.*, but also in the periodic publication *Onde Électrique* (Krylov and Bogolyubov 1935a,b, 1936a,b) (see *infra*).

11.7 The Notes in the *C.R.A.S.* of 1935: *Towards the Theory of Dynamical Systems*

On the 10th of December 1934, Hadamard introduced a note by Krylov and Bogolyubov (1935a) entitled: "On the study of the case of resonance in nonlinear mechanics problems", in which they explain the principle of the method they had developed. Again considering the equation (11.23), they reduced it to the equations system (11.12) which is the "first approximation". Then, they explain that while this system possesses a stable periodic solution (limit cycle), which is only possible if the characteristic exponent corresponding to this solution is negative:

> (…) then the equation (11.23), for the sufficiently small values of ε, admits a stable family (positively) of quasi-periodic solution with two fundamental frequencies. (Krylov and Bogolyubov 1935a, 114)

By comparing Krylov and Bogolyubov's approach with Andronov's (1929a), we see that they are based on the same stability criteria, which is the negativity of Lyapunov's exponent established by Poincaré (1892, vol. I, 177) in the first volume of his "Méthodes Nouvelles de la Mécanique Céleste", and implemented during the *Conférences sur la Télégraphie Sans Fil* given by Poincaré (1908, 391) to demonstrate the stability of the periodic solution to the differential equation characterizing the oscillations of a radio device of the same kind as the *singing arc* (see *supra* Part I).

The following year, Krylov and Bogolyubov (1936a) presented a second note entitled: "The general stationary motions in dynamical systems of the nonlinear mechanics", to the Academy of Science. It denoted a transition towards the theory of dynamical systems.

11.8 The Article in the *Onde Électrique* of 1936: *The Krylov-Bogolyubov Method*

In August 1936, an article written by Krylov and Bogolyubov (1936b) is published in the periodic *Onde Électrique*, entitled: "Application of Nonlinear mechanics to some problems of modern radio engineering". It is a report of the *commission de Radio-physique* of the General Assembly of the International Union of Radio Science (U.R.S.I.)[14] which was held in London between the 12th and 18th of September 1934, and to which also participated Balthazar Van der Pol, Leonid Mandelstam, Nikolai Papalexi, Aleksandr Andronov, Semen Khaikin and Aleksandr Witt (see *infra*). Krylov and Bogolyubov's report again started with a challenging of the Poincaré-Lindstedt methods:

> The critical statements apply mainly to theories based on the use of the rigorous Poincaré-Lyapunov methods. These famous methods hold a very specific importance in mathematical analysis, but are however hardly manageable for a practitioner, as they turn the study of oscillators, which have a somewhat complicated diagram, almost inaccessible from a practical point of view. Moreover, these methods, at least in their present state, hardly allow for more than the study of simply-periodic states, which means that the quasi-periodic states, of which we have established the existence, are apparently excluded from their application field. (Krylov and Bogolyubov 1936b, 509)

Though this time Krylov and Bogolyubov went much further, and explained the reason for such a questioning, and especially clarified that:

> These series, in the general case of the quasi-periodic oscillations, *are not convergent*, in the rigorous meaning of the term, and we showed that they will be semi-convergent, as in the famous Stirling series. The semi-convergent series of this kind are often used in Celestial Mechanics, and the experiment of the previous century has brilliantly confirmed the feasibility of their practical use. It must be noted however, that this kind of series, incidentally very convenient in regards to approximate calculus, nevertheless do not lend itself to the rigorous study of the properties of the structure with exact solutions. (Krylov and Bogolyubov 1936b, 509)

The *inadequation* of Poincaré-Lindstedt methods is therefore linked to the divergence of the series able to represent the solution. This divergence is caused for one part by the size of the oscillation parameter, which can, in the case of the relaxation phenomenon, become largely superior to the unity, and for another part the number of the system's degrees of freedom which imply, above one, the existence of quasi-periodic solutions, i.e. made of frequencies independent either mutually or immeasurable. Krylov and Bogolyubov then clarified the methods which they used to carry out their study.

> This study was carried out by using methods borrowed from topology, because it seems that it is actually the topology joined with the analytical theory of the numbers that must lead to the decisive breakthrough in the field of quasi-periodic solutions to differential equations. (Krylov and Bogolyubov 1936b, 510)

This sentence efficiently summarizes the transition happening towards the Theory of Dynamical systems from the *Analysis Situs* launched by Poincaré (1895) a few years before, and which is nowadays named Topology. By using symbolic calculus, Krylov and Bogolyubov present their method in detail (see *supra*) with which they studied the various phenomena evidenced by Van der Pol (see *supra*).

[14]For a brief history of the U.R.S.I. see Van Bladel (2006).

Chapter 12
The Mandel'shtam-Papalexi School: *The "Van der Pol-Poincaré" Method*

While Krylov and Bogolyubov developed their method, the Mandel'shtam-Papalexi School simultaneously developed its own technique of investigation of forced or coupled systems. Contrary to the followers of the Kiev School, who advocated the challenging of the Poincaré-Lindstetdt methods, Andronov and Witt (1930a), Mandel'shtam and Papalexi (1932), then Andronov and Witt (1934) continued to use them in a very specific framework.

12.1 Andronov's Second Note in the *C.R.A.S.*: *The Case of Two Degrees of Freedom*

In this note presented less than three months after the previous one (see *supra* Part II), Andronov and Witt (1930a) studies the case of a system with two degrees of freedom, which may be written with the form of an autonomous second order nonlinear differential equations system (see *supra*, (9.23)) which will be noted (12.1):

$$\begin{cases} \ddot{x} + \omega_1^2 x = \mu f\left(x, \dot{x}, y, \dot{y}, \mu\right) \\ \ddot{y} + \omega_2^2 y = \mu g\left(x, \dot{x}, y, \dot{y}, \mu\right) \end{cases} \tag{12.1}$$

From the start, Andronov and Witt (1930a, 256) clarify that they limited their analysis to the case of periodic solutions. They then searched for a pair of periodic solutions by using the Poincaré-Lindstedt method, in other words, by expanding these solutions "in an ordered series following the powers of μ, convergent for μ sufficiently small" (Andronov and Witt 1930a, 257).

$$\begin{cases} x\left(t, \mu\right) = R^* cos\left(\omega_1 t\right) + \mu A_1 + \mu^2 A_2 + \ldots \\ y\left(t, \mu\right) = 0 + \mu B_1 + \mu^2 B_2 + \ldots \end{cases} \tag{12.2}$$

J.-M. Ginoux, *History of Nonlinear Oscillations Theory in France (1880–1940)*, Archimedes 49, DOI 10.1007/978-3-319-55239-2_12

They then provide a condition necessary for the existence of periodic solutions.

$$\int_0^{2\pi} f\left(R\cos(\omega_1 t), -\omega_1 R \sin(\omega_1 t); 0, 0; 0\right) \sin(\omega_1 t)\, dt = 0 \tag{12.3}$$

This condition then allows them to determine the amplitude R of the periodic solutions. They then study what they call the "Lyapunov stability" of these periodic solutions. To this aim, they calculate, as in Andronov's previous note (1929a, 561), the *characteristic exponents* defined Poincaré (1892, 177) and Lyapunov (1907, 225) of this solution. The application of Routh's (1877) and Hurwitz's theorems (1895) leads them to the following conditions:

$$\begin{cases} \int_0^{\frac{2\pi}{\omega_1}} f'_x\left(R\cos(\omega_1 t), -\omega_1 R \sin(\omega_1 t); 0, 0; 0\right) dt < 0 \\ \int_0^{\frac{2\pi}{\omega_1}} g'_y\left(R\cos(\omega_1 t), -\omega_1 R \sin(\omega_1 t); 0, 0; 0\right) dt < 0 \end{cases} \tag{12.4}$$

By writing $g(x, \dot{x}, y, \dot{y}, \mu) = 0$ in the system (12.1) we therefore find the system (1.10) with one degree of freedom found by Andronov (1929a, 560) (see *supra* Part II). It then appears that the conditions (12.4) and (1.11) (see *supra* Part II) are perfectly identical.

Andronov and Witt (1930a, 258) then clarify that "the case where $\omega_2 = n\omega_1$ requires a dedicated discussion." They were of course referring to the drive and frequency demultiplication phenomena evidenced by Van der Pol (see *supra*). Indeed, by noting that the trigonometric function $\sin(\omega t)$ (resp. $\cos(\omega t)$) is the solution to the differential equation for a harmonic oscillator, it is feasible to transform[1] the non-autonomous differential equation (9.20) into a system with two degrees of freedom (12.1). The problem tackled by Andronov and Witt (1930a) thus becomes perfectly equivalent to the one of a forced oscillator which Van der Pol was faced with (1920).

Three years later, a more complete and detailed Russian version of this note written by Andronov and Witt is published (1934). The following year, the same article, entirely translated in French, is published in the *Journal of Technical Physics of the USSR* (1935). The analysis of the latter is very informative and confirms what came previously. Firstly, in the introduction, where Andronov and Witt explain:

> (...) the stationary motions may, it seems, be not only periodic, but also more complicated. (Andronov and Witt 1935, 249)

It is interesting to note that they did not use the expression "quasi-periodic", which would imply a reference to Krylov and Bogolyubov's work (see *supra*). Then, they present their approach, which is entirely based on Poincaré's (1892) and Lyapunov's (1907), and provides a definition of "Lyapunov stability":

[1]This type of transform had also been used by Andronov and Witt in this article in order to write the stability criteria of Routh and Hurwitz, which is deduced from the fourth order characteristic equation.

In the case where the initial differential equations do not depend explicitly on the time, one of the characteristic exponents is always zero.*** But in this case, in order to obtain the stability, in the meaning used by Lyapunov, all the other exponents simply must have negative real parts.*

*** Poincaré, Méthodes Nouvelles, t. I, p. 179.
*Andronov und Witt, Zur Stabilität nach Lyapunov, Phys. Zs. Der Sowjetunion, Bd. 4, H. 4, 1933. (Andronov and Witt 1935, 249)

It is then interesting to note that Andronov's demonstration (1929a) regarding the stability of the limit cycle of an oscillator with one degree of freedom (see *supra* Part I), just like Andronov and Witt's demonstration (1930a, 1934, 1935) in order to establish the stability of the periodic solutions of an oscillator with two degrees of freedom, is based for one part on the chapter entitled "Characteristic exponents" of Poincaré's (1892) "Méthodes Nouvelles de la Mécanique Céleste" ("New Methods of Celestial Mechanics"), in which he explains

The constants α are called characteristic exponents of the periodic solution. (...) Therefore if every α have their squares real and negative, the quantities[2] $\xi_1, \xi_2, \ldots, \xi_n$ will stay finite. The periodic solution $x_i = \varphi_i(t)$ is stable; contrariwise, this solution is unstable. (Poincaré's 1892, 177)

Another part on the first chapter of Lyapunov's book (1892, 1907, 208–245) on the "General problem of the motion stability" was published simultaneously in Russia. A few pages later, Andronov and Witt justify the terminology "Lyapunov stability":

We calculate the characteristic exponents from Poincaré's methods, although the theorems establishing the legitimacy of the equations with variations for the solution of the motion questions belong to Lyapunov. (Andronov and Witt 1935, 258)

Therefore, it appears that if Poincaré is responsible for this stability condition for the periodic solutions, it is Lyapunov who must be credited with the legitimacy of its application.

Andronov and Witt continue their demonstration without using Van der Pol's nor Krylov and Bogolyubov's methods, only using the Poincaré-Lindstedt method, allowing them to find the conditions (12.4). In the second part of their article, which is an application to the study of the stability of periodic solutions for coupled oscillators (see *supra*), they confirm what had previously been perceived regarding the phenomena evidenced by Van der Pol:

At the edge of this "trailing" field, the oscillatory regime, when we make one parameter vary, changes in a discontinuous manner, the system "jumps" from one frequency to the other. This "trailing" phenomenon has already been discussed theoretically by several authors,[3] but it was not given a rigorous mathematic theory, to our knowledge. (...) Another case requiring a specific study is the one where one frequency is a multiple of the other. (Andronov and Witt 1935, 265)

[2]These quantities result from "variations equations".

[3]Andronv and Witt refer among others, to Van der Pol (1922).

Andronov and Witt (1930a, 258) are indeed referring to the trailing phenomena, i.e. of drive and frequency demultiplication, in this note published in the *C.R.A.S.*

12.2 Mandel'shtam-Papalexi's Articles: *The "Van der Pol - Poincaré" Method"*

Between 1931 and 1933, many articles[4] were published in Russian on the study of resonance and frequency demultiplication phenomena by Mandel'shtam and Papalexi (1931, 1932, 1933, 1934). The next year, a summary of theses article was published in French in the report of the *commission de Radio-physique* of the General Assembly of the International Union of Radio Science (U.R.S.I.) which was held in London between the 12th and 18th of September 1934, and to which Leonid Mandel'shtam, Nikolai Papalexi, Aleksandr Andronov, Semen Khaikin and Aleksandr Witt participated (see *supra*). In it, they therefore presented Van der Pol's method and the various phenomena he observed:

> Although in the end, this method has been used for a long time in celestial mechanics, it was Van der Pol who was the first to systematically apply it to radio engineering problems, and he obtained a series of fundamental results on forced synchronization, "trailing" etc.[5] But until now, this method was not mathematically supported. Moreover, there still was some indetermination in the process of its application itself. The main difficulty on this level was eliminated[6] in a manner that we will explain using the example of the equation
>
> $$\ddot{x} + x = \mu f(x, \dot{x}, t) \tag{12.5}$$
>
> where the second member is a periodic function of t with a period 2π, and μ a "small parameter" from which will depend, as we will see, the order of approximation. (Mandel'shtam et al. 1935, 89)

It must first be noted that while Mandel'shtam et al. credited Van der Pol with the evidencing of several phenomena, they reproached that his method, as Krylov and Bogolyubov did (1933, 10) (see *supra*) lacked in of mathematical rigor, and also explained that it was already used in Celestial Mechanics (see *infra*). Secondly, it is very surprising that there is no mention of Krylov and Bogolyubov's work (1932a,b,c, 1933, 1934a,b,c,d), but on the contrary, of Mandel'shtam and Papalexi's work (1934). This may confirm the hypothesis that there was a competition between the two Schools, at least in this field. This hypothesis seems to be supported by Mandel'shtam and Papalexi's many publications during this time[7] and by the fact

[4]See also Mandel'shtam (1947–1955, vol. 1, 7–138).

[5]Mandel'shtam et al. refers to Van der Pol (1922) and Van der Pol and Van der Mark (1927a,b).

[6]Mandel'shtam et al. refers to Mandel'shtam and Papalexi (1934).

[7]See also Mandel'shtam (1947–1955, vol. 1, 7–138).

that Krylov and Bogolyubov are scarcely cited. In order to tackle this problem, Mandel'shtam et al. recalls the various steps of the implementation of Van der Pol's method (see *supra*). They then explain that by looking for a solution to (12.5) with the form (10.2), we transform the equation (12.5) into a system of two first order differential equations. They add:

> By replacing the second members with their means, we find van der Pol's equations, or "truncated equations". When we obtain them in this way, we can clearly enunciate a problem of approximation: we must establish when and how many (depending on μ) solutions to the truncated equations are close to those for the exact equations. This is a purely mathematical question, which has been studied by Fatou (1928) in an essay that we learned of only after we carried out our research on van der Pol's method. (Mandel'shtam et al. 1935, 90)

Mandel'shtam et al. thus show that Van der Pol's method, called "ingenious" and lacking in rigor by Krylov and Bogolyubov (1933, 10), was developed in order to solve a radio engineering problem and holds the mathematical foundations that it lacked, in the work of mathematician and astronomer Pierre Fatou. They then add:

> Moreover, Fatou's results allow to state that if μ is sufficiently small, to each balance point of the truncated system corresponds a periodic solution to the exact system, and that if this position this balance position is stable, the periodic solution is also stable. The question regarding van der Pol's mathematical basis is therefore cleared up. (Mandel'shtam et al. 1935, 90)

They therefore highlighted a bijective correspondence between the exact equations system, and the "truncated equations" one, and they explained that the existence of such a correspondence was established by Fatou (1928). Then, they describe their approach to the problem, which could be called the "Van der Pol-Poincaré method".

> Let us explain in a few sentences, the benefits of van der Pol's method. For an autonomous system with one degree of freedom, van der Pol's equations can be reduced to one same equation, which is easily solved with the quadrature. For the non-autonomous systems with one degree of freedom – the case which was just presented – van der Pol's equations are autonomous and yet, disputable through Poincaré's methods.[8] For more complicated systems, for instance the ones with two degrees of freedom (autonomous or subjected to external forces), van der Pol's equations are autonomous first order equations systems – with two equations for the more simple cases – which can be dealt with using Poincaré's methods. Van der Pol's method therefore allows the replacement of a system of nonlinear equations by another, simpler one. (Mandel'shtam et al. 1935, 90)

Thus, Mandel'shtam et al. suggest the application to autonomous systems with one degree of freedom, or non-autonomous ones with two degrees of freedom, a variation of Van der Pol's method, which consists in searching for a solution to the equations for forced or coupled oscillators with the form (10.2). The second member of the differential equations systems therefore obtained is then expanded in a Fourier series, then averaged. The final result, which is what (Mandel'shtam et al.

[8]Mandel'shtam et al. referred to Andronov and Witt (1930d).

1935, 90) call Van der Pol's "truncated equations", can then be dealt with by the means of the Poincaré-Lindstedt methods. Therefore, this process could be called "Van der Pol-Poincaré method". However, one must admit that this method is quite close to Krylov-Bogolyubov's. The only differences are firstly in the link, uncovered by Mandel'shtam, with Fatou's work, which represent the mathematical foundations of this method and also justify it, and secondly, in the study of the stability of the solutions to the truncated equations system by using the Poincaré-Lindstedt method. It must however be noted that the "Van der Pol-Poincaré method" does not apply to systems possessing periodic solutions, as the quasi-periodic case does not seem to have been addressed by the followers of Mandel'shtam's School[9]:

> We saw that the van der Pol method is entirely justified, at least for the simpler cases. But it only gives a "zero-th approximation". In some cases of radio engineering, this is enough. But there are others which require higher approximations: they are mainly questions of correction of the frequency, which only appears, in many problems, at the second approximation. A theory allowing the furthering of calculations' accuracy is thus necessary.Such a theory unfortunately exists for purely periodic phenomena. This is the "small parameter method", which we also owe to Poincaré. (Mandel'shtam et al. 1935, 91)

Again, Mandel'shtam et al. (1935) avoid making any reference to Krylov and Bogolyubov's work, and to their method, which they must be aware of. They kept to the periodic cases, and continued to apply Poincaré's method "of the small parameter", because they believed that it is the only one possessing a mathematical basis, just like the variation they develop and that it is justified by Fatou's work.

[9]There are no traces either in the original edition of Andronov and Khaikin's book (1937).

Chapter 13
From Quasi-periodic Functions to Recurrent Motions

In radio engineering devices, the existence of oscillations possessing at least two unmeasurable frequencies had therefore led Krylov and Bogolyubov to consider the representation of the solutions to the differential equations characterizing this kind of phenomena by quasi-periodic functions. These functions, discovered a few years before by Livonian mathematician Piers Bohl[1] (1865–1921) had been the subject of many works, in Copenhagen by Danish mathematician Harald Bohr (1887–1951) by Salomon Bochner (1899–1982), in the Soviet Union by Abram Besicovitch (1891–1970), Aleksandr Kovanko (1893–1975), Andrei Markov (1856–1922), Lev Pontrjagin (1908–1988) and Viacheslav Stepanov (1889–1950), in Germany by Hermann Weyl (1885–1955) and in France by a whole line of Astronomers, such as Ernest Esclangon (1876–1954), Pierre Fatou (1878–1929), Jean Favard (1902–1965) then Hervé Fabre (1905–1995), and Mathematicians such as Arnaud Denjoy (1884–1974), Jean Chazy (1882–1955) and Jacques Hadamard (1865–1963).

13.1 Ernest Esclangon's Work: *On Quasi-periodic Functions*

Born on the 17th of March 1876, in Mison, a small town near Sisteron (*Basses-Alpes*), he joined the *École Normale Supérieure* when he was nineteen, and passed the *agrégation des Sciences Mathématiques* in second place in 1898. The following year, he accepted the position of assistant-astronomer at the Bordeaux Observatory where, as it was customary for junior members, he was appointed to the *service méridien*, to study marker-stars. At the same time, he conducted mathematical research which led him to present a thesis at the Paris Faculty of Science, entitled: "Les fonctions quasi-periodiques" ("Quasi-periodic functions"), which he defended

[1]See Bohl (1893, 1900, 1910).

J.-M. Ginoux, *History of Nonlinear Oscillations Theory in France (1880–1940)*,
Archimedes 49, DOI 10.1007/978-3-319-55239-2_13

on the 29th of June 1904 before Misters Paul Appell (President), Henri Poincaré and Paul Painlevé (Examiners). This research was a continuation of Livonian mathematician Piers Bohl's own work (1893, 1900), in which he introduces the concept of quasi-periodic functions, although he did not give them this name. Indeed, it seems it was Esclangon (1902, 893) who was the first to formulate this neologism, in a note presented on the 22nd of November 1902 before the Academy of Science by Paul Painlevé, in which he defines this new type of function as follows:

> We find, in numerous problems, functions which can take the form of periodic functions with varying periods. It is possible to create, among those functions, a special classification, by showing that they do not belong to a more general class of functions, whose properties depend on a new extension of the notion of periodicity. These functions may be found in various problems where different periodic elements are in a way mixed together, and seem to play an important part. (Esclangon 1902, 891)

While of course, this astronomer's concern led him to considering quasi-periodicity in the field of Celestial Mechanics, it is impressive to note that as for wireless telegraphy, Mathematics have preceded (Radio) Engineering by almost twenty years. Two pages later, Esclangon unveils his new terminology by stating its definition:

> $f(x)$ is a *quasi-periodic* function if, given ε as small as we want, we can find δ such, that with the conditions
>
> $$|x + m_1 a_1 - x_1| < \delta, \ \ldots, \ |x + m_n a_n - x_n| < \delta$$
>
> we would have
>
> $$|f(x) - L(x_1, x_2, \ldots, x_n)| \leq \varepsilon \tag{13.1}$$
>
> and this for all the element of the total field. (Esclangon 1904, 4)

Two years later, he introduces his thesis dissertation with:

> After this publication, Mr. Bohl let me know that in his thesis,[2] he presented analogous considerations, albeit under a completely different form. (Esclangon 1904, 4)

He also gives a more accessible definition of the quasi-periodic functions:

> A periodic function $f(x)$ possesses the following property: given any interval $(\alpha, \ldots, \beta)$ and a number ε as small as we want, we can determine an infinity of intervals $(\alpha + h, \ldots, \beta + h)$, as removed as we want from the first one, in each of which the considered function *takes again the same series of value to less than* ε: hence the name *quasi-periodicity* given to this property. (Esclangon 1904, 3)

It therefore seems that to ε, the function is almost periodic. This was the terminology Bohr (1923a,b) chose twenty years later. Between 1902 and 1903, Esclangon was tasked with teaching Rational Mechanics at the Faculty of Science of Bordeaux. In 1905, he was appointed assistant-astronomer at the Bordeaux

[2]Esclangon refers to Bohl (1893).

Observatory, and went to the *service du grand équatorial* for the observation of small planets and comets. He was appointed *maître de conférences* the same year, then Assistant-Professor in 1908. He then taught the Differential Calculus, Infinitesimal Geometry and later, General Mathematics courses. During the First World War, Esclangon was assigned, as Jules Haag was (see *supra* Part II, 203), to the *polygone d'artillerie* in Gâvre under the presidency of General Prosper-Jules Charbonnier to study the sound phenomena caused by firing artillery. In a Note to the *Service Géographique* from the 22nd of September 1914, he was the first to suggest a method for spotting enemy battery through sound. His method was applied as early as 1916 with great success, before he extended it to the aerial acoustic and submarine detection.[3]

Right after the war, in 1919, he was appointed Director of the Strasbourg Observatory, as well as Astronomy Professor at the Faculty of Science.[4] The same year, he published in the *Annales de l'Observatoire de Bordeaux*, his "Nouvelles recherches sur les fonctions périodiques" ("New research on periodic functions") (Esclangon 1919). In the fifth part of his essay dealing with "linear differential equations with periodic coefficients", Esclangon writes:

> In Celestial mechanics, the quasi-periodic functions also intervene in the integration of differential equations for the theory of perturbation. In this field, many important results are yet to be obtained. In the second volume of his famous work: *Les Méthodes Nouvelles de la Mécanique Céleste*, H. Poincaré[5] studied the equation
>
> $$\frac{d^2x}{dt^2} + x_{\varphi}(t) = 0$$
>
> where
>
> $$\varphi(t) = 1 + \mu \sum a_i sin(\alpha_i t + \beta_i)$$
>
> the α_i are immeasurable constants two by two. By referring to methods he developed, Poincaré added: "In this case, the previous results are still applicable, but the series we

[3] See Esclangon (1925).

[4] When in 1929 Henri-Alexandre Deslandres (1853–1948) retired, Esclangon was given both the direction of the Meudon Observatory and Paris Observatory. In spite of his responsibilities, his productivity in science did not decrease. But the reason for his fame is most of all his invention of the very first speaking clock. Indeed, until now the observatories around the world were tasked with providing the exact time to the population. An agent was employed full time to this service, answering more or less precisely to the phone calls. To end this servitude, Esclangon used the process of sound film, recently invented. His clock had a "speaking" part, with sound tracks coiled around a cylinder and photoelectric pick-up heads moving automatically, and a clock part, made of the tops from a fundamental clock controlled by comparison with the astronomic determinations of the time. On the 14th of March 1932, the device was presented to the Academy of Science and on the 14th of February 1933 the speaking clock was implemented. People consulting the time by calling the number Odéon 84–00 (The speaking clock is still available nowadays), were far from compreheding the technical difficulties that the inventor had to overcome in order to provide the exact time at any moment, first to the tenth, then to the hundredth of a second.

[5] See Poincaré (1893, 277).

reach are not convergent anymore, so the processes have no other value than the one that can possess, from Chapter VIII, any method of formal calculus." A bit later, he considered the more general equation

$$\frac{d^2x}{dt^2} + x_f(t) = \varphi(t) \tag{13.2}$$

where $f(t)$ and $\varphi(t)$ functions developable in trigonometric series, in other words, ultimately, quasi-periodic functions. (Esclangon 1919, 193–194)

By comparing the equation (13.2) of Esclangon-Poincaré to the equation (13.3) of Krylov and Bogolyubov (1934e, 1592):

$$\ddot{x} + \omega_0^2 x = \sum_{n=1}^{\infty} \varepsilon^n f_n\left(t, x, \frac{dx}{dt}\right) \tag{13.3}$$

we see that on the one hand they are perfectly identical and, on the other, that they lead to the evidencing of quasi-periodic solutions.

This is incidentally confirmed by Esclangon on the next page:

To mention a simple case, we demonstrate for instance, as Bohl did,[6] that if the equation

$$\frac{d^2x}{dt^2} + x_f(t) = \varphi(t) \tag{13.4}$$

in which $f(t)$ and $\varphi(t)$ quasi-periodic functions, only admits bounded integrals, the general integral can be written with the form

$$x = c_1 \frac{\cos \int_0^t R dt}{\sqrt{R}} + c_2 \frac{\sin \int_0^t R dt}{\sqrt{R}} + T \tag{13.5}$$

in which R and T are quasi-periodic functions, c_1, c_2 are arbitrary constants. (Esclangon 1919, 195)

It therefore appears that the mathematical foundations required for the study of quasi-periodic solutions to the equation (13.2), i.e. (13.3), were already present in Poincaré's (1893), Bohl's (1906) and Esclangon's work (1919). A dozen years later, when the first *Conférence Internationale de Non linéaire* was held in Paris from the 28th to the 30th of January 1933 at the Henri Poincaré Institute, organized at the initiative of Balthasar Van der Pol and Nikolai Papalexi, Esclangon was invited, probably in order to discuss the applications of the quasi-periodic functions to the study of radio-engineering oscillators with Krylov and Bogolyubov who were supposed to attend (see *supra* Part II). Unfortunately, their absence prevented an exchange of points of views. Moreover, it proved impossible to establish whether Esclangon had managed to meet them during their second visit to Paris in 1935. The Ukranian mathematician Boris Levitan (1914–2004), who dedicated several

---

[6]Esclangon referred to Bohl (1906).

books to the theory of quasi-periodic functions,[7] writes in an article entitled: "A brief history of the theory of almost periodic functions":

> Общепризнано, что создателем теории почти-периодических функцины (в дал'неышем сокращенно п.п.ф.) являыецяя дацкиы математик Гарал'д Бор. В 1924–1926 гг. им были опубликованы в Аста Матхематжса три фундаментал'ных мемуара.[8] В первых двух мемуарах была изложена полная теория непрерывных п.п.ф. от одноы деыствител'ноы переменноы. Третиы посвящен теории аналитических п.п.ф.[9]

He then adds:

> Однако уже во втором мемуаре Бор упоминаыет о работах рижского математика П. Боля[10] и французского математика и астронома Е. Есклангона.[11]

Moreover, according to Gaiduk (1974), Esclangon had anticipated numerous results established later by Bohr. However, in their book entitled "The almost periodic functions and the differential equations", Levitan and Zhikov write:

> В дал'неышем (на протяжении 20–30-х годов) теория Бора получила существенноые развитиые в работах С. Бошера, Г. Веыля, А. Безиковича, Ж. Фавара, Дж. Неымана, В. В. Степанова, Н. Н. Боголюбова и др..[12]

13.2 Jean Favard's Work: *On Almost-Periodic Functions*

Born in Peyrat-la-Nonière in the *Creuse* in 1902, he joined the *École Normale Supérieure* at age nineteen, and passed the *agrégation* in Mathematical science in 1924. After his military service, he obtained a Rockefeller grant which allowed him to stay for one year in Copenhagen, with mathematician Harald Bohr. Upon his return to France, he presented a thesis[13] before the Faculty of Science of

---

[7]See Levitan (1953) and Levitan and Zhikov (1978).

[8]Levitan refers to Bohr(1924, 1925a, 1926).

[9]"It is more generally admitted that the creator of the theory of almost periodic functions (shortened a.p.f..) is Danish mathematician Harald Bohr. In the years 1924–1926, he published in Acta Mathematica three fundamental essays. In the first essays, he presented a complete theory of the a.p.f. continued from a real variable. The third addresses the analytical theory of the a.p.f." (Levitan 1980, 156).

[10]Levitan refers to Bohl (1893).

[11]"However, in his second essay, Bohr mentioned the work of mathematician Riga P. Bohl and French mathematician end astronomer E. Esclangon." (Levitan 1980, 156)

[12]"Later (in the 1920s and 1930s) Bohr's theory received important developments with the work of C. Bochner, H. Weyl, A. Besicovitch, J. Favard, J. von Neumann, V. V. Stepanoff, N. N. Bogolyubov etc." (Levitan and Zhikov 1978, 5)

[13]This document was published in the *Journal de Mathématiques Pures et Appliquées*. See Favard (1927a)

Paris "Sur les fonctions harmoniques presque périodiques" ("On almost-periodic harmonic functions") defended on the 11th of July 1927 before Misters Ernest Vessiot (President), Paul Montel and Arnaud Denjoy (Examiners), in which he recalls:

> In three Notes to the *Comptes rendus* and three Memoirs in the *Acta mathematica* ([1]) Mr. H. Bohr created the theory of almost-periodic functions ([2]). Various generalizations were noted by Misters Stepanoff ([3]) and Besicovitch ([4]), lastly, Mr. Bochner ([5]) gave new properties of Mr. Bohr's functions, as well as a method of summation of the developments analogous to Mr. Féjer's method for periodic functions.
>
> ---
>
> ([1]) H. BOHR, *Sur les fonctions presque périodiques* (*Comptes rendus*, t. 177, 1923. 737) ; *Sur l'approximation des fonctions presque périodiques par des sommes trigonométriques* (*Ibid.*, t. 177, 1923, 1090); *Sur les fonctions presque périodiques d'une variable complexe* (*Ibid.*, t. 180, 1925, 645) ; *Zur Theorie der fastperiodischen Funktionen*: I (*Acta mathematica*, t. 45, 29–127); II (*Ibid.*, t. 46, 1925, 101–214); III (*Ibid.*, t. 47. p. 237–274).
> ([2]) These functions contain, as a particular case, Bohl-Esclangon's quasi-periodic functions: BOHL, *Magister dissertation*, Dorpat, 1893; *Journal de Crelle*, t. 131, 1906, 268–321. ESCLANGON, *Thèse*, Paris, 1904; *Annales de l'Observatoire de Bordeaux*, 1919
> ([3]) STEPANOFF, *Sur quelques généralisation des fonctions presque périodiques* (*Comptes rendus*, t. 181, 1925, 90); *Über einige Verallgemeinerungen der fastperiodischen Funktionen* (*Math. Annalen*, t. 95, 1926, 473–498).
> ([4]) BESICOVITCH, *Sur quelques points de la théorie des fonctions presque périodiques* (*Comptes rendus*, t. 181, 1925, 394).
> ([5]) BOCHNER, *Sur les fonctions presque périodiques de Bohr* (*Comptes rendus* t. 180, 1925, 1156); *Beiträge zur Theorie der fastperiodischen funktionen* (*Math. Annalen* (sous presse)). (Favard 1927a, 229–230)

In this introduction it is remarkable that all the contributions from various mathematicians involved in the development of the theory of almost-periodic functions were published in French in the *C.R.A.S.* between 1923 and 1925.

In 1927, after a brief stay at the *lycée* of Amiens, Favard was a lecturer at the Faculty of Science of Grenoble.[14] The same year, he published in the periodic *Acta Mathematica* a lengthy article entitled "Sur les equations différentielles linéaires à coefficients presque-périodiques" ("On the linear differential equations with almost-

---

[14]In 1933 he was appointed *maître de conférences* at the Faculty of Science of Alger, but stayed at the one in Grenoble where he was appointed *Professeur de Mathématiques Générales* in 1935. He was mobilized in September 1939 as an artillery officer, and was captured and held prisoner until June 1940. He was sent to the oflag XVIII, in Lienz (Austria), where he created a Faculty of Science, of which he was the dean. Austrian mathematicians considered freeing him at the end of the war on the condition that he agreed to teach in Vienne; he refused. In 1941, he was appointed Professor at the Faculty of Science in Paris, but started teaching at the Sorbonne only after he was released in 1945. The research on the theory of almost periodic functions was carried out by mathematician Jean Delsarte (1903–1968). Author of numerous researches on medium-periodic functions (See Delsarte (1934, 1935a,b, 1938a, 1935b, 1939), he joined the *École Normale Supérieure* a year after Jean Favard (and at the same time as Yves Rocard and André Weil). It was also him who did the interim at the Faculty of Sciences of Grenoble when he was a prisoner.

periodic coefficients") (Favard 1927b), in which he analyzed the equation (13.5), which had initially been studied by Poincaré (1893, 283), Bohl (1893, 1906, 1910) and (Esclangon 1919, 195). He demonstrated, under specific conditions, that the solution to this equation which also has the form (13.5), is of an almost-periodic nature.[15] During the year 1929, supported by Gaston Julia, he became tenure holder at the Peccot Foundation, and gave lectures at the *Collège de France* which were published a few years later with the title "Leçons sur les fonctions presque-périodiques" ("Lessons on almost-periodic functions") (Favard 1933). In the preface of this book, Favard gives a glimpse of the possible applications for these functions:

> But the application fields are open to this theory: in Number Theory, some properties of the Dirichlet series that we encounter are consequences of the quasi-periodicity. In Dynamics, the motions neighboring a stable periodic motion are, in some cases, limit-periodic, as was shown by Mr. Birkhoff. Mr. Krylov's most recent Notes lead to predict applications to Mathematical physics. (Favard 1933, VII)

Favard implicitly referenced on the one hand Birkhoff's work (1912, 314; 1926, 518; 1927) and on the other, Krylov and Bogolyubov's Notes Krylov and Bogolyubov 1932a,b,c presented to the Academy of Science the year before the publication of his work. He then adds:

> There is therefore no fear that the theory may fall in outer obscurity: it might prove its worth, but the previous fields are barely weeded. (Favard 1933, VII)

Indeed, even if the Theory of almost-periodic functions was beginning to find application in the Theory of Nonlinear oscillations through the bridge of radio engineering, it seems this was only a first step (at least in France).

13.3 Arnaud Denjoy's Work: *Characteristics on the Surface of the Torus*

The problem of the quasi-periodicity had been considered by Poincaré in his memoir "Sur les courbes définies par une equation différentielle" ('On curves defined by a differential equation") (Poincaré 1885a, 220). Indeed, chapter XV is dedicated to a "Specific study of the torus". This genus 1 surface may be represented in cartesian coordinates by the following equation:

$$z^2 + \left(R - \sqrt{x^2 + y^2}\right)^2 = r^2 \qquad (13.6)$$

where R represents the radius of the generating circle, and r the radius of the meridian circle (see Fig. 13.1).

[15] See also (Favard 1926, 1124).

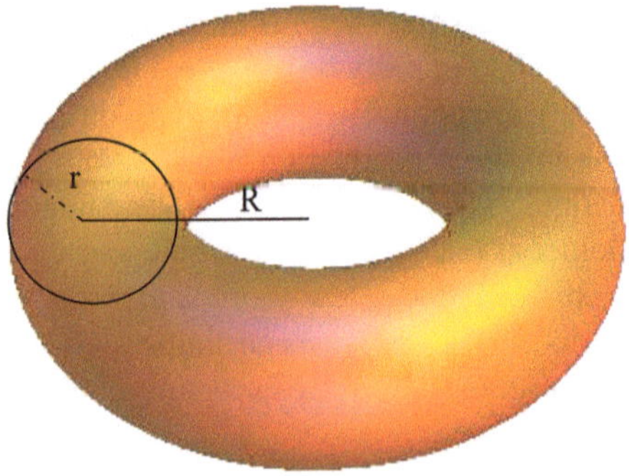

Fig. 13.1 Torus (genus 1 surface)

Poincaré then considers a first order differential equation for which the solution is constrained to evolve over the surface of the torus. By using the parametric representation of the torus:

$$\begin{cases} x = (R + rcos(\theta))\, cos(\varphi) \\ y = (R + rcos(\theta))\, sin(\varphi) \\ z = rsin(\theta) \end{cases} \tag{13.7}$$

He obtains the following differential equations system:

$$\begin{cases} \dfrac{d\theta}{dt} = \Omega(\varphi, \theta) \\ \dfrac{d\varphi}{dt} = \Phi(\varphi, \theta) \end{cases} \tag{13.8}$$

The ratio of these two equations led him to the following equation:

$$\frac{d\theta}{d\varphi} = A(\varphi, \theta)\,; A(\varphi, \theta) = \frac{\Omega(\varphi, \theta)}{\Phi(\varphi, \theta)} \tag{13.9}$$

He then considers a family of trajectories on the torus defined by the equation:

$$\frac{d\theta}{d\varphi} = \alpha \tag{13.10}$$

and thus defines the "rotation number" α which can be interpreted as the ratio of two frequencies of the rotation of the trajectory on the torus: the first one ω_1 depending on the generating circle, the second one ω_2 depending on the meridian circle. He then demonstrates that

- if α is rational, i.e. if ω_1 and ω_2 are rationally dependent, then any trajectory of (13.9) is closed (limit cycle) on the torus (see Fig. 13.2).

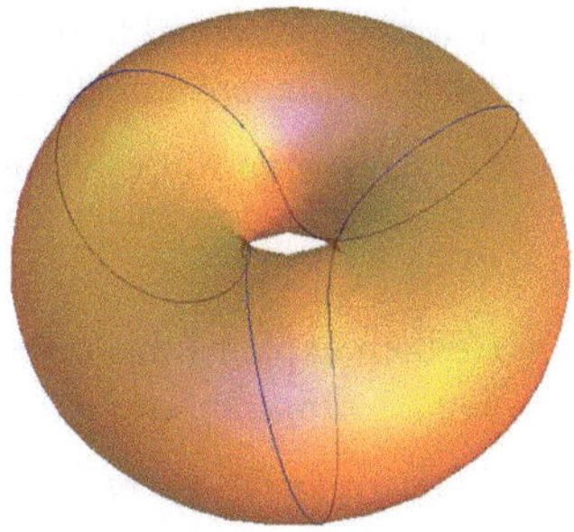

Fig. 13.2 Characteristics on the surface of the torus, by Ginoux (2009, 36)

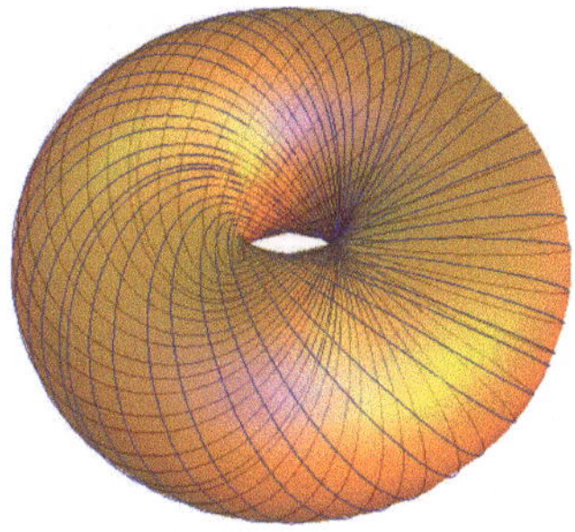

Fig. 13.3 Characteristics on the surface of the torus, by Ginoux (2009, 36)

- if α is irrational, i.e. if ω_1 and ω_2 are rationally independent, then any trajectory of (13.9) is dense on the torus[16] (see Fig. 13.3).

The solutions of (13.9) are then periodic in the first case, and quasi-periodic in the second case.

In the early 1930s, Arnaud Denjoy worked on the question of characteristics on the surface of the torus studied by Poincaré. Born on the 5th of January 1884 in Auch, he joined the *École Normale Supérieure* in 1902, where he was greatly influenced by his professors: Émile Borel, Paul Painlevé and Émile Picard, who helped him acquire solid knowledge in the fields of the theory of complex functions, continued fractions and differential equations. He was awarded a grant by the Thiers foundation, and dedicated his research to the study of infinite products. He presented his thesis at the Faculty of Science of Paris "Sur les produits canoniques d'ordre infini" ("On the infinite order standard products") defended on the 27th of November 1909 before Misters Henri Poincaré (President), Paul Painlevé and Émile Borel (Examiners). He was then appointed *maître de conférences* at the Montpellier University, where he taught until the First World War, in which he could not participate due to his faulty vision. He was then appointed associate professor at the Utrecht University in 1917 then Professor of Celestial Mechanics at the Sorbonne in 1922, where he stayed until he retired in 1955.

[16]See Poincaré (1885a, 220), Niemytski and Stepanoff (1960, 55), Pliss (1966, 120) and Arnold (1974, 213).

In a note presented to the *C.R.A.S.* by Jacques Hadamard on 7 March 1932, he completes Poincaré's result (1885a, 220) and demonstrates:

> Especially when A(θ, φ) is holomorphic (Poincaré's hypothesis) if α is irrational, all the characteristics pass indefinitely in the neighborhood of any point of the torus. (Denjoy 1932a, 833)

He then explains this result in more detail in an article published in the *Journal de mathématiques pures et appliquées*

> Therefore, *for the case of* $\frac{\partial A}{\partial \theta}$ (φ, θ) *being at* θ *with a total variation uniformly bounded in a field* $\varphi_0 \leq \varphi < \varphi_0 + 2\pi$, $\theta_0 < \theta < \theta_0 + 2\pi$ *and in the case of number* α *being irrational, the characteristics of the equation* (13.9) *cut any meridian* C *in a set that is dense everywhere.* Any characteristic passes an infinite number of times in the neighborhood of any point of the torus. From Poincaré's expression, the trajectories defined by the equation (13.9) are *stables*.
> In particular, this will be the case when $\frac{\partial^2 A}{\partial \theta^2}$ *is continuous* (or simply bounded) over the whole torus, and *a fortiori if* $A(\varphi, \theta)$ *is holomorphic* at φ and θ from Poincaré's hypothesis. (Denjoy 1932b, 375)

In 1935, when he analyzed Krylov and Bogolyubov's work on the "Fundamental problems of Nonlinear Mechanics" he recalls:

> The case that is the most remarkable, most perfect, but also most accidental, and yet least important in regards to physics or celestial mechanics applications, is the one of the periodic solutions. H. Poincaré and Lyapunov showed the fundamental role of these integral when they exist, and from this, founded the theory in essays that have now been classics for a long time. A more general case is the one of the quasi-periodic solutions, considered for the first time by P. Bohl, founder of the theory of functions of the same nature. But before N. Krylov's work, the existence of quasi-periodic solutions had been established with a rigor sufficient to only a very narrow class of differential equations. However, by studying in depth the remarkably simple case of the characteristics on the surface of a torus, H. Poincaré evidenced specific orders in fact likely to express themselves with an analogous form in much broader cases. In 1932, I myself completed the results of H. Poincaré on this matter, by establishing the quasi-periodicity in all the cases where there were no periodic solutions. G. Birkhoff studied, in the path opened by H. Poincaré, the systems with several degrees of freedom, and obtained fundamental results regarding the existence of quasi-periodic solutions and the properties of motions called *recurrent*. (Denjoy 1935, 390)

It is possible that Denjoy met Krylov during the first Congress of U.S.S.R. Mathematicians organized in Kharkiv in 1930, where was accompanying Hadamard. He also may have met him at the Henri Poincaré Institute during Hadamard's seminar in 1926–1927 to which Krylov had participated, or during the conferences given by Krylov in 1935 (see *supra*). It seems undeniable that the two men knew and respected each other a lot, as shown by Denjoy's previous quote, and confirmed this excerpt from Krylov and Bogolyubov's report n° 8 (1934d) in which we can read on page 108:

> Here, we must nonetheless note the deep research led by H. Poincaré himself on the characteristics on the surface of a torus, brilliantly completed in 1932 by Mr. A. Denjoy and pertaining to the equation
>
> $$\frac{d\Theta}{dt} = f(t, \Theta)$$

> where the function $f(t, \Theta)$ is periodic at t and Θ with the period 2π.
> Moreover, G. Birkhoff's research regarding the standard systems with two degrees of freedom shed stark light on the intimate relation between the existence of the invariant curve (depending on a specific isolated transform corresponding to given differential equations) and the existence of quasi-periodic solutions.
> We also owe to the illustrious American geometrician the deep results pertaining to the various properties of the motions called "recurrent". (Krylov and Bogolyubov 1934d, 101)

It thus appears that the theory of almost-periodic functions allowed the establishment of the existence of quasi-periodic solutions occurring in radio engineering problems. Moreover, this type of solutions can nowadays be described based on Birkhoff recurrent motions.

13.4 George Birkhoff's Work: *The Transition Towards Dynamical Systems*

Born on the 21st of March 1884 in Michigan, George David Birkhoff was an American mathematician, whose work on Dynamical systems and Ergodic theory had a notable impact in various fields. After studying at the Lewis Institute from 1886 to 1902, he joined the University of Chicago before joining Harvard University, where he was a student from 1903 to 1905. He then returned to Chicago to work on his Ph-D under the supervision of Eliakim Moore (1862–1932). He defended his thesis dissertation: "Asymptotic Properties of Certain Ordinary Differential Equations with Applications to Boundary Value and Expansion Problems" in 1907. He then taught at the University of Wisconsin in Madison from 1907 to 1909 before leaving for Princeton, where he became professor in 1911. In March 1912, Poincaré's last article was published (1912), with the title "Sur un théorème de géométrie" ("On a geometry theorem"), in which he shows that the existence of periodic solutions restricted three-body problem could be reduced using a very simple geometry theorem, but for which he could not establish the demonstration except for some specific cases.[17] After his passing away on the 17th of July 1912, Morse said that his article reached Princeton at the end of summer. Before the end of the year, Birkhoff managed to establish the demonstration that had remained forever unattainable to illustrious geometrician. In January 1913, this proof, published in the *Transactions of the American Mathematical Society* brought instant fame and Birkhoff became world-renowned (1913). The French translation of this article was published the

[17] The following year he went back to Harvard as assistant professor. He was appointed professor there in 1919, Perkins Professor in 1932 and became dean of the Arts and Science Faculty in 1936. According to Marston Morse (1892–1977), one of his students, Poincaré was indeed Birkhoff's supervisor. He said that during his stay at Princeton, when they discussed Birkhoff often referred to the "Méthodes Nouvelles de la Mécanique Céleste" written by Poincaré (1892) which he seemed to have studied in-depth.

following year (see Birkhoff 1914). Two years before, Birkhoff (1912) had shared, in an essay also published in the *Bulletin de la Société Mathématique de France* and entitled "Quelques théorèmes sur le mouvement des systèmes dynamiques" ("Some theorems on the motion of dynamical systems"), some of his preliminary results required in order to establish this demonstration. In it, he also introduced the concept of "recurrent motions":

> Among the motions of a dynamical system, there can exist some possessing the remarkable property of representing the whole motion with any order of approximation we may choose, during *any* time interval equal to T, where T only varies with the order of approximation. In the present Essay, these motions are called *recurrent* motions. They are a natural extension of the periodic motions. (Birkhoff 1912, 305)

By comparing this definition with Ernest Esclangon's (1904, 3) (see *supra*), it appears that the concept of "recurrent motion" covers, in a way, the definition of the *quasi-periodic* or *almost-periodic function*. Birkhoff then proposes a classification of the recurring motions depending on the dimension n. He explains that in the case of the planar motion ($n = 2$), the recurrent motions are periodic:

> (...) the curves representing a stable motion in the plane are closed or asymptotic with a closed curve.[18] (Birkhoff 1912, 317)

Regarding the case of the dimension n he expressed the solutions as complex exponential of the frequency $k_i$and the time "where the quantities $k_1, k_2, \ldots, k_{n-1}$ are real, positive and immeasurable.[19]" (Birkhoff 1912, 317)

He then considers the particular case of the three dimensional space:

> It is interesting to consider, in this regard, the case of $n = 3$ in which the trajectories of the Dynamic are the geodesics of a surface with a negative total curvature, with no singularities. Mr. Hadamard[20] dealt with this case. He divided the geodesics into three categories:
>
> 1° are closed or asymptotic to closed geodesics;
> 2° The ones that draw away on the infinite curved surface;
> 3° The remaining geodesics, which stay in the finite areas of the surface and have a finite number. (Birkhoff 1912, 317)

Birkhoff then demonstrates that:

> If there is a recurrent motion of this third category, Mr. Hadamard's research tends to show that it must be of discontinuous type. (Birkhoff 1912, 317)

This "third Hadamard category" is far from insignificant. Indeed, it is from this case that Hadamard established a generic property of the behavior of chaotic trajectories: the *sensitivity to initial conditions*.

[18]Birkhoff refers to Poincaré (1885a, 228). This is the chapter dedicated to " Étude particulière du torus " ('Particular study of the torus"). At page 228 there is a theorem ensuring the existence of a limit cycle.

[19]The case of immeasurable frequencies corresponds to the case of quasi-periodicity (see *supra*)

[20]Birkhoff refers to Hadamard (1898).

In short, as long as any geodesic which infinitely draws away is surrounded by a continuum of geodesics enjoying the same property, on the contrary, any change, as small as it may be, brought to the initial direction of a geodesic which remains at a finite distance is enough to bring a completely nondescript variation to the final appearance of the curve, as the perturbed geodesic may affect any of the previously listed forms. (Hadamard 1898, 71)

Indeed, according to Jean-Luc Chabert:

Here, we therefore have a theoretical model on which clearly appears, and in a proven manner, the phenomenon of sensibility to initial conditions, when the geodesics are interpreted in terms of trajectories. (Chabert 1992, 324)

A few pages later, he adds:

Later on, Hadamard's essay was addressed by G. D. Birkhoff, then by H. M. Morse[21] in their study on "recurrent motions". (Chabert 1992, 330)

Then, he concludes with:

The surfaces with negative curvatures are therefore the real paradigmatic example of the dynamical systems with sensibility to initial conditions. And Duhem expressed the essential, in his lovely description of Hadamard's theoretical model, when "he insisted on the fact that although deterministic, these systems do not allow for prediction". (Chabert 1992, 330)

It therefore appears that the Birkhoff's "recurrent motions" played a crucial role in the study of the behavior of trajectories of dynamical systems: periodic or quasi-periodic. They will also be the starting point of the "Symbolic Dynamics" under Morse's (1921) and his student Hedlund's impetus (1934).

In regards to the *quasi-periodicity*, Krylov and Bogolyubov established, in the middle of the 1930s, a condition of existence for the quasi-periodic motions based on Birkhoff's recurrent motions:

On another part, G. Birkhoff's research pertaining to standard systems with two degrees of freedom revealed the intimate link between the existence of the invariant curve (relating to a certain isolated transformation corresponding to given differential equations) and the existence of quasi-periodic solutions. We also owe to the illustrious American geometrician the profound results pertaining to the various properties of the motions called "recurrent". Let us incidentally note that based on Mr. Birkhoff's results, we can among other things establish that any recurrent motion, stable as per Lyapunov's meaning, will be almost periodic, and that we will obtain the same results, in a more general manner. Interesting results were obtained recently in this direction by Mr. Franklin and Mr. H. Bohr, founder of the modern theory of almost periodic functions. (Krylov and Bogolyubov 1934d, 100)

In the early 1960s, Niemytski and Stepanoff established in their work entitled: "Qualitative theory of differential equations" that:

Almost every periodic motion is recurrent (Niemytski and Stepanoff 1960, 384)

---

[21] Chabert refers to Birkhoff (1912) and Morse (1921).

Twenty years later, the various research carried out around the world on dynamical systems presenting a *sensitivity to initial conditions* and for which the trajectories have a behavior since then called *chaotic* will aim to determine by which transition or which road we reach this type of behavior. Several scenarios were considered, such as the ones proposed by Ruelle and Takens (1971) and then later clarified by Newhouse (1978). This approach can be summarized by considering a dynamical system in a stationary state which loses its stability by the increase of a control parameter. This situation, which corresponds mathematically to a Hopf bifurcation, is perfectly described by Pierre Collet:

> The Ruelle-Takens road is connected to Hopf's bifurcation. For a vector field, it starts from a fixed stable point, which bifurcates in a closed stable orbit. On this orbit, the motion is periodic. This orbit can **in** turn be destabilized by a Hopf bifurcation to generate an invariant attractor torus. We saw in the second chapter that it can occur with a Hopf bifurcation of the Poincaré map. On the torus, the motion may be quasi-periodic (two immeasurable frequencies) or periodic if there is a frequency clash. This clash phenomenon can also appear or disappear depending on the parameter. We may expect that a torus which presents a quasi-periodic motion will generally bifurcate to a triple torus with a quasi-periodic motion with three frequencies. If the generic bifurcation indeed leads to an invariant triple torus, the generic motion is not quasi-periodic. Ruelle and Takens's result is that on a triple torus, a quasi-periodic motion is not stable due to perturbation: in the whole neighborhood of the vector field (as small ; as this neighborhood may be), we can find another vector field which possesses a strange attractor, and in particular for which the attractor is not a triple torus. (Collet 1999, 35)

According to Bergé[22] the transition towards chaos by quasi-periodicity, i.e. the Ruelle-Takens road, can be described thus:

> For many physical examples, from the simple forced periodically pendulum to some hydrodynamical systems, the typical evolution of the asymptotic regime depending on the parameter, during an experiment of transition towards chaos, can be summarized in this way:
>
> $$\text{BALANCE} \xrightarrow{1} \text{PERIODIC BEHAVIOR} \xrightarrow{2} \text{BI-PERIODIC BEHAVIOR} \xrightarrow{3} \text{CHAOS}$$
>
> All these behaviors can be obtained by models described by ordinary differential equations. (Bergé et al. 1988, 117)

13.5 Marie Charpentier's Work: *Birkhoff's Legacy…*

At the time, there were few French scientists who took an interest in Birkhoff's work. However, in the early 1930s, Marie Charpentier (1903–1994) defended her thesis dissertation[23] entitled "Sur les points de Peano d'une equation différentielle du premier ordre" ("On Peano points of a first order differential equation") at

---

[22]See also Bergé et al. (1984, 171) and the work *L'ordre dans le chaos*, 1991, Pour la Science, Ed. Belin, Paris.

[23]See Charpentier (1931).

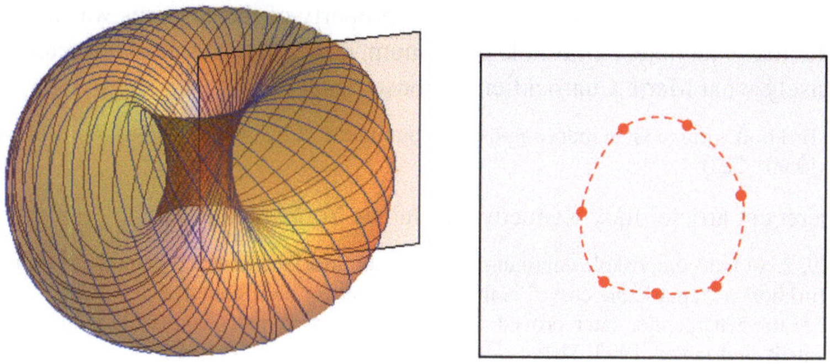

Fig. 13.4 Poincaré Map and Birkhoff's "remarkable curve"

the Faculty of Science of Poitiers under the Direction of Georges Bouligand (1889–1979). The president of the thesis committee was none other than the mathematician Paul Montel (1876–1975), who was very impressed by her results and recommended her for a grant from the Rockfeller Foundation. She therefore spent the following year studying in Harvard with George Birkhoff. Upon her return, Claude Chevalley[24] left from Princeton which allowed her to work at the Faculty of Science of Rennes.[25] In the meantime, Charpentier published many articles in the *C.R.A.S.* (Charpentier, 1934a), in the *Bulletin of the American Mathematical Society* (Charpentier 1932) then in the *Bulletin de la Société Mathématique de France* (Charpentier 1934b).

In the latter, Marie Charpentier demonstrated Birkhoff's result (1932) obtained two years before (probably during her stay at Harvard). Indeed, in an essay also published in the *Bulletin de la Société Mathématique de France,* Birkhoff (1932) had introduced the concept of "remarkable closed curves", which can be shown by using an example of coupled or forced oscillators, i.e. a dynamical system with two degrees of freedom[26] represented by the equation (9.21) or (9.23). In this case the trajectory which is a solution of one or the other equation can be described as a "characteristic on the surface of the torus" (see *supra*). If we consider a section of this torus with a plane passing through a meridian circle, the trajectory will intersect this plan with each (see Fig. 13.4).

We then observe (on the Fig. 13.4 on the right) that the "impacts" of the trajectory on this plane which is where the Poincaré map appears randomly, in such a way that it seems impossible to predict where the next impact will be. Gradually, as the trajectory intersects the plane, a "closed curve" is formed (dotted line on Fig. 13.4). This curve is Birkhoff's "remarkable closed curve". Actually, according to Patrice

[24]French mathematician and algebra specialist, Chevalley (1909–1984) was a member of the Bourbaki group.

[25]See Siegmund-Schultze (2001, 124–125).

[26]See Abraham and Shaw (1982–1988, vol. 2, 77).

Le Calvez (1988, 77) "the first remarkable property of these curves was to not be a curve (actually an indecomposable continuum, now called Birkhoff attractor)". This is precisely what Marie Charpentier demonstrates:

> Mr. Birkhoff's curve is an indecomposable continuum. (Charpentier 1934a, 703; Charpentier 1934b, 223).

In a recent article, Judy Kennedy and James Yorke recall that:

> In 1932, an indecomposable continuum arose as an invariant set in a dynamical system. G. D. Birkhoff's "remarkable curve" is the invariant set for a diffeomorphism on an annulus, and Marie Charpentier later proved that this curve is an indecomposable continuum.[27] (Kennedy and James 1995, 310)

A few years before, Patrice Le Calvez (1986) evidenced the existence of quasi-periodic orbits in the Birkhoff attractors.

On their side, Abraham and Shaw (1982–1988, vol. 2, 77–80) clarified that the Birkhoff's "remarkable closed curves" were only observed at the start of the 1980s by Shaw (1981) and that Birkhoff had thus discovered a chaotic attractor, which is a theoretical subject of study in discrete dynamical systems.

13.6 Hervé Fabre: *On the Recurrent Motions*

At the end of the 1930s, French astronomer Hervé Fabre[28] implemented Birkhoff's work in the field Celestial Mechanics. Born in 1905 in Aussillon (near Mazamet in the Tarn), Fabre became an orphan early in life. His father was killed during the First World War, and his mother passed away soon after. He was raised by his uncle, and after obtaining his *baccalauréat de Math Élém.*, a *licence* in Mathematics at the University of Montpellier when he was seventeen, as he was too young to register for the *agrégation*, he was self-taught when he became a Mathematics teacher. He then worked almost on his own on his thesis dissertation, entitled: "Les mouvements récurrents en mécanique céleste, et la variation des éléments des orbites" ("The recurrent motions in celestial mechanics, and the variation of the orbit elements"), presented to the Faculty of Science of Paris to obtain the rank of *Docteur ès Sciences Mathématiques*, and defended on the 24th of May 1938 before Misters Ernest Esclangon (President), Jean Chazy and Armand Lambert[29] (Examiners). According to astronomer Jean-Louis Pecker:

---

[27] Kennedy and Yorke refer to Birkhoff (1932) and Charpentier (1934b).

[28] This paragraph could not have been written without the collaboration and kind assistance of Mrs. Françoise Le Guet-Tully, astronomer at the *service Patrimoine* of the *Observatoire de la Côte d'Azur* and Mr. Luc Poirier, technician, who sent me biographic elements regarding Hervé Fabre (1905–1995).

[29] Armand Lambert (1880–1944) was chief of the *Service Méridien* at the Observatory of Paris. After June 1940 and the voting of the firsts anti-Semitic laws, he was forced to wear the yellow star. In spite of the risks, he did the interim at the direction of the Observatory of Paris, and ensured

This was actually an attempt at structuring celestial mechanics, based on anything but successive approximations processes. In the mathematical meaning of the term, Fabre stated that the stability of the solar system does not exist (a very fashionable idea nowadays!). In the ideal case where the number of degrees of freedom can be reduced to two (for example the case of the proof body with no notable mass situated in the gravitational field of a flattened spheroid), the orbits can maintain themselves inside a torus (a sign of a strange attractor?), which is a genuine stability. But then we must go back over the concept of "mean motion". The mean motion, as defined by astronomers, is slowly varying, whereas the rigorous, fixed, mathematical mean motion, depends on the filling of the torus when t varies from $-\infty$ to $+\infty$. The periodic and recurrent motions do not fill the torus. (Pecker 1996, 4)

Before his appointment as assistant-astronomer at the Nice Observatory, Fabre published two notes in the *C.R.A.S.* in 1937, which were presented by Jean Chazy (1882–1955) (see *infra*). In the first, Fabre (1937a) demonstrates that in stable motions of a force field with a revolution axis and equator, the longitudes of the node and periastron[30] (see Fig. 13.5) of an orbit are expressed:

(...) during a randomly long but finite time, like the sum of a linear term in t and a quasi-periodic function of t with two fundamental frequencies. This results from the replacing of the real orbit of the planet by another, called absolute orbit, for which the initial conditions may be chosen in such a way that the gap between absolute orbit and real orbit is randomly small during a randomly long time. (Fabre 1937b, 1215)

In his second contribution, Fabre continues his explanations expressed in the first, explaining that:

The idea of the *absolute orbits* is owed to Gyldèn, but could not be used because of some errors committed by this scientist.[31] (Fabre 1937b, 1215)

He then explains, based on Poincaré's work (1905a) "the convergence of Glydèn series" that consequently, the problem of the *absolute orbits* should be approached from another angle. He then formulates one of the main results found in his thesis (Fabre 1938):

Based on Mr. Birkhoff's theory of recurrent motions, and on the fact that any quasi-periodic motion is recurrent, we demonstrated the existence of absolute orbits. Fabre (1937b, 1216)

By expressing the longitudes of the node and periastron by using quasi-periodic functions and admitting, twenty years before it was established by Niemytski and Stepanoff (see *supra*) that "any quasi-periodic motion is recurrent", he demonstrates the existence of absolute orbits, i.e. of quasi-periodic solutions, thus reducing

the continuity of the work at the *Bureau international de l'heure*. He was arrested in 1943 at his home and never came back from Auschwitz.

[30]Or more generally the longitude of the *periapsis*, (noted ϖ), of a body in orbit is the longitude (measured from the vernal equinox point) for which the *periapsis* (point of the orbit closest to the central body) would be reached if the inclination of the body was null. The longitude of the *periastron* or *periapsis* is then expressed as the sum of the longitude of the ascending node and argument of periastron: $\varpi = \omega + \Omega$.

[31]Fabre refers to Poincaré (1905a) for a critical study of Glydèn's method.

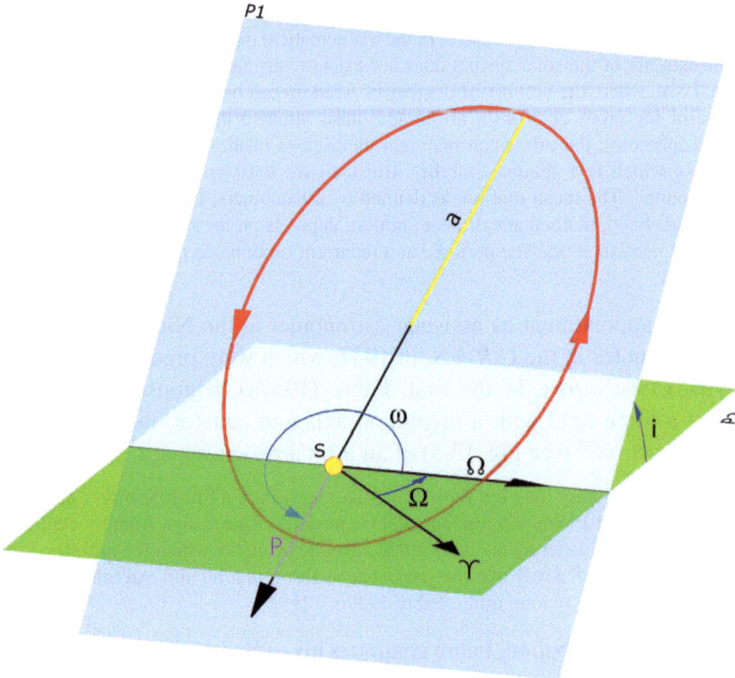

Fig. 13.5 Angular parameters of an elliptic orbit (Internet source), P_1 orbit plane, P_2 ecliptic plane, *P* periastron, *S* Sun, *a* semi-major axis, γ vernal point, *i* inclination, Ω longitude of the ascending node, ω argument of periastron

this Celestial Mechanics problem to a question that arises by radio engineering (see *supra*). In this context, it is not surprising to see the names of Krylov and Bogolyubov.

> Whatever the gravitational field we place ourselves in, *it is impossible, with the current state of our knowledge, to practically determine the Fourier series representing the coordinates on a quasi-periodic orbit*, and we must content ourselves with an approached representation, carried out for example by using Misters Krylov and Bogolyubov's methods. The approximations we obtained with these methods give the same appearance to the absolute orbits and real orbits. Let us add that the works of their authors provided criteria ([4]) to state the quasi-periodicity of the recurrent motions in relation to the longitude and time.
>
> ---
>
> ([4]) Les méthodes de la Mécanique non linéaire appliquées à la théorie des oscillations stationnaires, Kieff, Krylov and Bogolyubov (1934a, 88). (Fabre 1937b, 1216)

Here, Fabre confirms, on the one hand the impossibility of using Fourier's harmonic analysis to study the quasi-periodic solutions in the case of Celestial Mechanics, which incidentally can extend to Radio engineering, and mentions on the other hand, that it was thanks to the Krylov-Bogolyubov method that he managed to establish the quasi-periodicity of the recurrent motions. This clearly shows that contrary to what we could expect, there is not real airtight boundary between these various disciplines. Nevertheless, it is likely that the spreading was caused in this case by Arnaud Denjoy himself, whose work is precisely at the interface of these two fields. Fabre's thesis (1938) is a real synthesis of all the previous researches: Poincaré's, Birkhoff's, Esclangon's, Bohl's, Bohr's, Favard's, Denjoy's, etc.

In chapter V dedicated to the approached representation of the "recurrent motions" and "absolute orbits", he demonstrates that any absolute orbit corresponds to a recurrent motion, by first using Poincaré's work 1892. He recalls:

> By approaching the study of periodic solutions to the three body t problem, that Poincaré indeed called "very likely", the following proposition, which became famous from being quoted frequently:
> "With a given equations (analogous to the ones for the three body problem for the case where two of them have very small masses), and some specific solution to these equations, we can always find a periodic solution, whose period may, admittedly, be very long, such that the difference between the two solutions may be as small as we want it, for a time as long as we want it.[32]"
> This role which Poincaré attributes, without demonstration, to periodic motions, surely belongs to the more general motions of which Mr. Birkhoff demonstrated the existence, and which he called *recurrent*.[33] (Fabre 1938, 82)

It is interesting to note once again the importance of the analogy in the reasoning used by Fabre. Indeed, he made an analogy between the role played by the motions that Poincaré saw as periodic, with the motions that Birkhoff called recurrent. He then clarifies the definition of this concept introduced by Birkhoff (1912, 1920):

> The necessary and sufficient condition for a motion to be recurrent is: "*given a positive number δ, as small as we want it, there exists a positive number T large enough for the arc of the trajectory described by the traveling objects between the instants t and $t + T$ (any one) to possess points distant of less than δ from any point of the trajectory* ([2]). "
>
> ---
>
> ([2]) Therefore, any quasi-periodic, or almost periodic motion, is recurrent, the reverse is not true. (Fabre 1938, 83)

This shows that he did not actually demonstrate Niemytskii and Stepanoff's theorem (see *supra*), but that he admitted as "likely" the conjecture he formulated in the footnote, which will allow him to demonstrate however, that any absolute orbit corresponds to a recurrent motion. In order to ensure this, he uses the concepts of the Poincaré map and Birkhoff's remarkable closed curves:

> A periodic motion corresponds to a finite set of points on the map surface[34]: $P_1, P_2, \ldots, P_n$, such that each coincides with its nth transform (…). The non-periodic recurrent motions, in a system with two degrees of freedom, form families, and each family is connected to the whole of the points of a closed curve, or the whole of the points of several closed curves, or,

[32]This slightly reworked quote from Fabre is taken from Chapter III entitled "Periodic solutions", of the first volume of the "Méthodes Nouvelles de la Mécanique Céleste" ("New Methods of Celestial Mechanics"). See Poincaré (1892, 82).

[33]Fabre refers to Birkhoff (1912).

[34]This is a Poincaré map (see *supra* Fig. 13.5).

> if there is only one family, connected to the whole of the points of the whole map surface, "a single minimal set of recurrent motions[35]". (Fabre 1938, 88)

Then, he applies the Krylov and Bogolyubov method (1934d) which allowed him to transform the differential equations for the motion, and established, based on the work of Poincaré (1881, 1882, 1885a,b, 1892, 1893, 1899), Birkhoff (1912, 1920, 1932), Denjoy (1932a), Favard (1933), Levi-Civita (1900a,b)[36] that any quasi-periodic motion is recurrent (Fabre 1938, 82–114).

> An exact quasi-periodic solution corresponds to a recurrent motion. (Fabre 1938, 113)

Fabre's research, which reflects his personality – profound, and discreet– seems to represent one of the first applications, in France, of the Krylov-Bogolyubov methods, i.e. of the Nonlinear Mechanics to a field other than Radio engineering.

However, his stance seems marginal, as many astronomers such as Fatou or Chazy continued to use the Poincaré-Lindstedt methods to study the *three-body problem*.

[35]Fabre refers to Birkhoff (1920), probably paragraph 42 entitled: "Existence of closed invariant curves".

[36]See also Levi-Civita (1906a,b, 1911).

Chapter 14
Hadamard and His Seminary: *At the Crossroads of Ideas and Theories*

Throughout this study, Jacques Hadamard has been mentioned several times, notably as he shared numerous notes with the Academie des sciences, in Paris: those of Andronov (1929a), Andronov and Witt (1930a,b,c,d), Krylov and Bogolyubov (1932a,c, 1934e, 1935a), etc. Hadamard also belongs to the group of leading scientists invited to participate to the first *Conférence Internationale de Non-linéaire* (*International Conference on Nonlinear Oscillations*) which took place in January 1933 at the Henri Poincaré Institute. By his training and position at the time of Poincaré's demise in 1912, Hadamard appears to be the mathematician most suited to apprehend his work.

In November 1897, he became deputy to Maurice Lévy (1838–1910) at the chair of *Mécanique Analytique et Mécanique Céleste du Collège de France* (Analytical Mechanics and Celestial Mechanics at the Collège de France). It is at this time that he published two dissertations entitled: *Sur certaines propriétés des trajectoires en Dynamique* (On certain properties of dynamics trajectories) (Hadamard 1897) and *Sur les surfaces à courbures opposées et les lignes géodésiques* (On surfaces having opposite curvatures and geodesic lines) (Hadamard 1898).[1] In 1909, he becomes chair of the Analytic mechanics and Celestial mechanics at the Collège de France. Three years later, he succeeds Henri Poincaré at the Académie des sciences and he also succeeds Camille Jordan (1838–1922) at the École polytechnique. In the period between the wars, he played an essential role in the French and international mathematics community.[2] He notably organised a seminary at the Collège de France in which many participants' testimonies came through to us.[3] However, here can

[1]See the article by Jean-Luc Chabert (1992).

[2]See H. Gispert and J. Leloup (2009, 70–75).

[3]See, for instance, in Maz'ya and Shaposnnikova (1998/2005), where it has been stated that the following speakers were participating: Émile Borel, Paul Montel, Henri Lebesgue, Paul Lévy, Maurice Fréchet, Jean Chazy, Henri Villat, Gaston Julia, Arnaud Denjoy, Georges Valiron, Élie Cartan, André Weil, Vito Volterra, Tulio Lévi-Civita, George Birkhoff.

J.-M. Ginoux, *History of Nonlinear Oscillations Theory in France (1880–1940)*,
Archimedes 49, DOI 10.1007/978-3-319-55239-2_14

Fig. 14.1 Jacques Hadamard (Harcourt Studio, internet source)

be used a source which was of particular interest for this subject. That interest has been crystallised by J.-L. Chabert and C. Gilain in their work on Hadamard during the First World War: the *Annuaire du Collège de France* (the *Collège de France yearbook*), which contains, from 1900, the "lessons summaries" of each scholar year, written by the professor. Hadamard has transformed, as soon as the year 1913–1914, but especially from the year 1920–1921,[4] a growing part of his courses at the Collège de France in "analysis of scientific memoirs", presented by young or confirmed mathematicians, be they French or foreign, with great space for discussion. This "seminary" part will then take over all of Hadamard's lessons. These documents contain information that is directly linked to the theme of this thesis. The lecture part of Hadamard's lessons will be presented first, followed by the seminary part (Fig. 14.1).

14.1 Spreading the Legacy of Poincaré

14.1.1 Lessons at the Collège de France

Hadamard's lessons were mainly based on his own research or recent work by other mathematicians, also influenced by Poincaré's work which constituted his main interest. Some lessons are also dedicated to Lyapunov's work from 1907–1908, as his thesis had just been translated to French. In 1910–1911, his lessons relate

[4]Until 1937, the year he retired.

to "Quasi-periodic functions and their application to analytical mathematics and to celestial mathematics". He also outlined both Bohl and Ernest Eclangon's work.[5] However, from 1919–1920 until 1924–1925, Hadamard's lessons were dedicated to the systematic presentation of Poincaré's work. For instance, he wrote in 1924:

> The lesson from 1923–1924 has been, as it has been the previous years, composed of two distinct parts. The first lesson was dedicated to the first years of Poincaré's work. The last dissertations that were studied were those that dealt with arithmetic and the theory of functions; this time, dissertations on the theory of differential equations were studied, starting with those which treated these dissertations from a purely analytical point of view, that is, from the perspective of singularities as real as they are complex, and the development of proper series, able to offer solutions. The opposite view point, that is the one which is real and qualitative, is represented in four famous dissertations *Sur les courbes définies par les équations différentielles* (On curves defined by differential equations) (Journal de Math., 1880–1885) was discussed towards the end of the lesson. Its study will be continued this year. (*ACF*, 1923–1924, 38)

Here we can seen, on Hadamard's part, a willingness to transmit to the new generation of mathematicians a heritage that could be lost, at least partially, since Poincaré's premature demise in 1912, followed by a long period of war.

14.1.2 Hadamard's Lectures Abroad

Through his important contribution in the issue *Acta mathematica*, dedicated to Poincaré and published in 1921 before the war, many other elements show Hadamard's "militant" willingness. These are lectures he gave abroad concerning Poincaré's work in 1920–1925. In spring 1920, Hadamard presented during a series of lectures given at the Rice Institute (Houston, Texas) "The early scientific works of Henri Poincaré" (Hadamard 1922). The following year, his intervention at the *Institut d'Études Catalanes*, in Barcelona,[6] was entitled: "Poincaré I la Teoria de les equacions diferencials[7]" (Hadamard 1923). In 1925, he delivers a second series of lectures at the Rice Institute entitled "The Later Scientific Work of Henri Poincaré" (Hadamard 1933). However, Poincaré's legacy can also be revived by promoting the development of new research prolonging his work, which will be one of the functions of his seminary at the Collège de France.

---

[5]See Esclangon (1902, 1904) and Bohl (1893, 1904, 1906).

[6]Professor Emma Sallent del Colombo, epistemologist from the physics department at the University of Barcelona, has pointed out that Hadamard also traveled to Madrid, as he was invited to the *Junta de Ampliación de Estudios* (Graduate studies Council) to present a cycle of lectures on Poincaré's work.

[7]Poincaré and the theory of differential equations.

14.2 The Seminary Part of Hadamard's Lectures *a Few Subjects Addressed*

14.2.1 The Work of George Birkhoff

The work carried out by Birkhoff will often be part of the seminary's program. In 1922–1923, the complete demonstration of Poincaré's latest theorem, which he stopped working on and shortly after died, was presented. This demonstration allowed an opportunity to take over this theorem. This work allows "the establishment of the existence of a limited number of periodic solutions in a case with three bodies that is more general than all those considered until then".[8] Published in January 1913 in the *Transactions of the American Mathematical Society* (Birkhoff 1913), it is published in French the next year (see Birkhoff 1914).

The seminary also studies, the same year, Birkhoff's work addressing "recurrent motions" (Birkhoff 1912), then his dissertation on the punctual transformation of surfaces, in 1925–1926. Birkhoff himself will attend the seminary to deliver lectures in 1930–1931, as he is invited to the Collège de France[9] under the Michonis lecturer title.[10] Thus, the complements brought to the unsettled issues left by Poincaré were the subject of presentations, discussions and debated during the Hadamard seminaries. This will also be the case with the problem enquired by Poincaré (1885a, 220) in his "Particular Study of the Torus".

14.2.2 Nonlinear Phenomena

Alike Poincaré, Hadamard insists many times in his lectures' summaries on the importance of staying attentive when it comes to links that can establish themselves between pure mathematics and other disciplines. Thus, in 1931:

> Mentions must be made, writes Hadamard, about the relation between pure mathematics and other disciplines: sessions dedicated to same ideas have been particularly interesting this year and this is how applications that were recently received, in purely practical and professional domains, were some of the most beautiful ideas issued by Poincaré in his dissertations on curves defined by differential equations. (*ACF*, 1930–1931, 39)

---

[8]See Darboux (1914, 145).

[9]A young mathematician, Marie Charpentier (1903–1904), after her thesis, will work at Harvard with Birkhoff, with the support of the Rockfeller grant. She improved many of his results. (see Charpentier 1934a,b).

[10]In 1903, M. G. Michonis donated to the Collège de France, a sum that served to "have happen, whenever possible, by a scientist or a foreign thinker appointed by a professor or an administrator at the Collège de France, and which, when the circumstances allow it, will include a philosopher or a historian in religious sciences, a series of conferences.". His wishes were executed from 1905.

One can infer that the previous extract is linked with the fact that Hadamard had just presented to the Science Academy a note by Andronov, on the 14th of October 1929, then the one by Andronov and Witt, on the 20th of January 1930 (see *supra*, Part II).

14.2.2.1 Van der Pol's Research

During the school year 1931–1932, Philippe Le Corbeiller was invited to present Van der Pol's work:

> Twice, writes Hadamard, we've been fortunate enough to be shown results, and most of all we've had the chance to ask questions which call for more research: one by M. Léon Brillouin (*Atomic mechanics of crystalline network*) and one by M. Le Corbeiller (*Generalization of Van der Pol's research to systems of differential equations*). (*ACF*, 1931–1932, 39)

The fact that a specific theme linked with the non-linear domain, taking off at the time, was discussed during the Hadamard seminary shows the appeal that these new researches had on the French scientific community. It is necessary to remind ourselves that during the years following Le Corbeiller's first intervention at the Collège de France, Van der Pol had been invited twice in Paris to present his work, in 1928 and 1930 (see *supra* Part II). Finally, Le Corbeiller (1931a) had given three lectures on these themes in 1931 and 1932. It should be mentioned that, even though he obtained the rank of Doctor of Mathematical Sciences at the Science Faculty of Paris, Le Corbeiller belonged to the field of engineering (X 1910 and E.S.P.T. 1914) and more specifically to the engineers of the T.S.F. It is therefore surprising to find him in such a crowd. This shows the disciplinary open-mindedness wanted by Hadamard, and that the demand introduces new mathematical means to deal with Industry and Engineering issues. During the school year of 1934–1935, Le Corbeiller is once again invited to the Collège de France:

> With M. LE CORBEILLER, we have discovered how an electro-technical application problem , particularly a modern one, raises analytical problems that apply to various orders of analysis. (*ACF*, 1934–1935, 63)

Between these two interventions by Le Corbeiller, several major events, as we know, happened in Paris: the International Electrical Congress in July 1932 during which a great number of international scientific figures in the field of T.S.F. and non-linear oscillations met, such as Van der Pol, Appleton, Blondel, Gutton, Bethenod, Krylov and Bogolyubov; the first International Conference on Non-linear Oscillations also greeted Hadamard and Philippe Le Corbeiller. It is therefore conceivable that this second invitation to the seminary following these events shows Hadamard's willingness to establish a link between science and technology.

14.2.2.2 Kostitzin's Work

Vladimir Kostitzin was invited as early as 1929 to present his work at the Henri Poincaré Institute. His research was at the convergence of the work of Poincaré and Vito Volterra, i.e. at the interface of the domains of qualitive theory of differential equations and "hereditary physics".[11] At the beginning of the 1930s, Hadamard presented Kostitzin's notes (1932, 1933) at Science Academy. He was then invited to present his work during the Hadamard seminary in the 1934–1935 school year:

> The work on hereditary elasticity phenomena by M. KOSTITZIN, shows how beneficial mathematics can be to shed a light on a physics problem: in this case, on the interpretation of recent experiences, it was proven that we could no longer continue to "linearize" the problem. (ACF, 1934–1935, 62)

This last sentence, which declares the end of the linearization methods, appears to be out of sync compared to numerous research that tend to demonstrate the contrary. Many, such as Blondel, Van der Pol, Le Corbeiller and Andronov, to quote only the most important scientists, had already completed this stage before 1934. However, Kostitzin and Le Corbeiller's interventions allow the demonstration that, on the one hand, non-linear phenomena interests mathematicians greatly, and on the other hand, that the necessity to investigate possible applications for different concepts and theories is being felt increasingly acutely. Thus, it appears that the Hadamard seminary, intellectual centre of mathematical life between the two wars, didn't remain indifferent to the development of non-linear oscillations theory, and its first fundamental results were published in France during this period.

14.2.3 The Problem of the Characteristics on the Surface of the Torus

14.2.3.1 Denjoy's Work

In an article written shortly after Poincaré's demise and entitled "L'œuvre mathématique de Poincaré" (Poincaré's mathematical achievement), Hadamard reminds the problem:

> After studying the sphere, Poincaré attempts, from the same viewpoint, to study the torus, he deduces that this second case can result in a great amount of new circumstances that the first could not foresee. However, what happens must first be determined. There are also new difficulties, hence many questions must be asked to resolve these. The questions raise difficult arithmetical problems which have so far remained without answers. (Hadamard 1921, 247)

[11] Which implied the use of the integral kernel by Volterra to show a certain delay between the effect and the cause in a predator-prey type of relationship. See Volterra (1931, 141 and following).

Arnaud Denjoy presents his work on these issues at the Hadamard seminary in 1931–1932, as indicated in the following summary written by Hadamard himself:

> Finally, the exchanges that happened during our reunions have had a direct scientific repercussion, provoking, under short intervals, the work dedicated to Minkowski's function (x) by M. Denjoy, which has also presented during our sessions the great discovery through which he resolved the question, dealing with the defined curve by a differential equation on the surface of a torus, which Poincaré left with no answer. (*ACF*, 1931–1932, 39)

14.2.3.2 Weil's Work

During this period, French mathematician André Weil, one of the founders of the Bourkabi group, also showed interest in the problem of the curves on the torus. In 1932, he presented a method allowing the study of this question, which was published in the *Journal of the Indian Mathematical Society* (Weil 1932). The following year, the Annuary of the Collège de France reveals he made several presentations during the Hadamard seminary, but it can not be confirmed whether its subject was the same, as Weil was active participant, and talked about subjects. However, several years later, Weil (1936) published an article entitled "Les familles de courbes sur le tore" (Families of curves on the torus) in which he reminds:

> The problem has been dealt with for the first time by Poincaré, in his differential equations theory: he supposed that the family is defined by a differential equation
>
> $$\frac{dx}{X(x,y)} = \frac{dy}{Y(x,y)},$$
>
> where X, Y are periodical in x and y of period 1, satisfy Lipschitz conditions and do no cancel each-other simultaneously. The hypotheses, from the viewpoint of the study of the family, are not really restrictive; however Poincaré assumed that X never vanished, which implies that on the basis (x, y), which is the universal cover of the torus studied, every curve cuts every vertical $x = C$ in one and only point: it is said, in that case, that the meridians $x =$ are sectional curves, and we can easily observe that the issue stems back to the study of topological invariants of a homomorphic transformation of a circle itself, Poincaré demonstrates that there is, for a certain transformation, three possible cases: or they form from a same point (whatever this point may be) an ensemble that is uniformly dense, the transformation is then equivalent to the rotation of an irrational angle: that angle ("rotation number") is then the unique invariant, or the transformations have as an accumulative point a perfect non-dense ensemble; or there exists at least one point from which the transformed are in a finite number, that is (going back to the torus) a shut curve belonging to the family. Following Poincaré's footsteps, many authors were preoccupied with the same problem: the most interesting result is the one found by Denjoy, which shows the impossibility of the second case when the function defining the transformation satisfies at only certain very large conditions of the regularity. (Weil 1936, 779)

It can be noted that later in his article "The future of mathematics", Weil (1948) states:

> (…) the study of Van der Pol's equations, and the said-to-be relaxation oscillations, is one of the rare interesting questions derived to mathematicians by contemporary physics; as this study of nature, formerly one of the main sources of the greatest mathematical problems,

seems to have, in the last several years, borrowed to us much more than it has given back. (Weil 1948, 316–317)

Although this affirmation is not alike Hadamard's usual point of view, it is most likely related to what was heard during the Collège de France seminary on non-linear oscillations.

Conclusion of Part III

In this third and last part, we analyzed the various mathematical methods developed in order to study the oscillatory behaviors of the solutions of forced or coupled systems, omnipresent in Radio engineering. What played a crucial part here is the property of quasi-periodicity of the solutions. In this case, the methods employed in order to obtain approximations for the amplitude and period of the periodic solutions were not suitable anymore. In the early 1920s, Van der Pol started working on quasi-periodic oscillations for a triode subjected to a forcing or a coupling. When faced with problems of convergence in the Fourier or Poincaré-Lindstedt asymptotic expansions, he was led to offer another method, nowadays called the "Van der Pol method", which allowed him to evidence a whole series of new phenomena such as the forced synchronization, hysteresis resonance, frequency demultiplication, or the trailing. Our analysis of this section of Van der Pol's body of researches, show that this may be his most important contribution to the development of the nonlinear oscillations theory, while in historiography, his contribution is often reduced to the discovery of relaxation oscillations. In the early 1930s, the scientists from the Kiev School Nikolai Krylov and Nikolai Bogolyubov developed, then formalized, the "Van der Pol method", at the same time stressing that the Poincaré-Lindstedt method was inadequate for the study of Radio engineering problems, where quasi-periodic solutions occurred. By systematically going back over the phenomena discovered by Van der Pol, and applying a rigorous mathematical process to them, they managed to lay the foundations of a new field, which they called Nonlinear mechanics. In a series of publications in France, and in French, Krylov and Bogolyubov clarified the reasons why the "Poincaré-Lindstedt methods" did not allow the study of the quasi-periodic solutions of Radio engineering devices, and introduced the various steps of the development of what they called the "new methods of Nonlinear mechanics". These methods were synthesized in their work entitled "Introduction to Nonlinear mechanics", published in Russian in 1937, the same year Andronov and Khaikin's famous work Theory of oscillations stemmed from the Gorki School. At the same time, Mandelstam and Papalexi wrote an article in French concerning a hybrid method which could be called the "Van der Pol-Poincaré method", and they noted that the "Van der Pol method" held its mathematical founding principles in the research lef by French astronomer and mathematician Pierre Fatou. It also appeared that the mathematical framework likely to describe the behavior of Radio engineering solutions had been developed as early as the beginning of the century by another French scientist, Ernest Esclangon, who authored theoretical contributions on the quasi-periodic functions. These types of functions were the subject of numerous researches, especially by Favard or Denjoy in France. Around the late 1930s, another French astronomer, Hervé Fabre, linked this concept of quasi-periodicity with the concept of "recurrent motions" previously introduced by Birkhoff. At this point, the theory of nonlinear oscillations reached its maturity, and a transition occurred towards the development of a new theory named Dynamical systems.

General Conclusion

Until now, historiography's reconstruction of the events showed a "jump" of almost half a century between the introduction of the limit cycle concept by Poincaré (1882) and the moment when this concept was implicitly applied by Van der Pol (1926d) and the correspondence was established by Andronov (1929a).

The aim of this study was therefore to analyze the period ranging from 1880 to 1940, in order to measure the importance, on the one hand, of the contribution brought by the French scientific community, and on the other hand, of the role played by France, which thus appeared as a "crossroads" where various participants to the development of a theory of nonlinear oscillations met.

The first part of this work, entitled: "From sustained oscillations to relaxation oscillations" allowed the unveiling of the existence of four devices which were analogous, from a mathematical standpoint: the series-dynamo machine, the singing arc, the triode, and Abraham and Bloch's multivibrator, inside of which an oscillatory phenomenon of a new kind was produced, then studied, which was called *sustained oscillations*.

The first series-dynamo machine, allowed the evidencing of a component's presence displaying a nonlinear behavior responsible for these oscillations.

The second one led to the discovery of lectures on wireless telegraphy given by Poincaré in May and June 1908, which had been apparently forgotten, in which he established that the necessary and sufficient condition for the establishment of a stable regime of sustained waves in the singing arc was the existence of a stable limit cycle (although he did not use this term).

From the third – the triode, it was established that it was André Blondel who carried out in 1919 – one year before Van der Pol – a first modeling which *de facto* led him to the writing of the equation for the oscillations sustained by a triode.

The fourth- Abraham and Bloch's multivibrator, showed that it was at the source of the concept of *relaxation oscillations* introduced by Van der Pol to describe this new oscillatory phenomenon. In 1925, probably unaware of Poincaré's work, Janet submitted the problem of the study of the solutions to the nonlinear

J.-M. Ginoux, *History of Nonlinear Oscillations Theory in France (1880–1940)*,
Archimedes 49, DOI 10.1007/978-3-319-55239-2

differential equation representing the oscillations of the series-dynamo machine to mathematicians Élie and Henri Cartan, who demonstrated the existence of a periodic solution, and determined its characteristics.

The next year, Van der Pol published his famous article entitled "On relaxation-oscillations", in which he gave the prototype equation characterizing the oscillations of a triode, and consequently, of the singing arc and series-dynamo machine. Although he represented it by using a graphical integration of the solution to this equation with a closed curve form, he did not recognize it was Poincaré's limit cycle.

In 1928, the engineer and mathematician Alfred Liénard considered a second order differential equation which was more general than Van der Pol's, and established, under specific conditions, the existence and uniqueness of a periodic solution. However, he did not associate it with Poincaré's concept of limit cycle either.

Although the fundamental result established by Poincaré in 1908 did not seem to be used until it was found independently by Andronov in 1929, the whole of the French work carried out in this field played an important part in the conceptualization of relaxation oscillations by Van der Pol, and the introduction of new methods in nonlinear analysis.

The aim of the second part of this work, entitled: "From relaxation oscillations to self-oscillations", was to study the emergence of the concepts of relaxation oscillations and self-oscillations, their initial spreading in France during the late 1920s and early 1930s, as well as their reception by the French scientific community. This allowed the establishment that faced with the conceptualizations of nonlinear oscillations, the French scientific community reacted in various ways. While Andronov's scientific work caused relative indifference, a great enthusiasm rose in favor of Van der Pol's ideas and had caused a true "hunt for the relaxation effect".

Moreover, Van der Pol's various presentations in Paris, the great number of publication from the Russian School in the *C.R.A.S.*, and the holding of the very first *Conférence Internationale sur les Oscillations Non Linéaires* at the Henri Poincaré Institute in 1933 allowed the evidencing of the role played by France, which therefore appears as a "crossroads" where various movements participating in the development of the theory of nonlinear oscillations met. However, the failure of this conference highlights the inability of the French scientific community to federate its own results around a common thematic, prerequisite to the development of a theory.

In this third and last part entitled: "From self-oscillations to quasi-periodic oscillations", the various mathematical methods developed in order to study the oscillatory behavior of the solutions to forced or coupled systems, omnipresent in Radio engineering, were analyzed. Again, the role of France as a venue for presentations, debates, and discussions regarding these methods was also connected to Poincaré's work, but Fatou's work was also stressed as being crucial. Moreover, the analysis of the part played by the French contribution to the development of the theory of quasi-periodic functions and the solving of the tricky problem of the characteristics on the surface of a torus left pending by Poincaré showed the importance of his work in the development of the theory of nonlinear oscillations, at the same time as it reached its maturity and a transition began towards the development of a new theory, that is Dynamical systems.

This study, which led to reconsider the historiography by taking into account, on the one hand, the contribution of the French scientific community, and on the other, the role of France as a "crossroads" where various movements participating in the development of the theory of nonlinear oscillations met, also poses again the problem, widely discussed in the past few years, of the complexity of Poincaré's scientific legacy in this field, as already stated by David Aubin and Amy Dahan-Dalmedico:

> As we shall see, however, various results found by Poincaré would be picked up by generations of successors, studied, developed, and extended in a great variety of directions. Nonetheless it can be safely said that is was never truly *lost* or *forgotten*. What is true, on the other hand, is that Poincaré's results, which we have been summarized above, were not, until much later, mobilized in an integrated manner. Comprising dozens of books and hundreds of articles, his lifework was never, so to speak, digested by successors. In fact, whether they even appeared as such to Poincaré himself is not clear. The dynamical systems synthesis is a *post facto* construction that has to be accounted for on its own terms. As a matter of fact, Poincaré's published papers and books have constantly been revisited until today, and only through this process could the dynamical system theory develop. And it is this very persistent back-and-forth motion between the master and later contributors that makes the *longue-durée* history an unavoidable, but especially slippery, task for the chaos historian. (Aubin and Dahan Dalmedico 2002, 10)

Bibliography

Abréviations

Bulletin de la Société Mathématique de France: *B.S.M.F.*
Comptes rendus hebdomadaires des séances de l'Académie des sciences: *C.R.A.S.*
Journal de mathématiques pures et appliquées: *J.M.P.A.*
Journal de Physique et le Radium: *J. Phys. Radium*
Journal de Physique Théorique et Appliquée: *J. Phys. Théor. Appl.*
Revue générale des sciences pures et appliquées: *R.G.S.P.A.*

Sources Imprimées

Annuaire du Collège de France (ACF), Paris: Ernest Ledoux éd.

J.-M. Ginoux, *History of Nonlinear Oscillations Theory in France (1880–1940)*, Archimedes 49, DOI 10.1007/978-3-319-55239-2

References

Abarbanel, Henry, Mikhail Rabinovich, and Mikhail Suschik. 1993. *Introduction to nonlinear dynamics for physicists*, Lecture notes in physics, vol. 53. Singapore: World Scientific.

Abelé, Jean. 1942a. Système d'entretien à amplitude autostabilisée. *Comptes rendus hebdomadaires des séances de l'Académie des sciences* 214(18 mai 1942): 841–842.

Abelé, Jean. 1942b. Application du système d'entretien à amplitude autostabilisée à un pendule élastique. *Comptes rendus hebdomadaires des séances de l'Académie des sciences* 214(1er juin 1942): 875–877.

Abelé, Jean. 1943. *Étude d'un système oscillant entretenu à amplitude autostabilisée et application à l'entretien d'un pendule élastique*. Thèse de sciences physiques, Faculté des sciences de Paris. Toulouse: Privat.

Abelé, Jean. 1945a. Définition cinématique des oscillations de relaxation. *Comptes rendus hebdomadaires des séances de l'Académie des sciences* 220(9 avril 1945): 511–513.

Abelé, Jean. 1945b. Définition cinématique des oscillations de relaxation. *Journal de Physique* 9(1945): 96–103.

Abelé, Jean. 1945c. Définition cinématique des oscillations de relaxation discontinues. *Comptes rendus hebdomadaires des séances de l'Académie des sciences* 221(26 novembre 1945): 656–658.

Abelé, Jean. 1946. Les oscillations de relaxation et le problème de leur définition analytique. *Revue générale des sciences pures et appliquées* 53(1946): 68–77.

Abelé, Jean. 1947. Construction d'oscillateurs non linéaires sinusoïdaux par la méthode de l'axe mobile. *Comptes rendus hebdomadaires des séances de l'Académie des sciences* 225(26 novembre 1945): 1270–1271.

Abraham, Henri, and Eugène Bloch. 1919a. Sur la mesure en valeur absolue des périodes des oscillations électriques de haute fréquence. *Comptes rendus hebdomadaires des séances de l'Académie des sciences* 168(2 juin 1919): 1105–1108.

Abraham, Henri, and Eugène Bloch. 1919b. Sur l'entretien des oscillations mécaniques au moyen des lampes à trois électrodes. *Comptes rendus hebdomadaires des séances de l'Académie des sciences* 168(16 juin 1919): 1197–1198.

Abraham, Henri, and Eugène Bloch. 1919c. Mesure en valeur absolue des périodes des oscillations électriques de haute fréquence. *Journal de Physique Théorique et Appliquée* 9(1919): 211–222.

Abraham, Henri, and Eugène Bloch. 1919d. Entretien des oscillations d'un pendule ou d'un diapason avec un amplificateur à lampes. *Journal de Physique Théorique et Appliquée* 9(1919): 225–233.

Abraham, Henri, and Eugène Bloch. 1919e. Mesure en valeur absolue des périodes des oscillations électriques de haute fréquence. *Annales de Physique* 9(1919): 237–302.

J.-M. Ginoux, *History of Nonlinear Oscillations Theory in France (1880–1940)*,
Archimedes 49, DOI 10.1007/978-3-319-55239-2

Abraham, Henri, and Eugène Bloch. 1931. Peut-on maintenir une salle à température constante? *Comptes rendus hebdomadaires des séances de l'Académie des sciences* 193(21 juin 1931): 1402–1403.

Abraham, Henri, Eugène Bloch, and Léon Bloch. 1920. Appareil à lecture directe pour les mesures en courants alternatifs: Voltmètre-amplificateur. *Journal de Physique et le Radium* 1, 2(1920): 44–57.

Abraham, Ralph Herman, and Christopher David Shaw. 1982–1988. *Dynamics: The geometry of behavior*, 4 vol., Santa Cruz: Aerial Press, Inc.

Acot, Pascal, and Jean-Marc Drouin. 1997. L'introduction en France des idées de l'écologie scientifique américaine dans l'entre-deux-guerres. *Revue d'histoire des sciences* 50(1997): 461–480.

Adrian, Edgar Douglas. 1930. The effects of injury on Mammalian nerve fibres. *Proceedings of the Royal Society* 106(1930): 596–618.

Airy, George Biddell. 1840. On the regulator of the clock-work for effecting uniform movement of Equatoreals. *Memoirs of the Royal Astronomical Society* 11(1840): 249–267.

Alembert, Jean Le Rond d'. 1748. Suite des recherches sur le calcul intégral, quatrième partie: Méthodes pour intégrer quelques équations différentielles. *Hist. Acad. Berlin pour 1748* IV(1750): 275–291; *Œuvres complètes de d'Alembert*, vol. I/4a, 215–245. Paris: CNRS Editions, 2007.

Andronov, Aleksandr Aleksandrovich. 1928. Предельные циклы Пуанкаре и теория колебаний, in *IVs'ezd ruskikh fizikov* (5–16.08: 23–24). Moscow: N.-Novgorod, Kazan, Saratov; *Œuvres complètes*: 32–33. [Les cycles limites de Poincaré et la théorie des oscillations, Ce rapport a été lu lors du IVe congrès des physiciens russes à Moscou du 5 au 16 août 1928: 23–24].

Andronov, Aleksandr Aleksandrovich. 1929a. Les cycles limites de Poincaré et la théorie des oscillations auto-entretenues. *Comptes rendus hebdomadaires des séances de l'Académie des sciences* 189(14 octobre 1929): 559–561.

Andronov, Aleksandr Aleksandrovich. 1929b. Предельные циклы Пуанкаре и теория автоколебаний. *Œuvres complètes*, 41–43 [Les cycles limites de Poincaré et la théorie des autooscillations].

Andronov, Aleksandr Aleksandrovich. 1933. Математические проблемы теории автоколебаний, Эти доклад прочитан А. А. Андроновым на Всесоюзной конференция по колебанням в 1933 г. Опубликобано в кн.: *I*. Всесоюзной конференция по колебанням. М-Л. ГТТН, 1933, с. 32. [Problèmes mathématiques de la théorie des auto-oscillations. Ces rapports ont été lus par A. A. Andronov à la Conférence de l'Union Soviétique sur les oscillations en 1933 et publiés dans le tome I: Conférence de l'Union Soviétique sur les oscillations, M-L. GTTI, 1933: 32].

Andronov, Aleksandr Aleksandrovich. 1943. Андронов Александр Александрович: Автобиография Аан Ссср, Ф. 411, Ох. 25, Д. 12, Л. 3–6. Подлинник. [Aleksandr Aleksandrovich Andronov: Autobiographie U.R.S.S. AAN, folio 411, 25, 12: 3–6].

Andronov, Aleksandr Aleksandrovich. 1956. Собрание трудов, [*Œuvres complètes*], Moscow: Izd. AN SSSR.

Andronov, A.A., and Aleksandr Adol'fovich Witt. 1930a. Sur la théorie mathématique des auto-oscillations. *Comptes rendus hebdomadaires des séances de l'Académie des sciences* 190(20 janvier 1930): 256–258.

Andronov, A.A., and Aleksandr Adol'fovich Witt. 1930b. Unstetige periodische Bewegungen und Theorie des Multivibrators von Abraham und Bloch. *Bulletin de l'Académie des Sciences de l'U.R.S.S.* 8: 189–192. [Mouvements périodiques discontinus et la théorie du multivibrateur d'Abraham et Bloch].

Andronov, A.A., and Aleksandr Adol'fovich Witt. 1930c. К математической теории захватывания. Журнал технишеской Физики, 1930, т. 7, №4, с. 3–20. *Journal de Physique Technique de l'URSS* 7(1930): 3–20. [Sur la théorie de l'entraînement].

Andronov, A.A., and Aleksandr Adol'fovich Witt. 1930d. Zur theorie des Mitnehmens von Van der Pol. *Archiv für Elektrotechnik* XXIV(1930): 99–110.

Andronov, A.A., and Aleksandr Adol'fovich Witt. 1933. Об устоЪичивости по ЛЪпунову, Ъурнал ЪкспериментталЪноЪи и теоретическоЪи Физики 3: 373–374. [Sur lastabilité au sens de Lyapunov]. [*OEuvres complètes*, 140–141].

Andronov, A.A., and Aleksandr Adol'fovich Witt. 1934. К математической теории автоколебательных систем с двумя степенями свободы, Журн. Техн. физики, т. 4, №1, с. 122–143. [La théorie mathématique des systèmes auto-oscillants à deux degrés de liberté].

Andronov, A.A., and Aleksandr Adol'fovich Witt. 1935. Sur la théorie mathématique des systèmes auto-oscillatoires à deux degré de liberté. *Journal de Physique Technique de l'URSS* 1: 249–271. Trad. fr. de [1934].

Andronov, A.A., Nikolaï Nikolaevich Bautin, and G.S. Gorelik. 1945. Auto-oscillations d'un schéma simplifié contenant une hélice à pas variable automatique. *Comptes rendus de l'Académie des sciences de l'U.R.S.S.* XLVII(1945): 263–267.

Andronov, A.A., G.S. Gorelik, and Sergei Mikhailovich Rytov. 1947. Некоторые исследования в области нелинейных колебаний, проведённые в сссп, начиная с 1935 г., Успехи Физических Хаук, т. 33, вып. 3, с. 335–352. [Les recherches sur les oscillations non linéaires menées en URSS, à partir de 1935].

Andronov, A.A., and Semen Emanuilovich Khaikin. 1937. ТеориЪ колебаниЪи [Théorie des oscillations], ОбЪединное Хаучно-Техническое ИздателЪство, НКТП, СССП, главнаЪ редакциЪ техно-теоретическоЪи литературы, Москва 1937. Ленинград. Trad. angl., *Theory of oscillations*. Princeton: Princeton University Press, 1949.

Andronov, A.A., and S.E. Khaikin. 1949. Theory of oscillations (Transl. and adapted by S. Lefschetz). Princeton: Princeton University Press.

Andronov, A.A., A.A. Witt, and S.E. Khaikin. 1959. Теория колебаний [Théorie des oscillations], Государственное издательство физико-математической литературы, Москва 1959. Trad. angl., *Theory of oscillators*. Oxford: Pergamon Press, 1966.

Andronov, A.A., and Lev Semenoviéc Pontrjagin. 1937. Systèmes grossiers. *Comptes rendus de l'Académie des sciences de l'URSS* XIV(2 janvier 1937): 247–250.

Andronov, A.A., and Evgenya Aleksandrovna Leontovich. 1938. Sur la théorie de la variation de la structure qualitative de la division du plan en trajectoires. *Comptes rendus de l'Académie des sciences de l'URSS* XXI(31 octobre 1938): 423–426.

Appleton, Edward Victor. 1922. The automatic synchronization of triode oscillators. *Proceedings of the Cambridge Philosophical Society* 21(1922): 231–248.

Appleton, Edward Victor, and Balthasar Van der Pol. 1921. On the form of free triode vibrations. *Philosophical Magazine* 42(1921): 201–220.

Appleton, Edward Victor, and Balthasar Van der Pol. 1922. On a type of oscillation-hysteresis in a simple triode generator. *Philosophical Magazine* 43(1922): 177–193.

Appleton, Edward Victor, and William Michael Herbert Greaves. 1923. On the solution of the representative differential equation of the triode oscillator. *Philosophical Magazine* 45(1923): 401–414.

Arnold, Engelbert. 1904. *La machine dynamo à courant continu*. Paris: Ch. Béranger éditeur.

Arnold, Vladimir. 1974. *Équations différentielles ordinaires*, Moscou: Éditions Mir.

Atten, Michel, François Du Castel, and Marie Pierre. 1999. *Les télécoms: Histoire des Écoles supérieures des télécommunications*. Paris: Hachette.

Aubin, David. 2004. Analyses d'ouvrages: Nicolas Bouleau. *Philosophie des mathématiques et de la modélisation: Du chercheur à l'ingénieur* (L'Harmattan, 1999) et Giorgio Israël, *La Mathématisation du réel: essai sur la modélisation mathématique* (Seuil, 1996), in *Revue d'histoire des sciences* 57(2004): 451–454.

Aubin, David, and Amy Dahan Dalmedico. 2002. Writing the history of dynamical systems and chaos: *Longue Durée* and revolution, disciplines and cultures. *Historia Mathematica* 29(2002): 273–339.

Auger, Léon. 1932. Sur les mouvements des anches battantes pour tuyaux d'orgue. *Comptes rendus hebdomadaires des séances de l'Académie des sciences* 195(17 août 1932): 516–518.

Auger, Léon. 1937. Sur les fréquences des sons donnés par un tuyau cylindrique à anche battante. *Comptes rendus hebdomadaires des séances de l'Académie des sciences* 207(5 avril 1937): 1169–1171.

Auger, Léon. 1938. Sur l'accord des tuyaux sonores à anche battante considéré comme phénomène de relaxation. *Comptes rendus hebdomadaires des séances de l'Académie des sciences* 208(4 juillet 1938): 216–217.

Auger, Léon. 1939. Sur la stabilité des sons émis par un tuyau à anche soumis à une pression constante. *Comptes rendus hebdomadaires des séances de l'Académie des sciences* 209(18 septembre 1939): 508–217.

Auger, Léon. 1945. *Oscillations auto-entretenues dans les tuyaux sonores à anches battantes*. Thèse de sciences physiques, Faculté des sciences de Paris (1943), Publications scientifiques et techniques du ministère de l'Air. Paris: Gauthier-Villars, 1945.

Ayrton, Hertha. 1895. Variation of potential difference of the electric arc with current, size of carbons, and distance apart. *The Electrician*, 1895, vol. XXXIV, pp. 335, 364, 399, 471, 541, 610; 1895, vol. XXXV, pp. 418, 635, 743; 1895, vol. XXXVI, pp. 36, 225; 1896, vol. XXXVI, p. 539.

Ayrton, Hertha. 1902. *The electric arc*. London: "The Electrician" Printing and Publishing Company.

Ayrton, William Edward, and John Perry. 1882. The resistance of the electric Arc. *Proceedings of the Physical Society of London* 5(1882): 197–204.

Barbillion, Louis, and Paul Cayère. 1918. Note sur une nouvelle solution du problème de la régulation des groupes électrogènes. *Revue Générale de l'Électricité* tome III, vol. 21, p. 747–764, and tome III, vol. 22, p. 787–796 (1918).

Barkhausen, Heinrich. 1907. *Das Problem der Schwingungserzeugung mit besonderer Berücksichtigung schneller elektrischer Schwingungen*. Leipzig: S. Hirzel.

Barkhausen, Heinrich. 1920. Die Vakuumröhre und ihre technischen Anwendungen. *Jahrbuch d. drahtlosen Telegraphie u. Telephonie* 16: 82–114.

Barkhausen, Heinrich. 1929. *Lechbuch der Elektronen-Röhren und ihrer technischen Anwendungen*, 3. Band. Rückkoplung, Leipzig: S. Hirzel.

Bedeau, François. 1928. Les différentes méthodes de détermination de la condition des oscillations dans les émetteurs à lampe. *Onde électrique* (juillet 1928): 265–286.

Bedeau, François. 1931. Cours élémentaire de télégraphie et téléphonie sans fil. *Ed. Librairie Vuibert*, Paris.

Bedeau, François, and Jehan-George De Mare. 1928. Stabilisation des oscillations de relaxation. *Comptes rendus hebdomadaires des séances de l'Académie des sciences* 187(17 juillet 1928): 209–210.

Bénard, Henri. 1908. Formation de centres de giration à l'arrière d'un obstacle en mouvement. *Comptes rendus hebdomadaires des séances de l'Académie des sciences* 147(2 novembre 1908): 839–842.

Bendixson, Ivar. 1901. Sur les courbes définies par des équations différentielles. *Acta Mathematica* 24(1901): 1–88.

Bendrikov, G.A., and G.A. Sidorova. 1981. Aleksandr Adolfovich Witt, Источник: Г. А. Бендриков, Г. А. Сидорова. Александр Адольфович Витт, История и методология естественных наук. Вып. *XXVI*, Физика. Изд-во МГУ, 1981, p. 150–168.

Bennett, Stuart. 1979. *A history of control engineering 1800–1930*. London: Peter Peregrinus Ltd.

Bennett, Stuart. 1984. Nicolas Minorsky and the automatic steering of ships. *IEEE Control Systems* 4(1984): 10–15.

Bergé, P., Y. Pomeau, and C. Vidal. 1984. *L'ordre dans le chaos*. Paris: Hermann.

Bergé, P., Y. Pomeau, and C. Vidal. 1988. *Le Chaos, Théorie et Expériences*. Paris: Eyrolles.

Besicovic, Abram. 1925. Sur quelques points de la théorie des fonctions presque périodiques. *C.R.A.S.* 181: 394–397.

Besicovic, Abram. 1931. H. Bohr: Almost periodicity and general trigonometric series. *Acta Mathematica* 57: 203–292.

Besicovic, Abram. 1932a. Analysis of conditions of almost periodicity. *Acta Mathematica* 58: 217–230.
Besicovic, Abram. 1932b. *Almost periodic functions*. Cambridge: Cambridge University Press.
Bethenod, Joseph. 1937. Étude de la décharge d'un condensateur à travers un tube à gaz. *Comptes rendus hebdomadaires des séances de l'Académie des sciences* 205(26 octobre 1937): 727–730.
Bethenod, Joseph. 1938a. Sur l'entretien du mouvement d'un pendule au moyen d'un courant alternatif de fréquence élevée par rapport à sa fréquence propre. *Comptes rendus hebdomadaires des séances de l'Académie des sciences* 207(2 novembre 1938): 847–849.
Bethenod, Joseph. 1938b. André Blondel. *Revue Générale de l'Electricité* 44(1938): 751–756.
Bethenod, Joseph. 1939. Détermination du régime de fonctionnement d'un arc à courant alternatif. *Comptes rendus hebdomadaires des séances de l'Académie des sciences* 208(20 février 1939): 562–565.
Birkhoff, George D. 1912. Quelques théorèmes sur le mouvement des systèmes dynamiques. *Bulletin de la Société Mathématique de France* 40(1912): 305–323.
Birkhoff, George D. 1913. Proof of Poincaré's geometric theorem. *Transactions of the American Mathematical Society* 14(1913): 14–22.
Birkhoff, George D. 1914. Démonstration du dernier théorème de géométrie de Poincaré. *Bulletin de la Société Mathématique de France* 42(1914): 1–12.
Birkhoff, George D. 1920. Surface transformations and their dynamical applications. *Acta Mathematica* 43:1–119.
Birkhoff, George D. 1926. Sur la signification canonique des équations de la dynamique. *Comptes rendus hebdomadaires des séances de l'Académie des sciences* 183(30 août 1926): 516–519.
Birkhoff, George. 1927. *Dynamical systems*, vol. IX. New York: American Mathematical Society Colloquium Publications.
Birkhoff, George D. 1932. Sur quelques courbes fermées remarquables. *Bulletin de la Société Mathématique de France* 60:1–26.
Bissell, C.C. 1994. Postcards from a forgotten man. *The Times Higher Educational Supplement* 15.
Bissell, Chris. 2009. A history of automatic control. In *Springer handbook of automation*, ed. Y. Nof Shimon. Berlin/Heidelberg: Springer.
Blondel, André. 1891a. Études expérimentales sur l'arc à courants alternatifs. *La Lumière Électrique* XLII(1891): 551–560, 618–623.
Blondel, André. 1891b. Études expérimentales sur l'arc à courants alternatifs. *La Lumière Électrique* XLIII(1891): 51–61.
Blondel, André. 1893a. Oscillographes: Nouveaux appareils pour l'étude des oscillations électriques lentes. *Comptes rendus hebdomadaires des séances de l'Académie des sciences* 116(6 mars 1893): 502–506.
Blondel, André. 1893b. Conditions générales que doivent remplir les instruments enregistreurs ou indicateurs; problème de la synchronisation intégrale. *Comptes rendus hebdomadaires des séances de l'Académie des sciences* 116(10 avril 1893): 748–752.
Blondel, André. 1893c. Nouvelles recherches sur l'arc à courants alternatifs. *La Lumière Électrique* XLIX(1893): 502–508, 557–569, 608–617.
Blondel, André. 1905a. Sur les phénomènes de l'arc chantant. *Comptes rendus hebdomadaires des séances de l'Académie des sciences* 140(13 juin 1905): 1680–1682.
Blondel, André. 1905b. Sur les phénomènes de l'arc chantant. *Revue générale des sciences pures et appliquées* 16(1905): 708; *Journal de Physique Théorique et Appliquée* (IV) 5(1906): 77–97.
Blondel, André. 1905c. Sur les phénomènes de l'arc chantant. *Éclairage Électrique* XLIV(1905): 41–58, 81–104.
Blondel, André. 1912. Henri Poincaré. *La Lumière Électrique* (2e série) XIX(1912): 99–101.
Blondel, André. 1915a. Systèmes oscillants à amortissement discontinu. *La Lumière Électrique*, (2e série) XXXI(1915): 193–196; *La Lumière Électrique* (2e série) XXXI(1915): 217–222. *La Lumière Électrique* (2e série) XXXI(1915): 241–245.
Blondel, André. 1915b. Systèmes oscillants à amortissement discontinu. *Comptes rendus hebdomadaires des séances de l'Académie des sciences* 161(26 octobre 1915): 546–552.

Blondel, André. 1915c. Sur les oscillations forcées d'un système oscillant à amortissement discontinu. *Comptes rendus hebdomadaires des séances de l'Académie des sciences* 161(22 novembre 1915): 625–630.

Blondel, André. 1916. Remarques sur les applications du courant continu à haute tension dans la télégraphie et la téléphonie sans fil. *La Lumière Électrique*, (2e série) XXXIII(1916): 49–54 *La Lumière Électrique*, (2e série) XXXIII(1916): 97–105.

Blondel, André. 1919a. Caractéristiques d'oscillations des lampes à trois électrodes utilisées comme générateurs d'oscillations entretenues. *Comptes rendus hebdomadaires des séances de l'Académie des sciences* 169(20 octobre 1919): 676–680.

Blondel, André. 1919b. Amplitude du courant oscillant produit par les audions générateurs. *Comptes rendus hebdomadaires des séances de l'Académie des sciences* 169(17 novembre 1919): 943–948.

Blondel, André. 1919c. Étude graphique du fonctionnement des audions à circuit résonant comme récepteurs sensibilisés ou comme désamortisseurs. *Comptes rendus hebdomadaires des séances de l'Académie des sciences* 169(8 décembre 1919): 1377–1382.

Blondel, André. 1919d. Sur les systèmes à oscillations persistantes, et en particulier sur les oscillations entretenues par auto-amorçage. *Journal de Physique Théorique et Appliquée* (5e série) IX(1919): 117–151, 153–161.

Blondel, André. 1920. Théorie graphique des audions générateurs et calcul de l'amplitude des oscillations. *Radioélectricité* 1(1920): 7–13, 63–72.

Blondel, André. 1923a. Sur les conditions de rendement des lampes-valves génératrices ayant une caractéristique d'arc électrique chantant et sur la définition de leur puissance. *Comptes rendus hebdomadaires des séances de l'Académie des sciences* 177(9 juillet 1923): 87–92.

Blondel, André. 1923b. Méthode rationnelle pour les essais de spécification des lampes triodes destinées à fonctionner en clapet. *Comptes rendus hebdomadaires des séances de l'Académie des sciences* 177(23 juillet 1923): 376–382.

Blondel, André. 1926. Contribution à la théorie de l'arc chantant musical. *Comptes rendus hebdomadaires des séances de l'Académie des sciences* 182 (12 avril 1926): 900–903.

Blondel, André, and Fernand Carbenay. 1916. Remarques complémentaires sur les oscillations forcées des systèmes à amortissement discontinu. *La Lumière Électrique* (2e série) XXXIV(1916): 98–105.

Blondel, A., and Charles Lavanchy. 1919. Contribution à la théorie des audions générateurs, condition d'amorçage et degré d'amortissement des oscillations de faible amplitude obtenue par ces appareil. *Revue Générale de l'Electricité* 25(1919): 875–893, 923–930.

Bochner, Salomon. 1925. Sur les fonctions presque périodiques de Bohr. *C.R.A.S.* 180: 1156–1158.

Bochner, Salomon. 1927. Beitrage zur Theorie der fastperiodischen Funktionen. *Math. Annalen* 96: 119–147 et 383–409.

Bochner, Salomon. 1929. Uber gewisse Differential und allgemeinere Gleichungen, deren Losungen fastperiodisch sind - I Teil - Der Existaezsatz. *Math. Ann.* 102: 489–504.

Bochner, Salomon. 1930. Uber gewisse Differential und allgemeinere Gleichungen, deren Losungen fastperiodisch sind - II Teil - Der Beschrankteitssatz. *Math. Ann.* 103: 588–597.

Bochner, Salomon. 1931. Uber gewisse Differential und allgemeinere Gleichungen, deren Losungen fastperiodisch sind - III Teil - Systeme von Gleichungen. *Math. Ann.* 104: 579–587.

Bochner, Salomon. 1933. Abstrakte fastperiodische Funktionen. *Acta Mathematica* 61: 149–184.

Bogolyubov, Nicolas, and Nicolas Krylov. 1928. Sur les méthodes des différences finies pour la résolution approchée des problèmes fondamentaux de la Physique mathématique. *Comptes rendus hebdomadaires des séances de l'Académie des sciences* 186(13 février 1928): 422–425.

Bogolyubov, Nicolas, and Nicolas Krylov. 1928. Sur la théorie mathématique des oscillographes. *Comptes rendus hebdomadaires des séances de l'Académie des sciences* 187(5 novembre 1928): 938–940.

Bohl, Piers. 1893. Über die Darstellung von Funktionen einer Variabeln durch trigonometrische Reihen mit mehreren einer Variabeln proportionalen Argumenten, Thèse, Dorpat.

Bohl, Piers. 1900. О некоторых дифференциальных уравнений общего характера, применимых в механике. Thèse, Institut Polytechnique de Riga, Dorpat.

Bohl, Piers. 1904. Über die Bewegung eines mechanischen Systems in der Nähe einer Gleichgewichtslage. *Journal für die reine und angewandte Mathematik* 127(1904): 179–276.
Bohl, Piers. 1906. Über eine Differentialgleichung der Störungstheorie. *Journal fur die Reine und Angewandte Mathematik* 131(1906): 268–321.
Bohl, Piers. 1910. Sur certaines équations différentielles d'un type général utilisables en mécanique. *Bulletin de la Société Mathématique de France* 38(1910): 5–138.
Bohr, Harald. 1923a. Sur les fonctions presque périodiques. *C.R.A.S.* 177: 737–739.
Bohr, Harald. 1923b. Sur l'approximation des fonctions presque périodiques par des sommes trigonométriques. *C.R.A.S.* 177: 1090–1092.
Bohr, Harald. 1924. Zur Theorie der fastperiodischen Funktionen I. Eine Verallgemeinerung der Theorie der Fourierreihen. *Acta Mathematica* 45: 29–127.
Bohr, Harald. 1925a. Zur Theorie der fastperiodischen Funktionen II. Zusammenhang der Fastperiodischen Funktionen mit Funktionen von unendlich vielen Variablen; gleichmäßige Approximation durch trigonometrische Summen. *Acta Mathematica* 46: 101–214.
Bohr, Harald. 1925b. Sur les fonctions presque périodiques d'une variable complexe. *C.R.A.S.* 180: 645–647.
Bohr, Harald. 1926. Zur Theorie der fastperiodischen Funktionen. III. Dirichletentwicklung analytischer Funktionen. *Acta Mathematica* 47: 237–281.
Bose, Jagadish Chandra. 1919. Life movements in plants. *Transactions of the Bose Institute* 2(1919): 255–597.
Bouasse, Henri. 1924. *Les oscillations électriques*. Paris: Librairie Delagrave.
Bouasse, Henri. 1926. *Acoustique Générale*. Paris: Librairie Delagrave.
Boucherot, Paul. 1904. Observations d'ordre général sur les moteurs. *Bulletin de la Société internationale des Électriciens* 21(1904): 71–77.
Boulanger, Julien Auguste. 1885. *Sur les progrès de la science électrique et sur les nouvelles machines d'induction*. Paris: Gauthier-Villars.
Boulanger, Julien Auguste, and Gustave Ferrié. 1899. *La Télégraphie sans fil*. Paris: Berger-Levrault et Cie.
Bouligand, Georges. 1932a. Progrès récents dans l'intégration approchée des équations de la physique mathématique. *Revue Générale des Sciences* 43(1932): 77–81.
Bouligand, Georges. 1932b. Sur l'intégration qualitative. *Revue Générale des Sciences* 43(1932): 329–336.
Bouligand, Georges. 1961. Jean Abelé. *Revue Générale des Sciences* 68(1961): 195–196.
Bourrières, François-Joseph. 1932. Sur les oscillations libres des extrémités de tubes élastiques parcourus par un courant uniforme de fluide. *Comptes rendus hebdomadaires des séances de l'Académie des sciences* 194(28 décembre 1931): 49–51.
Bourrières, François-Joseph. 1937a. Sur les oscillations auto-entretenues des extrémités de tubes élastiques déversant un courant continu de fluides et celles d'anches libres encastrées. *Comptes rendus hebdomadaires des séances de l'Académie des sciences* 204(28 février 1937): 654–656.
Bourrières, François-Joseph. 1937b. Sur les auto-oscillations de jets réels en milieu identique par auto plissements. *Comptes rendus hebdomadaires des séances de l'Académie des sciences* 204(5 mai 1937): 111–113.
Bourrières, François-Joseph. 1939. *Sur un phénomène d'oscillation auto-entretenue en mécanique des fluides*. Publications Scientifiques et Techniques du Ministère de l'Air vol. 147 (1939), E. Blondel La Rougery. Paris: Gauthier-Villars.
Bouthillon, Léon. 1915. Sur la charge des condensateurs au moyen d'une force électro motrice constante et leur décharge dans un circuit à étincelle. *Comptes rendus hebdomadaires des séances de l'Académie des sciences* 160(14 juin 1915): 800–802.
Bouthillon, Léon. 1915. Sur l'application à la télégraphie sans fil à étincelles du procédé de charge des condensateurs au moyen de dynamos à force électro motrice constante en combinaison avec un éclateur tournant. *Comptes rendus hebdomadaires des séances de l'Académie des sciences* 161(15 novembre 1915): 583–586.
Bouthillon, Léon. 1919–1921. *La Théorie et la Pratique des Radiocommunications*, 2 tomes. Paris: Librairie Delagrave.

Bouthillon, Léon. 1926. Radiogoniomètre et radiophares à maximum accentué *Comptes rendus hebdomadaires des séances de l'Académie des sciences* 183(22 novembre 1926): 955–957.

Bouthillon, Léon. 1927. Inclinaison des ondes et système dirigé. *Comptes rendus hebdomadaires des séances de l'Académie des sciences* 184(24 janvier 1927): 190–192.

Boyko, E. C. 1983. Школа академика А. А. Андронова [L'École de l'académicien A. A. Andronov], Moskva: Nauka. Reed., sous la direction de A. T. Grigoryan.

Braun, Karl Ferdinand. 1897. Über ein Verfahren zur Demonstration und zum Studium des zeitlichen Verlaufes variabler Ströme. *Annalen der Physik und Chemie* 60(1897): 552–559.

Brillouin, Marcel. 1897. Théorie d'un alternateur auto-excitateur. *Éclairage Électrique* XI(1897): 49–59.

Busch, Hans. 1913. *Stabilität, Labilität und Pendelungen in der Elektrotechnik*. Leipzig: S. Hirzel.

Camichel, Charles. 1927. Sur les tourbillons provoqués par un obstacle immergé dans un courant liquide. *Comptes rendus hebdomadaires des séances de l'Académie des sciences* 184(7 juin 1927): 1509–1512.

Camichel, C., Pierre Dupin, and Max Teissié-Solier. 1927. Sur l'application de la loi de similitude aux périodes de formation des tourbillons alternés de Bénard-Karman. *Comptes rendus hebdomadaires des séances de l'Académie des sciences* 185(19 décembre 1927): 1556–1559.

Camichel, C., Pierre Dupin, and Max Teissié-Solier. 1928. Sur l'existence dans l'écoulement d'un fluide autour de cylindres immergés, d'un phénomène périodique en régime de Poiseuille. *Comptes rendus hebdomadaires des séances de l'Académie des sciences* 186(9 janvier 1928): 203–205.

Camichel, C., and Max Teissié-Solier. 1935. Influence d'une perturbation sur le sillage en régime de Poiseuille d'un corps immergé. *Comptes rendus hebdomadaires des séances de l'Académie des sciences* 200(28 janvier 1935): 704–707.

Carrière, Z. 1924. Entretien des vibrations d'une anche libre. *Journal de Physique et le Radium* 5, 11(1924): 338–344.

Cartwright, Marie Lucy, and John Edensor Littlewood. 1945. On non-linear differential equations of the second order, I: The Equation $\ddot{y} - k\left(1 - y^2\right)\dot{y} + y = b\lambda kcos\,(\lambda t + a)$, k large. *Journal of the London Mathematical Society* 20(1945): 180–189.

Cartwright, Marie Lucy, and John Edensor Littlewood. 1947. On nonlinear differential equations of the second order, II: the equation $\ddot{y} - kf(y)\dot{y} + g(y,k) = p(t) = p_1(t) + kp_2(t)$; $k > 0$, $f(y) \geq 1$. *Annals of Mathematics* 48(1947): 472–494.

Cartwright, Marie Lucy, and John Edensor Littlewood. 1948. Errata. *Annals of Mathematics* 49(1948): 1010.

Cartwright, Marie Lucy, and John Edensor Littlewood. 1949. Addendum. *Annals of Mathematics* 50(1949): 504–505.

Cartwright, Marie Lucy. 1948. Forced oscillations in nearly sinusoidal systems. *Journal of the Institution of Electrical Engineers* 95 (1948): 88–96.

Cartwright, Marie Lucy. 1960. Balthasar Van der Pol. *Journal of the London Mathematical Society* 35(1960): 367–376.

Cartwright, Marie Lucy. 1965. Jacques Hadamard. *Biographical Memoirs of Fellows of the Royal Society* 11(1965): 65–98.

Castagnetto, Louis. 1939. *Contribution à l'étude des tourbillons alternés de Bénard-Karman*. Thèse, Faculté des Sciences de Toulouse.

Castagnetto, Louis. 1958. Sur l'intégration numérique de l'équation de stabilité. *Comptes rendus hebdomadaires des séances de l'Académie des sciences* 246(13 janvier 1958): 255–257.

Castagnetto, Louis, and Jean-Claude Matheau. 1961a. Sur une généralisation de la méthode de Kron aux réseaux actifs. *Comptes rendus hebdomadaires des séances de l'Académie des sciences* 252(4 janvier 1961): 85–86.

Castagnetto, Louis, and Jean-Claude Matheau. 1961b. Sur le problème de l'adaptation. *Comptes rendus hebdomadaires des séances de l'Académie des sciences* 252(30 janvier 1961): 689–691.

Cavaillès, Jean. 1946. Sur la logique et la théorie de la science. Paris: P.U.F.

Chabert, Jean-Luc. 1992. Hadamard et les géodésiques des surfaces à courbure négative. In *Chaos et déterminisme*, 306–330. Paris: Seuil (coll. Points Sciences).

Champeix, Robert. 1980. Grande et petite histoire de la lampe T.M. *La liaison des Transmissions* n° 126 (Nov.-Déc. 1980).
Chandrashekaran, Maroli Krishnayya, and R. Subbaraj. 1996. J. C. Bose's view on biological Rythms. *Indian Journal of History of Science* 31(1996): 375–382.
Charpentier, Marie. 1931. *Sur les points de Peano d'une équation différentielle du premier ordre*. Thèse, Faculté des sciences de Poitiers, 1931. Cluj, Institutul de Arte Grafice "Ardealul".
Charpentier, Marie. 1932. On certain dynamical systems with points of Peano. *Bulletin of the American Mathematical Society* 38(12)(1932): 849–854.
Charpentier, Marie. 1933. Sur la semi-continuité d'inclusion des trajectoires de la Dynamique. *Comptes rendus hebdomadaires des séances de l'Académie des sciences* 196(6 juin 1933): 1771–1773.
Charpentier, Marie. 1934a. Sur quelques propriétés des courbes de M. Birkhoff. *Comptes rendus hebdomadaires des séances de l'Académie des sciences* 198(19 janvier 1934): 701–703.
Charpentier, Marie. 1934b. Sur quelques propriétés des courbes de M. Birkhoff. *Bulletin de la Société Mathématique de France* 62(1934): 193–224.
Chenciner, Alain, 1985. Systémes dynamiques différentiables. *Encyclopédia Universalis* XX: 594–630.
Child, Clement Dexter. 1909. Counter Electromotive Force in the Electric Arc. *Physical Review* (Series I) 29(1909): 229–242.
Cliquet, Édouard. 1949. *Émetteurs de petite puissance sur ondes courtes*. 2e éd., 2 vol. Paris: Éditions Technique & Vulgarisation.
Colin, J. 1893. Sur la rigidité des liquides. *Comptes rendus hebdomadaires des séances de l'Académie des sciences* 116(1893): 1251–1253.
Collet, Pierre. 1999. Leçons sur le chaos, Cours donné au Centre de Physique Théorique de l'École Polytechnique.
Cornu, Alfred. 1887a. Sur la condition de stabilité du mouvement d'un système oscillant soumis a une liaison synchronique pendulaire. *Journal de Physique Théorique et Appliquée* 6(1887): 445–452.
Cornu, Alfred. 1887b. Sur la synchronisation d'une oscillation faiblement amortie indicatrice de synchronisation représentant le régime variable. *Journal de Physique Théorique et Appliquée* 6(1887): 452–464.
Curie, Pierre. 1891. Équations réduites pour le calcul des mouvements amortis. *La Lumière Électrique* (1re série) 41 (1891): 201–209. *La Lumière Électrique* (1re série) 41 (1891): 270–276. *La Lumière Électrique* (1re série) 41(1891): 307–362.
Daan, Serge. 2010. A history of chronobiological concepts. In *Protein reviews*, The circadian clocks, vol. 12. New York: Springer.
D'adhémar, Robert. 1934. *La balistique extérieure*, Mémorial des sciences mathématiques, fasc. 65. Paris: Gauthier-Villars.
Dahan Dalmedico, Amy. 1996. Le difficile héritage de Henri Poincaré en systèmes dynamiques. In *Henri Poincaré: Science et philosophie*, éd. J.-L. Greffe, G. Heinzmann, and K. Lorenz, 13–33. Paris: Blanchard/Berlin: Akademie Verlag.
Dahan Dalmedico, Amy. 2004a. L'École d'Andronov à Gorki, profil d'un centre scientifique dans la Russie soviétique (avec la collaboration d'Irina Gouzévitch). In *Les Sciences pour la guerre, 1940–1960*, éd. A. Dahan and D. Pestre, 275–316. Paris: Editions de l'EHESS.
Dahan Dalmedico, Amy. 2004b. Early developments of nonlinear science in Soviet Russia: The Andronov School at Gor'kiy (in collaboration with Irina Gouzévitch). *Science in Context* 17(2004): 235–265.
Dahan Dalmedico, Amy, J.L. Chabert, and Karine Chemla, éd. 1992. *Chaos et déterminisme*. Paris: Éditions du Seuil (coll. Points Sciences).
Darboux, Gaston. 1878. Sur les équations différentielles algébriques du premier ordre et du premier degré. *Bulletin de la Société Mathématique de France* 2(2): 60–96, 123–143, 151–200.
Darboux, Gaston. 1914. Éloge historique d'Henri Poincaré. *Mémoires de l'Académie des sciences* 52(1914): 81–148.

Darrieus, Georges. 1929. Les modèles mécaniques en électrotechnique, leur application aux problèmes de Stabilité. *Bulletin de la Société Française des électriciens* 9: 794–809.

Darrieus, Georges. 1968. Notice nécrologique sur Hippolyte Parodi (1874–1968). membre de la division des applications de la Science à l'Industrie. *Comptes rendus hebdomadaires des séances de l'Académie des sciences* 267(18 novembre 1968): 81.

David, Pierre. 1927. Les oscillations par décharges (Relaxation Oscillations). *Onde Électrique* 69(6)(sept. 1927): 38A.

Decaux, Bernard, and Philippe Le Corbeiller. 1931. Sur un système autoentretenu utilisant un tube à néon. *Comptes rendus hebdomadaires des séances de l'Académie des sciences* 193(26 octobre 1931): 723–725.

Dejussieu-Pontcarral, Pierre. 1961. *L'épopée du tube électronique*. Paris: Les éditions du XXe siècle.

Delsarte, Jean. 1934. Les fonctions moyenne-périodiques. *C.R.A.S.* 198 (22 janvier 1934): 330–332.

Delsarte, Jean. 1935a. Application de la théorie des fonctions moyenne-périodiques àă la résolution des équations de Fredholm-Nörlund. *C.R.A.S.* 200 (28 janvier 1935): 371–372.

Delsarte, Jean. 1935b. Les fonctions moyenne-périodiques. textitJournal de mathématiques pures et appliquées 14(9): 403–453.

Delsarte, Jean. 1938a. Sur une extension de la formule de Taylor. *Journal de mathématiques pures et appliquées* 17(9): 213–231.

Delsarte, Jean. 1938b. Sur une extension nouvelle de la notion de presque-périodicité. *C.R.A.S.* 206(21 février 1938): 573–575.

Delsarte, Jean. 1939. Une extension nouvelle de la théorie de fonctions presque-périodiques de Bohr. *Acta Mathematica* 69: 257–317.

Denjoy, Arnaud. 1932a. Sur les caractéristiques à la surface du tore. *Comptes rendus hebdomadaires des séances de l'Académie des sciences* 194(7 mars 1932): 830–833, 2014–2016.

Denjoy, Arnaud. 1932b. Sur les courbes définies par les équations différentielles à la surface du tore. *Journal de mathématiques pures et appliquées* (IX) 11(1932): 333–376.

Denjoy, Arnaud. 1935. Les nouvelles méthodes de la Mécanique d'après le Professeur N. Krylov. *Revue Générale des Sciences* 46(1935): 389–391.

Deprez, Marcel. 1881. Distribution de l'énergie par l'électricité. *Comptes rendus hebdomadaires des séances de l'Académie des sciences* 92(28 novembre 1881): 892–895, 952–955.

Diner, Simon. 1992. Les voies du chaos déterministe dans l'École russe. In *Chaos et déterminisme*, ed. A. Dahan, J.-L. Dalmedico, J.-L. Chabert, and K. Chemla, 331–368. Paris: Éditions du Seuil (coll. Points Sciences).

Dorodnitsyn, Anatoly Alekseevich. 1947. Асимптотическое решение уравнения Вандер-Поля [Solution asymptotique de l'équation de Van der Pol]. Прикладная математические и механика 11: 313–328.

Du Bois Duddell, William. 1897. Oscillographs. *The Electrician* 39(1897): 636–638.

Du Bois Duddell, William. 1900a. On rapid variations in the current through the direct-current arc. *Journal of the Institution of Electrical Engineers* 30(1900): 232–283.

Du Bois Duddell, William. 1900b. On rapid variations in the current through the direct-current arc. *The Electrician* 46(1900): 269–273, 310–313.

Du Bois Duddell, William. 1901a. On the resistance and electromotive forces of the electric arc. *Proceedings of the Royal Society of London* 68(1901): 512–518.

Du Bois Duddell, William. 1901b. Sur la résistance et les forces électromotrices de l'arc électrique. *Revue Générale des Sciences* 12(1901): 986.

Du Bois Duddell, William. 1901c. The musical arc. *The Electrician* 52(1901): 902.

Du Bois Duddell, William. 1904. On the resistance and electromotive forces of the electric arc. *Philosophical Transactions of the Royal Society of London Series A* 203(1904): 305–342.

Duffing, Georg. 1918. *Erzwungene Schwingungen bei veränderlicher Eigenfrequenz und ihre technische Bedeutung*. Braunschweig: F. Vieweg u. Sohn.

Dufour, A., and René Mesny. 1923. Étude oscillographique de quelques émetteurs à triodes. *Onde Électrique* 23(novembre 1923): 620–633. *Onde Électrique* 24(décembre 1923): 692–705.

Eccles, William Henry, and Winifred A. Leyshon. 1927. Some new methods linking mechanical and electrical vibrations. *Proceedings of the Physical Society* 40(1927): 229–233.

Eck, Jean-Louis. 1936. Oscillations de relaxation à période stable obtenues avec une triode à gaz. *Journal de Physique et le Radium* 7, 5(1936): 227–232.

Eck, Jean-Louis. 1947. Sur les discontinuités de potentiel entre un métal et un semi-conducteur liquide. *Comptes rendus hebdomadaires des séances de l'Académie des sciences* 225(22 décembre 1947): 1294–129.

Edlund, Erik. 1867. Untersuchung über den galvanischen Lichtbogen. *Annalen der Physik* 207(1867): 586–607.

Enns, Richard H., and George C. McGuire. 2001. *Nonlinear physics with mathematica for scientists and engineers*. Boston: Birkhäuser.

Esclangon, Ernest. 1902. Sur une extension de la notion de périodicité. *Comptes rendus hebdomadaires des séances de l'Académie des sciences* 135(22 novembre 1902): 891–894.

Esclangon, Ernest. 1904. *Les fonctions quasi-périodiques*. Thèse de sciences mathématiques, Faculté des Sciences de Paris. Paris: Gauthier-Villars.

Esclangon, Ernest. 1919. *Nouvelles recherches sur les fonctions quasi-périodiques*. Annales de l'observatoire de Bordeaux. Paris: Gauthier-Villars.

Esclangon, Ernest. 1925. *L'acoustique des canons et des projectiles*. Mémorial de l'Artillerie française, T. 4, 3e fascicule. Paris : Gauthier-Villars.

Esnault-Pelterie, Robert. 1946. *L'analyse dimensionnelle*. Paris: Gauthier-Villars.

Fabre, Hervé. 1937a. Sur les déplacements des nœuds et des apsides dans les systèmes planétaires. *C.R.A.S.* 205: 1042–1044.

Fabre, Hervé. 1937b. Sur les orbites absolues de Gylden. *C.R.A.S.* 205: 1214–1216.

Fabre, Hervé. 1938. Les mouvements récurrents en mécanique céleste, et la variation des éléments des orbites. Thèse de sciences mathématiques, Faculté des Sciences de Paris, Paris: Gauthier-Villars.

Fabry, Charles. 1903. La période dans l'arc électrique chantant. *Éclairage Électrique* XXXIV(1902): 375–376.

Fatou, Pierre. 1927a. Sur le mouvement des nœuds de certaines orbites. *C.R.A.S.* 184: 1535–1537.

Fatou, Pierre. 1927b. Sur la recherche des orbites périodiques. *C.R.A.S.* 185: 35–36.

Fatou, Pierre. 1928. Sur le mouvement d'un système soumis à des forces à courte période. *B.S.M.F.* 56: 98–139.

Favard, Jean. 1926. Sur les équations différentielles à coefficients presque périodiques. *C.R.A.S.* 182: 1122–1124.

Favard, Jean. 1927a. Sur les fonctions harmoniques presque périodiques. *Thèse, reproduite dans Journal de mathématiques pures et appliquées*, (IX) 6: 229–336.

Favard, Jean. 1927b. Sur les équations différentielles linéaires à coefficients presque-périodiques. *Acta Mathematica* 51:31–81.

Favard, Jean. 1933. *Leçons sur les fonctions presque périodiques*. Paris: Gauthier-Villars (Cahiers scientifiques, fasc. 13).

Feinberg, Eugenii L'vovich. 2006. The development of radio in Russia. *Herald of the Russian Academy of Sciences* 76(2006): 64–70.

Ferrié, Gustave. 1921. *Notice sur les Travaux Scientifiques et Techniques de M. Gustave Ferrié*. Paris: Gauthier-Villars.

Ferrié, Gustave, and Raymond Jouaust. 1925. L'emploi des cellules photo-électriques à l'observation et à l'entretien des pendules astronomiques. *Comptes rendus hebdomadaires des séances de l'Académie des sciences* 180(14 avril 1925): 1145–1148.

Fessard, Alfred. 1931. Les rythmes nerveux et les oscillations de relaxation. *L'Année psychologique* 32(1931): 49–117.

Fessard, Alfred. 1936. *Les propriétés rythmiques de la matière vivante*. Paris: Hermann.

Flanders, Donald A., and James Johnston Stoker. 1946. A limit case of relaxation oscillations. In: *Studies in nonlinear vibration theory*, 50–64. New York: New York University.

Flemming, John Ambrose. 1916. *The principle of electric wave telegraphy and telephony*, 3e éd. New York: Longman Green and Co.

Foch, Adrien. 1935. *Acoustique*. Paris: Armand Colin.
Fortrat, René. 1927–1929. *Introduction à la Physique Théorique*, 7 fascicules. Paris: Hermann.
Fortrat, René. 1941. *Leçons de Radio-Électricité*. Paris: Masson.
Fortrat, René. 1945. Louis Barbillion – 11 mai 1873–27 mars 1945. *Annales de l'université de Grenoble* 21(1945): 21–25.
Fortrat, René. 1951. Oscillations pendulaires et de relaxation. *Journal de Physique et le Radium* 12(1951): 41–50.
Fortrat, R., and A. Caravel. 1939. Oscillations radioélectriques par triode à grille. *Comptes rendus hebdomadaires des séances de l'Académie des sciences* 209(25 septembre 1939): 554–558.
Fremont, Charles. 1915. Un échappement d'horloge au treizième siècle. *Comptes rendus hebdomadaires des séances de l'Académie des sciences* 161(6 décembre 1915): 690–692.
Friedrichs, K., P. Le Corbeiller, N. Levinson, and J.J. Stoker. 1942–1943. *Nonlinear mechanics*. Brown University.
Friedrichs, Kurt, and J.J. Stoker. 1943. Forced vibrations of systems with nonlinear restoring force. *Quarterly of Applied Mathematics* 1(1943): 97–115.
Friedrichs, Kurt, and Wolfgang R. Wasow. 1946. Singular perturbations of non-linear oscillations. *Duke Mathematical Journal* 13(1946): 367–381.
Frith, Julius, and Charles Rodgers. 1895. On the resistance of the electric arc. *Proceedings of the Physical Society of London* 14(1895): 307–327.
Gaiduk, Ju D. 1974. On the history of the theory of almost periodic functions. *Narisi Istor. Prirodoznav. i Tehn. Vyp.* 18(111): 35–42.
Gaponov-Grekhov, Andreï Viktorovich, and Mikhail I. Rabinovich. 1979. L. I. Mandel'shtam and the modern theory of nonlinear oscillations and waves. *Soviet Physics Uspekhi* 22(1979): 590–614.
Gaponov-Grekhov, Andreï Viktorovich, and Mikhail I. Rabinovich. 1992. *Nonlinearities in action*. New York: Springer.
Gau, Émile. 1925. *Calculs numériques et graphiques*. Paris: Armand Colin.
Gause, Georgii Frantsevich, and Wladimir Wladimirovich Alpatov. 1931. Die logistische Kurve von Verhulst-Pearl und ihre Anwendung im Gebiet der quantitativen Biologie. *Biol. Zentralbl.* 51(1931): 1–14.
Gause, Georgii Frantsevich. 1934. *The struggle for existence*. Baltimore: Williams & Wilkins.
Gause, Georgii Frantsevich. 1935. *Vérification expérimentale de la théorie mathématique de la lutte pour la vie*. Paris: Hermann (Actualités scientifiques et industrielles, vol. 277).
Gause, G.F., O.K. Nastukova, and W.W. Alpatov. 1934. The Influence of Biologically Conditioned Media on the Growth of a Mixed Population of Paramecium caudatum and P. aureliax. *Journal of Animal Ecology* 3(1934): 222–230.
Gause, G.F., and A.A. Witt. 1934. On the periodic fluctuations in the numbers of animals. A mathematical theory of the relaxation interaction between predators and prey and its application to a population of protozoa. [en russe, avec un résumé en anglais] *Bulletin de Leningrad*, 1551–1559.
Gause, G.F., and A.A. Witt. 1935. Behavior of mixed populations and the problem of natural selection. *The American Naturalist* 69(1935): 596–609.
Gause, G.F., N.P. Smaragdova, and A.A. Witt. 1936. Further studies of interaction between predators and prey. *Journal of Animal Ecology* 5(1936): 1–18.
Gerard-Lescuyer, Jean-Marie-Anatole. 1880a. Sur un paradoxe électrodynamique. *Comptes rendus hebdomadaires des séances de l'Académie des sciences* 168(16 juillet 1880): 226–227.
Gerard-Lescuyer, Jean-Marie-Anatole. 1880b. On an electrodynamical paradox. *Philosophical Magazine* (V) 10(1880): 215–216.
Gilain, Christian. 1977. *La théorie géométrique des équations différentielles de Poincaré et l'histoire de l'analyse*, Thèse de 3e cycle de l'Université de Paris I.
Gilain, Christian. 1991. La théorie qualitative de Poincaré et le problème de l'intégration des équations différentielles. In *La France mathématique. La Société mathématique de France (1872–1914)*, éd. dans H. Gispert. Paris: SFHST et SMF.

Ginoux, Jean-Marc. 2009. *Differential geometry applied to dynamical systems*, Series on nonlinear science, A, vol. 66. Singapore: World Scientific.

Ginoux, J.M., and Loïc Petitgirard. 2010. Poincaré's forgotten conferences on wireless telegraphy. *International Journal of Bifurcation & Chaos* 20(11): 3617–3626.

Gispert, Hélène, and Juliette Leloup. 2009. Des patrons des mathématiques en France dans l'entre-deux-guerres. *Revue d'histoire des sciences* 62(2009): 39–117.

Gleick, James. 1989. *La théorie du chaos*. Paris: Albin Michel.

Goodwin, Richard M. 1949. The business cycle as a self-sustaining oscillation. *Econometrica* 17(1949): 184–185.

Goodwin, Richard M. 1951. The nonlinear accelerator and the persistence of business cycles. *Econometrica* 19(1951): 1–17.

Graffi, Dario. 1940. Sopra alcune equazioni differenziali non lineari della fisica-matematica. *Memorie della Accademia delle scienze dell'Istituto di Bologna* (IX) 7(1940): 121–129.

Graffi, Dario. 1942. Sopra alcune equazioni differenziali della radiotecnica. *Memorie della Accademia delle scienze dell'Istituto di Bologna* (IX) 9(1942): 145–153.

Graffi, Dario. 1951. Forced oscillations for several nonlinear circuits. *Annals of Mathematics* (II) 54(1951): 262–271.

Granier, Jean. 1936. *Les systèmes oscillants*. Paris: Dunod, 1936. Résumé dans *Revue Générale des Sciences* 46(1935): 693.

Grassman, J. 1987. *Asymptotic methods for relaxation oscillations and applications*. Applied mathematical sciences, vol. 63. New York: Springer.

Greaves, William Michael Herbert. 1924. On the Stability of the Periodic States of the Triode Oscillator. *Mathematical Proceedings of the Cambridge Philosophical Society* 22(1924): 16–23.

Gutton, Camille. 1901. Sur la propagation des oscillations hertziennes dans l'eau. *Journal de Physique Théorique et Appliquée* (IX) 1(1901): 752–756.

Gutton, Camille. 1919a. La lampe-valve à trois électrodes. *Revue générale de l'Electricité* 5 (1919): 629–640.

Gutton, Camille. 1919b. Entretien des oscillations électriques par une lampe-valve à trois électrodes. *Revue générale de l'Electricité* 6(1919): 14–24.

Gutton, Camille. 1919c. La télégraphie et la téléphonie sans fil par les lampes-valves à trois électrodes. *Revue générale de l'Electricité* 6(1919): 365–376.

Gutton, Camille. 1922. Sur l'entretien simultané d'un circuit oscillant et de circuits harmoniques. *Comptes rendus hebdomadaires des séances de l'Académie des sciences* 174(3 avril 1922): 941–943.

Gutton, Camille. 1929. *La lampe à trois électrodes*. Paris: P.U.F.

Gutton, Camille. 1937. *Leçons de radioélectricité*. Paris: Eyrolles.

Haag, Jules. 1915a. Sur un système de formules différentielles concernant les éléments de tir d'un projectile soumis à une résistance quadratique de l'air. *Comptes rendus hebdomadaires des séances de l'Académie des sciences* 161(27 septembre 1915): 379–381.

Haag, Jules. 1915b. Sur la méthode d'Otto. *Comptes rendus hebdomadaires des séances de l'Académie des sciences* 161(2 novembre 1915): 524–526.

Haag, Jules. 1915c. Sur le calcul du temps. *Comptes rendus hebdomadaires des séances de l'Académie des sciences* 161(22 novembre 1915): 633–634.

Haag, Jules. 1921. Sur le calcul des trajectoires et de leurs altérations. *Journal de l'École Polytechnique* (II) 21e cahier (1921): 3–52.

Haag, Jules. 1929. Théorie générale de la synchronisation. *Comptes rendus hebdomadaires des séances de l'Académie des sciences* 189(23 décembre 1929): 1244–1246.

Haag, Jules. 1934a. Sur le calcul des oscillations mécaniques. *Comptes rendus hebdomadaires des séances de l'Académie des sciences* 198 (19 février 1934): 693–695.

Haag, Jules. 1934b. Sur les oscillations auto-entretenues. *Comptes rendus hebdomadaires des séances de l'Académie des sciences* 199(5 novembre 1934): 906–909.

Haag, Jules. 1936. Sur l'étude asymptotique des oscillations de relaxation. *Comptes rendus hebdomadaires des séances de l'Académie des sciences* 202(13 janvier 1936): 102–104.

Haag, Jules. 1937. Sur la théorie des oscillations de relaxation. *Comptes rendus hebdomadaires des séances de l'Académie des sciences* 204(22 mars 1937): 932–934.

Haag, Jules. 1938. Formules asymptotiques concernant les oscillations de relaxation. *Comptes rendus hebdomadaires des séances de l'Académie des sciences* 204(20 avril 1938): 1235–1237.

Haag, Jules. 1943. Étude asymptotique des oscillations de relaxation. *Annales scientifiques de l'École Normale Supérieure* (III) 60(1943): 35–64, 65–111 et 289 (errata).

Haag, Jules. 1944. Exemples concrets d'études asymptotiques d'oscillations de relaxation. *Annales scientifiques de l'École Normale Supérieure* (III) 61(1944): 73–117.

Haag, Jules. 1952–1955. *Les mouvements vibratoires*, 2 vol. Paris: P.U.F.

Hadamard, Jacques. 1897. Sur certaines propriétés des trajectoires en Dynamique. *Journal de mathématiques pures et appliquées* (V) 3(1897): 331–388.

Hadamard, Jacques. 1898. Sur les surfaces à courbures opposées et les lignes géodésiques. *Journal de mathématiques pures et appliquées* (V) 4(1898): 27–73.

Hadamard, Jacques. 1921. L'œuvre mathématique de Poincaré. *Acta Mathematica* 38(1921): 203–287.

Hadamard, Jacques. 1922. The early scientific work of Henri Poincaré: Lectures delivered at the Rice Institute in March, 1920. *Rice Institute Pamphlet* 9(1922): 111–183.

Hadamard, Jacques. 1923. Poincaré I la Teoria de les equacions diferencials. *Publicacions de l'Institut de Cienciès* (Col'lecció de Cursos de Fisica i Matem'atica): 5–87.

Hadamard, Jacques. 1933. The later scientific work of Henri Poincaré: Lectures delivered at the Rice Institute in May, 1925. *Rice Institute Pamphlet* 20(1933): 1–86.

Hadamard, Jacques. 1968. [*Œuvres*] *Œuvres de Jacques Hadamard*, 4 vol. Paris: Éditions du C.N.R.S.

Hamburger, Ludwig. 1930. Een Nieuwe Weg Voor Conjunctuur-Onderzoek, Een Nieuwe Richt lijn Voor Conjunctuur-Politiek, [Une nouvelle approche pour la recherche économique, Une nouvelle directive pour le dévelop-pement économique et politique]. *De Economist* LXXIX(1930):1–38.

Hamburger, Ludwig. 1931. Analogie des Fluctuations économiques et des Oscillations de relaxation, Institut de Statistique de l'Université de Paris. *Supplément aux Indices du Mouvement des Affaires*, 1–35. Janvier.

Hamburger, Ludwig. 1934. Note on economic cycles and relaxation oscillations. *Econometrica* 2(1934): 112.

den Hartog, Jacob Pieter. 1934. *Mechanical vibrations*. New York: MacGraw Hill Book, 1934; 2e éd., 1940.

den Hartog, Jacob Pieter. 1936. *Vibrations et mouvements vibratoires*. Paris: Dunod.

Heegner, Kurt. 1924. Über Selbsterregungserscheinungen bei Systemen mit gestörter Superpositionn. *Zeitschrift für Physik* 29(1924): 91–109.

Heegner, Kurt. 1927. Über Schwingungserzeugung mittels eines Elektronenröhrensystems, welches keine Selbstinduktion enthält. *Zeitschrift für Physik* 42(1927): 773–778.

Henri, and Cartan Élie. 1925. Note sur la génération des oscillations entretenues. *Annales des P.T.T.* 14: 1196–1207.

Herz, Cornelius. 1884. Le Comte Th. Du Moncel. *La Lumière Électrique* XI(1884): 381–385.

Hilbert, David. 1900. Mathematische Probleme, Vortrag, gehalten aufdem internationalen Mathematiker Kongress zu Paris 1900. *Nachrichten von der Königlichen Gesellschaft der Wissenschaften zu Göttingen. Mathematischen Physiques Klasse* (1900): 253–297.

Hochard, Étienne. 1933. Sur des oscillations entretenues. *Comptes rendus hebdomadaires des séances de l'Académie des sciences* 196(27 mars 1933): 905–906.

Hospitalier, Édouard. 1881. Lampe électrique et veilleur automatique de M. Anatole Gérard. *La Nature* n° 418(4 juin 1881): 220–222.

Hospitalier, Édouard. 1881. *Les principales applications de l'électricité*. Paris: Masson, 1881; 3e éd., 1884.

Hospitalier, Édouard. 1885. Sur la mesure des courants redressés. *Comptes rendus hebdomadaires des séances de l'Académie des sciences* 100(1885): 1456–1458.

Hurwitz, Adolf. 1895. Über die Bedingungen unter welchen eine Gleichung nur negatives reellen Theilen besitzt. *Mathematische Annalen* 46: 273–284.

Israel, Giorgio. 1996. *La Mathématisation du réel*. Paris: Seuil.

Israel, Giorgio. 2004. Technological innovations and new mathematics: Van der Pol and the birth of non-linear dynamics. In *Technological concepts and mathematical models in the evolution of modern engineering systems*, éd. M. Lucertini, A.M. Gasca, and F. Nicolo, 52–78. Basel: Birkhäuser.

Janet, Paul. 1893. Sur les oscillations électriques de période moyenne. *Journal de Physique Théorique et Appliquée* (II) 1 (1893): 337–352.

Janet, Paul. 1893. *Cours d'Électricité Industrielle*, 2e année, Grenoble: Alexandre Gratier & Cie.

Janet, Paul. 1894a. *Électricité industrielle*, Grenoble: Alexandre Gratier et Cie.

Janet, Paul. 1894b. *Premiers principes d'électricité industrielle*. Paris: Gauthier-Villars.

Janet, Paul. 1900. *Leçons d'électrotechnique générale*, 3 vol. Paris: Gauthier-Villars.

Janet, Paul. 1902a. Application de l'arc chantant de Duddell à la mesure des faibles coefficients de self-induction. *Comptes rendus hebdomadaires des séances de l'Académie des sciences* 134(24 février 1902): 462–463.

Janet, Paul. 1902b. Quelques remarques sur la théorie de l'arc chantant de Duddell. *Comptes rendus hebdomadaires des séances de l'Académie des sciences* 134(14 avril 1902): 821–823.

Janet, Paul. 1902c. L'arc voltaïque: Propriétés générales; expériences nouvelles. *Revue Générale des Sciences* 13(30 avril 1902): 416–422.

Janet, Paul. 1919. Sur une analogie électrotechnique des oscillations entretenues. *Comptes rendus hebdomadaires des séances de l'Académie des sciences* 168(14 avril 1919): 764–766.

Janet, Paul. 1925. Note sur une ancienne expérience d'électricité appliquée. *Annales des P.T.T.* 14(1925): 1193–1195.

Jordan, Camille. 1916. Notice nécrologique sur M. Léauté. *Comptes rendus hebdomadaires des séances de l'Académie des sciences* 163(6 novembre 1916): 501–502.

Jouaust, Raymond, and B. Decaux. 1928. Note sur quelques perfectionnements des amplificateurs à courant continu. *Onde électrique* (juillet 1928): 306–308.

Jouguet, Émile. 1943. Application de la méthode de variation des constantes à l'étude des oscillations non-linéaires. *Comptes rendus hebdomadaires des séances de l'Académie des sciences* 217(30 août 1943): 218–220.

Kennedy, Judy, and James Yorke. 1995. Bizarre topology is natural in dynamical systems. *Bulletin (New Series) Of The American Mathematical Society* 32(3): 309–316.

Kolmogorov, Aleksandr. 1936. Sulla Teoria di Volterra della lotta per l'Esisttenza. *Giorni dell'Instituto Italiano degli Attuari* 7(1936): 74–80.

Korowine, J. Th. 1921. A propos d'une analogie électrotechnique des oscillations entretenues. Revue Générale de l'Électricité II: 586.

Kostitzin, Vladimir. 1930. Sur quelques applications des équations intégrales. *Annales de l'institut Henri Poincaré* 1(1930): 177–203.

Kostitzin, Vladimir. 1932. Sur application géologique des équations différentielles biologiques. *Comptes rendus hebdomadaires des séances de l'Académie des sciences* 195(19 décembre 1932): 1219–1222.

Kostitzin, Vladimir. 1933. Sur quelques phénomènes quasi-périodiques dans les bassins fermés. *Comptes rendus hebdomadaires des séances de l'Académie des sciences* 196(16 janvier 1933): 214–215.

Kostitzin, Vladimir. 1934. *Symbiose, Parasitisme et Évolution*, Actualités Scientifiques et Industrielles, vol. 96. Paris: Hermann.

Kostitzin, Vladimir. 1936. Sur les solutions asymptotiques d'équations différentielles biologiques. *Comptes rendus hebdomadaires des séances de l'Académie des sciences* 203(16)(30 novembre 1936): 1124–1126.

Kostitzin, Vladimir. 1937. *Biologie Mathématique*. Paris: Armand Colin.

Kovanko, Alexandr'. 1928a. Sur une généralisation des fonctions presque-périodiques. *C.R.A.S.* 186: 354–355.

Kovanko, Alexandr'. 1928b. Sur quelques généralisations des fonctions presque-périodiques. *C.R.A.S.* 186: 729–730.
Kovanko, Alexandr'. 1929a. Sur l'approximation des fonctions presque périodiques généralisées. *C.R.A.S.* 188: 142–143.
Kovanko, Alexandr'. 1929b. Sur une classe de fonctions presque périodiques qui engendre les classes de fonctions p.p. de W. Stepanoff, H. Weyl et Bezicovitch. *C.R.A.S.* 189: 393–394.
Krylov, Nicolas. 1925. Sur l'estimation de l'erreur commise dans l'application de la méthode de W. Ritz pour l'intégration approchée des équations différentielles. *Comptes rendus hebdomadaires des séances de l'Académie des sciences* 180(4 mai 1925): 1316–1318.
Krylov, Nicolas. 1925. Sur différents procédés d'intégration approchée en physique mathématique. *Annales de la facultés des sciences de Toulouse* (III) 17(1925): 153–186.
Krylov, Nicolas. 1927. Sur différents procédés d'intégration approchée en physique mathématique. *Annales de la facultés des sciences de Toulouse* (III) 19(1927): 167–199.
Krylov, Nicolas. 1931. *Les méthodes de solution approchée des problèmes de la physique mathématique*, Mémorial des sciences mathématiques, fasc. 49. Paris: Gauthier-Villars.
Krylov, Nicolas, and Nicolas Bogolyubov. 1932a. Quelques exemples d'oscillateurs non-linéaires. *Comptes rendus hebdomadaires des séances de l'Académie des sciences* 194(22 février 1932): 957–960.
Krylov, Nicolas, and Nicolas Bogolyubov. 1932b. Sur le phénomène de l'entraînement en radiotechnique. *Comptes rendus hebdomadaires des séances de l'Académie des sciences* 194(22 février 1932): 1064–1066.
Krylov, Nicolas, and Nicolas Bogolyubov. 1932c. Les phénomènes de démultiplication de fréquence en radiotechnique. *Comptes rendus hebdomadaires des séances de l'Académie des sciences* 194(22 février 1932): 1119–1122.
Krylov, Nicolas, and Nicolas Bogolyubov. 1933. Problèmes fondamentaux de la mécanique non-linéaire. *Revue Générale des Sciences* 44(1933): 9–19.
Krylov, Nicolas, and Nicolas Bogolyubov. 1934a. *Méthodes nouvelles de la Mécanique Non-Linéaire dans leur application à l'étude du fonctionnement de l'oscillateur à lampe*, Moscou: Édit. techn.-théor. d'État.
Krylov, Nicolas, and Nicolas Bogolyubov. 1934b. L'application des méthodes de la mécanique non linéaire à la théorie des perturbations des systèmes canoniques. Всеукринская Академия наук, Институт строительной механики, n° 4, Киев.
Krylov, Nicolas, and Nicolas Bogolyubov. 1934c. Sur quelques développement formels en série dans la mécanique non linéaire. Всеукринская Академия наук, Институт строительной механики, n° 5, Киев.
Krylov, Nicolas, and Nicolas Bogolyubov. 1934d. Méthodes de mécanique non linéaire appliquées à la théorie des oscillations stationnaires. Всеукринская Академия наук, Институт строительной механики, n° 8, Киев.
Krylov, Nicolas, and Nicolas Bogolyubov. 1934e. Sur les solutions quasi périodiques des équations de la mécanique non linéaire. *Comptes rendus hebdomadaires des séances de l'Académie des sciences* 199(26 décembre 1934): 1592–1593.
Krylov, Nicolas, and Nicolas Bogolyubov. 1935a. Sur l'étude du cas de la résonance dans les problèmes de la mécanique non-linéaire. *Comptes rendus hebdomadaires des séances de l'Académie des sciences* 200(7 janvier 1935): 113–115.
Krylov, Nicolas, and Nicolas Bogolyubov. 1935b. Méthodes de mécanique non linéaire appliquées à la théorie des oscillations stationnaires. *Èasopis pro pìstování matematiky a fysiky* 64(1935): 107–115.
Krylov, Nicolas, and Nicolas Bogolyubov. 1936a. Les mouvements stationnaires généraux dans les systèmes dynamiques de la mécanique non-linéaire. *Comptes rendus hebdomadaires des séances de l'Académie des sciences* 202(20 janvier 1936): 200–201.
Krylov, Nicolas, and Nicolas Bogolyubov. 1936b. Application de la Mécanique Non-Linéaire à quelques problèmes de la radiotechnique moderne. *Onde Électrique* 15(1936): 508–531.

Krylov, Nicolas, and Nicolas Bogolyubov. 1937. *Introduction à la Mécanique non-linéaire* (en russe), Izd. Akad. Nauk. Sssr, 1937. Trad. angl. par S. Lefschetz. *Introduction to non-linear mechanics*. Princeton University Press, 1943.

Kuhn, Thomas. 1972. *Structure des révolutions scientifiques*, trad. fr. Paris: Flammarion.

Labrique, J.P. 1968. Oscillations de relaxation enregistrées à l'extrémité de cellules végétales. *Bulletin de la Société Royale de Belgique* 10(22 Nov. 1967): 205–211.

Lallemand, Charles. 1926. Allocution prononcée en prenant possession du fauteuil de la présidence. *Comptes rendus hebdomadaires des séances de l'Académie des sciences* 182(1er février 1926): 295–297.

Lasalle, J. 1949. Relaxation oscillations. *Quarterly of Applied Mathematics* 7(1): 1–19.

Latour, Marius. 1912. La génération industrielle des courants à haute fréquence. *La Lumière Électrique* (2e série) XVIII: 355–359.

Latour, Marius. 1916. Étude théorique sur l'audion. *La Lumière Électrique* (2e série) XXXV: 289–292.

Lawden, Derek F. 1954. *Mathematics of engineering systems*. London: University Paperbacks.

Léauté, Henry. 1885. Mémoire sur les oscillations à longue période dans les machines actionnées par des moteurs hydrauliques et sur les moyens de prévenir ces oscillations. *Journal de l'École Polytechnique* 55(1885): 1–126.

Léauté, Henry. 1887. Sur la caractéristique cinématique d'un système mécanique en mouvement. *Journal de mathématiques pures et appliquées* (IV) 4(1887): 465–476.

Lebon, Ernest. 1912. *Henri Poincaré, biographie, bibliographie analytique des écrits*. Paris: Gauthier-Villars (Coll. Savants du Jour).

Leblanc, Maurice. 1899. Étude des propriétés des machines d'induction. *Bulletin de la Société internationale des Électriciens* 16(1899): 54–65.

Leblanc, Maurice. 1921. Sur les lampes à trois électrodes, anode, cathode et grille intermédiaire, où le courant est transporté par des ions et leurs applications. *Comptes rendus hebdomadaires des séances de l'Académie des sciences* 175(10 juillet 1921): 132–137.

Lebon, Ernest. 1912. *Henri Poincaré, biographie, bibliographie analytique des écrits*. Paris: Gauthier-Villars (Coll. Savants du Jour).

Le Calvez, Patrice. 1986. Existence d'orbites quasi-périodiques dans les attracteurs de Birkhoff. *Communications in Mathematical Physics* 106: 383–394.

Le Corbeiller, Philippe. 1926. Sur un type d'oscillations entretenues par réaction. *Annales des P.T.T.* 8(1926): 1–24.

Le Corbeiller, Philippe. 1931a. *Les systèmes auto-entretenus et les oscillations de relaxation*, Actualités scientifiques et industrielles, vol. 144. Paris: Hermann.

Le Corbeiller, Philippe. 1931b. Sur les oscillations des régulateurs. In *Proceedings of the Third International Congress of Applied Mechanics* (Stockholm, 24–29 août 1930), vol. 3, 205–212.

Le Corbeiller, Philippe. 1932. Le mécanisme de la production des oscillations. *Annales des P.T.T.* 21(1932): 697–731.

Le Corbeiller, Philippe. 1933a. Le mécanisme de la production des oscillations. *Onde Électrique* 12(1933): 116–148.

Le Corbeiller, Philippe. 1933b. Les systèmes auto-entretenus et les oscillations de relaxation. *Econometrica* 1(3): 328–332.

Le Corbeiller, Philippe. 1934. Le mécanisme de la production des oscillations. *Annales des P.T.T.* 23(1934): 893–907.

Le Corbeiller, Philippe. 1936. Non-linear theory of maintenance of oscillations. *Journal of Institution of Electrical Engineers* 79(1936): 361–378.

Lefschetz, Solomon. 1919. Sur l'analyse situs des variétés algébriques. *Comptes rendus hebdomadaires des séances de l'Académie des sciences* 168(31 mars 1919): 672–674.

Lefschetz, Solomon. 1930. Les transformations continues des ensembles fermés et leurs points fixes. *Comptes rendus hebdomadaires des séances de l'Académie des sciences* 190(13 janvier 1930): 99–100.

Lefschetz, Solomon. 1943. Existence of periodic solutions for certain differential equations. *Proceedings of the National Academy of Sciences of the United States of America* 29(1943): 29–32.

Lefschetz, Solomon. 1957. *Differential equations: Geometric theory*. New York: Interscience Publishers.

Lemoine, Raoul. 1890. *L'Électricité dans l'Industrie*. Rouen: Impr. de Benderitter.

Levi-Civita, Tullio. 1900a. Sur l'instabilité des trajectoires périodiques. *C.R.A.S.* 131: 173–176.

Levi-Civita, Tullio. 1900b. Sur le problème restreint des trois corps. *C.R.A.S.* 131: 236—–239.

Levi-Civita, Tullio. 1906a. Sur la résolution qualitative du problème restreint des trois corps. *Acta Mathematica* 30: 305–327.

Levi-Civita, Tullio. 1906b. Sur la recherche des solutions particulières des systèmes différentiels et sur les mouvements stationnaires. *Prace Matematyczno-fizyczne* 17: 1–40.

Levi-Civita, Tullio. 1911. Sur les équations à coefficients périodiques et sur le moyen mouvement du nIJud Lunaire. Annales scientifiques de l'École Normale Supérieure (III) 325–376. http://www.numdam.org:80/numdam-bin/feuilleter?id=ASENS_1911\3_2828. (http://www.numdam.org:80/numdam-bin/recherche?h=py&format=short&py=1911&j=ASENS1911)

Levinson, Norman, and Oliver K. Smith. 1942. A general equation for relaxation oscillations. *Duke Mathematical Journal* 9(1942): 382–403.

Levitan, Boris Moisseevich. 1953. Почти-периодические функции, [Fonctions presque périodiques], функции государственное издательство технико-теоретической литературы, москва.

Levitan, Boris Moisseevich. 1980. Очерк истории теории почти-периодических функций, [Une brève histoire de la théorie des fonctions presque périodiques]. *Istoriko-Matematiceskie Issledovanija* 25: 156–166.

Levitan, B.M., et V.V. Zhikov. 1978. Почти-периодические функции и дифференциальные уравнения, [Les fonctions presque périodiques et les équations différentielles], Издателрьство Московского Университета.

Leyshon, Winifred A. 1923. Forced Oscillations in self-maintained oscillating circuits. *Philosophical Magazine* (Series 6) 46(1923): 686–698.

Leyshon, Winifred A. 1930. Characteristics of discharge tubes under "flashing" conditions as determined by means of the cathode ray oscillograph. *Proceedings of the Physical Society* 42(1930): 157–169.

Leyshon, Winifred A. 1931a. Note on a frequency recorder for use with a neon lamp excitation circuit. *Proceedings of the Physiological Society in Journal of Physiology* LXXI(november 15, 1930).

Leyshon, Winifred A. 1931b. Note on the use of the flashing neon lamp circuit for physiological purposes, and on a time-marker for use with the circuit. *Journal of Scientific Instruments* 8(1931): 202–204.

Leyshon, Winifred A. 1932. On periodic movements of the negative glow in discharge tubes. *Proceedings of the Physical Society* 44(1932): 171–189.

Leyshon, Winifred A. 1935. Some experiments on electronic oscillations. *Proceedings of the Physical Society* 47(1935): 277–286.

Leyshon, Winifred A. 1936. On a new type of electronic oscillator tube with parallel plane grids. *Proceedings of the Physical Society* 48(1936): 469–476.

Leyshon, Winifred A. 1938. Demonstrations: A plane-electrode demountable thermionic tube used as a generator of ultra-high-frequency oscillations. *Proceedings of the Physical Society* 50(1938): 146–148.

Leyshon, Winifred A. 1941. On ultra-high-frequency oscillations generated by means of a demountable thermionic tube having electrodes of plane form. *Proceedings of the Physical Society* 53(1941): 141–156, 490–491.

Lyapunov, Aleksandr. 1892. *Obshchaya zadacha ob ustoichivosti dvisheniya*. Kharkoff: Kharkov Mathematical Society.

Lyapunov, Aleksandr. 1907. Problème général de la stabilité du mouvement (traduit du russe par M. Édouard Davaux, Ingénieur de la Marine à Toulon). *Annales de la faculté des sciences de Toulouse* (II) 9(1907): 203–474.
Liénard, Alfred. 1928. Étude des oscillations entretenues. *Revue générale de l'Electricité* 23(1928): 901–912, 946–954.
Liénard, Alfred. 1931. Oscillations auto-entretenues. In *Proceedings of the Third International Congress of Applied Mechanics* (Stockholm, 24–29 août 1930), vol. 3, 173–177.
Liénard, Alfred. 1934. *Notice sur les Travaux Scientifiques de M. A Liénard.* Paris: Gauthier-Villars.
Lindsay, R. Bruce. 1981. Le Corbeiller, Philippe (1891–1980). *The Journal of the Acoustical Society of America* 69(1981): 1524.
Lotka, Alfred. 1925. *Elements of physical biology.* Baltimore: Williams & Wilkins Company.
Lotka, Alfred. 1932. The growth of mixed populations: Two species competing for a common food supply. *Journal of the Washington Academy of Science* 72(1932): 461–469.
Lorenz, Edward Norton. 1963. Deterministic nonperiodic flow. *Journal of Atmospheric Science* 20(1963): 130–141.
Luggin, Hans. 1888. Versuche und Bemerkungen über den galvanischen Lichtbogen. *Centralblatt für Elektrotechnik* X(1888): 567–581.
Maisel, S. 1903. The frequency of the singing arc. *Physikalische Zeitschrift* 4(1903): 532.
Maisel, S. 1904. The frequency of the singing arc. *Physikalische Zeitschrift* 5(1904): 550.
Maisel, S. 1905. The frequency of the singing arc. *Physikalische Zeitschrift* 6(1905): 38.
Mandel'shtam, L. I. 1947–1955. (Complete Works), Polnoe sobranie trudov, Moscou : Izd. AN SSSR, 1947–1955, Vol. 1Ű4. (vol. 1Ű3 edited by S. M. Rytov ; vol. 4 edited by M. A. Leontovich, 1955).
Mandel'shtam, Leonid Isaakovich, and Nikolaï Dmitrievich Papalexi. 1931. О резонансных явлениях при делении частоты. [Sur le phénomène de résonance et de démultiplication de fréquence]. *Zeitschrift für Physik* 73(1931): 223–248.
Mandel'shtam, L.I., and N.D. Papalexi. 1932. О явлениях резонанса л-го рода. [Sur le phénomène de résonance de la *n*-ième espèce] Журн. Техн. Физики 2: 775–811. [*Journal de Physique Technique de l'URSS* 2: 775–811].
Mandel'shtam, L.I., and N.D. Papalexi. 1933. О возбуждении колебаний в электрической колебательной системе при помощи периодического изменения емкости. [Sur l'excitation paramétrique d'oscillations électriques par variation périodique de la capacité d'un circuit] Журн. Техн. Физики 3: 1141–1144. [*Journal de Physique Technique de l'URSS* 3: 1141–1144].
Mandel'shtam, L.I., and N.D. Papalexi. 1934. Об обосновании одного метода приближенного решения дифференциальных уравнений, [Justification d'une méthode d'ap- proximation des solutions d'équations différentielles]. Журн. Техн. Физики 4: 117–122. [*Journal de Physique Technique de l'URSS* 4: 117–122].
Mandel'shtam, L.I., N.D. Papalexi, A.A. Andronov, S.E. Chaikin, and A.A. Witt. 1935. Exposé des recherches récentes sur les oscillations non-linéaires. urn. Tehn. Fiziki 2: 81–134. *Journal de Physique Technique de lŠURSS* 2: 81–134.
Markoff, Andreï. 1929. Sur les mouvements oscillations presque périodiques. *Comptes rendus hebdomadaires des séances de l'Académie des sciences* 189(4 novembre 1929): 732–734.
Markov, Andreï. 1929. Sur les mouvements oscillations presque périodiques. *C.R.A.S.* 189: 732–734.
Maxwell, James Clerk. 1867. On the dynamical theory of gases. *Philosophical Transactions of the Royal Society* 157(1867): 49–88.
Maxwell, James Clerk. 1868. On governors. *Proceedings of the Royal Society of London* 16(1868): 270–283.
Maz'ya, Vladimir, and Tatyana Shaposhnikova. 1998. Jacques Hadamard, a Universal Mathematician, History of Mathematics, 14. Providence/London: American Mathematical Society and London Mathematical Society. There are also two revised and expanded editions: the

French translation Maz'ya, Vladimir; Shaposhnikova, Tatyana (January 2005) [1998], Jacques Hadamard, un mathématicien universel, Sciences & Histoire (in French), Paris: EDP Sciences.
Maxwell, James Clerk. 1890. *Scientific papers*. Cambridge: Cambridge University Press.
Mercier, Jean. 1929. Le mécanisme de la stabilisation des oscillations dans un oscillateur à lampes. *Onde Électrique* 85(1929): 29–36, 60–67.
Mercier, Jean. 1934a. Sur les différents genres possibles d'oscillations électriques. *Comptes rendus hebdomadaires des séances de l'Académie des sciences* 198(22 janvier 1934): 349–351.
Mercier, Jean. 1934b. Sur les différents genres d'oscillations électriques. *Journal de Physique et le Radium* 5(1934): 126–131.
Mercier, Jean. 1934c. *Les circuits oscillants*. Paris: Librairie Delagrave.
Merger, Michèle. 1997. Hippolyte Parodi (1874–1868): Un des pionniers de l'électrification ferroviaire en France. In *Revue d'Histoire des Chemins de Fer*, Hors Série n°5(juillet 1997), Collection Histoire de l'Électricité n° 10: 147–161.
Merton, Robert K. 1968. The Matthew effect. *Science* 159(3810): 56–63.
Mesny, René. 1922. Génération par tubes électroniques d'oscillations polyphasées de haute fréquence. *Comptes rendus hebdomadaires des séances de l'Académie des sciences* 175(23 octobre 1922): 691–692.
Mesny, René. 1923a. Les déviations des ondes électromagnétiques. *Journal de Physique et le Radium* 4(1923): 129–143.
Mesny, René. 1923b. *Génération d'oscillations polyphasées au moyen de tubes électroniques*, tome 2, *Fonctionnement des lampes émission réception*. Paris: Étienne Chiron.
Mesny, René, and Pierre David. 1923. Les ondes très courtes en radiotélégraphie. *Comptes rendus hebdomadaires des séances de l'Académie des sciences* 177(26 novembre 1923): 1106–1107.
Mestraud, André. 1921. Sur une analogie électrotechnique des oscillations entretenues. *Revue Générale de l'Électricité* I: 363–368.
Migulin, Vladimir Vasil'evitch. 1981. N. D. Papalexi (100th anniversary of his birth). *Soviet Physics Uspekhi* 24: 614–618.
Minorsky, Nicholas. 1947. *Introduction to non-linear mechanics*, Originally published 1944–1946, as Restricted reports by David W. Taylor Model Basin, U.S. Navy.
Minorsky, Nicholas. 1949. Sur l'oscillateur de Van der Pol. *Comptes rendus hebdomadaires des séances de l'Académie des sciences* 228(20 décembre 1948): 60–61.
Minorsky, Nicholas. 1951a. Sur l'oscillateur non linéaire de Mathieu. *Comptes rendus hebdomadaires des séances de l'Académie des sciences* 232(11 juin 1951): 2179–2180.
Minorsky, Nicholas. 1951b. Sur une équation différentielle de la physique. *Comptes rendus hebdomadaires des séances de l'Académie des sciences* 232(18 décembre 1951): 1060–1062.
Minorsky, Nicholas. 1954. Sur l'interaction des oscillateurs non-linéaires. *Seminario Matematico e Fisico di Milano* XXV(1954): 145–163.
Minorsky, N. 1962. *Nonlinear oscillations*. Princeton/Toronto/New York: D. Van Nostrand.
Minorsky, Nicholas. 1967. *Théorie des oscillations*, Mémorial des sciences mathématiques, fasc. 163. Paris: Gauthier-Villars.
Mira, Christian. 1997. Some historical aspects of nonlinear dynamics – Possible trends for the future. *Journal of the Franklin Institute-Engineering and Applied Mathematics* 334B(1997): 1075–1113.
Mitropolski, Yurii Alekseevich. 1964. *Problems of the Asymptotic Theory of Nonstationary Oscillations*, (en russe) Mockba: Nauka.
Mitropolski, Yurii Alekseevich. 1966. *Problèmes de la théorie asymptotique des oscillations non stationnaires*. Paris: Gauthier-Villars.
Mitropolski, Yurii Alekseevich. 1999. On the role of N. N. Bogolyubov in the development of the theory of nonlinear oscillations. *Ukrainian Mathematical Journal* 51(1999): 1143–1165.
Moncel, Théodose du. 1858. *Étude sur le magnétisme et l'électro-magnétisme au point de vue de la construction des électro-aimants*. Paris: L. Hachette et Cié/Mallet-Bachelier.
Moncel, Théodose du. 1872–1878. *Exposé des applications de l'électricité*, 5 vol. Paris: Librairie Scientifique, Industrielle et Agricole, Eugène Lacroix.

Moncel, Théodose du. 1878. *Le Téléphone, le Microphone et le Phonographe*. Paris: Librairie Hachette et Cie.

Moncel, Théodose du. 1879. *L'Éclairage électrique*. Paris: Librairie Hachette et Cie.

Moncel, Théodose du. 1880. Réactions réciproques des machines dynamo-électriques et magnétoélectriques. *La Lumière Électrique* II(1880): 352.

Moncel, Théodose du. 1882. *Le Microphone, le Radiophone et le Phonographe*. Paris: Librairie Hachette et Cie.

Morched-Zadeh, R. 1936. *Étude des oscillations de relaxation et des différents modes d'oscillations d'un circuit comprenant une lampe au néon*. Thèse de sciences physiques, Faculté des sciences de Toulouse.

Morse, Marston. 1921. A one-to-one representation of geodesics on a surface of negative curvature. *American Journal of Mathematics* 43: 33–51.

Mosseri, Rémy. 1999. *Léon Brillouin, 1889–1969: à la croisée des ondes*. Paris: Belin.

Moussiegt, J. 1949. Les oscillations de relaxation dans les tubes à décharge; application à l'étude de l'allumage. Paris: Masson and Co.

Mutabazi, Innocent, José Éduardo Wesfreid, and Étienne Guyon. 2005. *Dynamics of spatio-temporal cellular structures: Henri Bénard centenary review*, Tracts in modern physics series, vol. 207. New York: Springer.

Nasmyth, George W. 1908. The frequency of the singing arc. *Physical Review* (Series I) 27(1908): 117–140.

Neimark, Yurii Isaakovich. 2001. Создатель нелинейной динамики и физики: к 100-летию со дия рождения академика А. А. Андронова, Вестник Российской Академии Наук, Tom 71, № 4, S. 231–236. [Le 100e anniversaire de l'académicien A. A. Andronov: Créateur de la physique et de la dynamique non-linéaire]. *Bulletin de l'Académie russe des sciences* 71(2001): 231–236.

Newhouse, S., D. Ruelle, and F. Takens. 1978. Occurrence of strange Axiom A attractors near quasiperiodic flows on Tm, m≥3. *Communications in Mathematical Physics* 64(1): 35–40.

Niemytski, Viktor Vladimirovich. 1934. Sur les systèmes dynamiques instables. *Comptes rendus hebdomadaires des séances de l'Académie des sciences* 199(2 juillet 1934): 18–20.

Niemytski, V.V., and Vyacheslaw Vassilievich Stepanoff. 1960. *Qualitative theory of differential equations*. Princeton: Princeton University Press.

Ocagne, Maurice d'. 1893. *Le Calcul simplifié par les procédés mécaniques et graphiques*. Paris: Gauthier-Villars.

Odani, Kenzi. 1995. The limit cycle of the Van der Pol equation is not algebraic. *Journal of Differential Equations* 115(1995): 146–152.

Panc-Tcheng, Kao. 1930. Oscillations de relaxation produites par un oscillateur à quartz piézoélectrique. *Comptes rendus hebdomadaires des séances de l'Académie des sciences* 191(17 novembre 1930): 932–934.

Papalexi, N.D. 1934. Me dunarodna nelineinym konferenci. [Conférence internationale de non linéaire] *Zeitschrift für Technische Physik* 4: 209–213.

Parodi, Hippolyte. 1909. Les méthodes de calculs des projets de traction. *Bulletin de la Société française des Électriciens* 3(1909): 151–191.

Parodi, Hippolyte. 1931. Méthode d'intégration par arcs successifs permettant d'obtenir simplement, dans le calcul de l'arc élémentaire, une approximation aussi élevée qu'on le désire. *Comptes rendus hebdomadaires des séances de l'Académie des sciences* 193(26 octobre 1931): 816–818.

Parodi, Hippolyte. 1942a. Traction électrique automatique et équations de relaxation. *Comptes rendus hebdomadaires des séances de l'Académie des sciences* 215(20 juillet 1942): 169–171.

Parodi, Hippolyte. 1942b. Système simple d'équations analogue à une équation de relaxation unique. *Comptes rendus hebdomadaires des séances de l'Académie des sciences* 215(3 août 1942): 125–127.

Parodi, Hippolyte. 1942c. Détermination graphique du cycle limite solution des équations de relaxation. *Comptes rendus hebdomadaires des séances de l'Académie des sciences* 215(18 août 1942): 196–198.

Parodi, Hippolyte, and Maurice Parodi. 1942. Méthode d'intégration par arcs successifs des équations de relaxation. *Comptes rendus hebdomadaires des séances de l'Académie des sciences* 215(31 août 1942): 268–270.

Parodi, Hippolyte, and Maurice Parodi. 1943. Les équations de la relaxation. Cas particulier des équations de la marche du train. *La revue scientifique* 81(1943): 110–120.

Pearl, Raymond, and Lowel J. Reed. 1920. On the rate of growth of the population of the United States since 1790 and its mathematical representation. *Proceedings of the National Academy of Science (U.S.A.)* 6(1920): 275–288.

Pechenkin, Alexander A. 1999. The importance of the Strasbourg period in L.I. Mandelstam's life for his further work in science. *NTM: International Journal of History and Ethics of Natural Sciences, Technology and Medicine* 7(1999): 93–104.

Pechenkin, Alexander A. 2002. The concept of self-oscillations and the rise of synergetics ideas in the theory of nonlinear oscillations. *Studies in History and Philosophy of Modern Physics* 33(2002): 269–295.

Pecker, Jean-Claude. 1996. Hervé Fabre, 27 November 1905–1995. *Journal des Astronomes Français* 50: 4–6.

Pellat, Henri. 1901. Sur un phénomène d'oscillation électrique. *Journal de Physique Théorique et Appliquée* 10(1901): 471–475.

Petit, Gaston Émile, and Léon Bouthillon. 1910. *La Télégraphie sans fil, la Téléphonie sans fil.* Paris: Librairie Delagrave.

Pétrovitch, Michel. 1901. Les Analogies Mathématiques et la Philosophie Naturelle. *Revue Générale des Sciences* 12(1901): 626–632.

Pétrovitch, Michel. 1907. *La Mécanique des Phénomènes fondée sur les Analogies*. Collection Scientia n° 27. Paris: Gauthier-Villars.

Pétrovitch, Michel. 1921.*Mécanismes communs aux phénomènes disparates*. Paris: Librairie Félix Alcan.

Pétrovitch, Michel. 1931. *Intégration qualitative des équations différentielles*, Mémorial des sciences mathématiques, vol. 48. Paris: Gauthier-Villars.

Peychés, Yvan. 1970. Notice sur sa vie et l'œuvre de Hippolyte Parodi. *Notices et discours* t. 5(1963–1972), Institut de France, Académie des sciences, 611–628.

Picard, Émile. 1891–1894. *Traité d'Analyse*, 3 tomes. Paris: Gauthier-Villars.

Pliss, Viktor Aleksandovich. 1966. *Nonlocal problems of the theory of oscillations*. New York/London, Academic Press.

Poincaré, Henri. 1879. *Sur les propriétés des fonctions définies par les équations aux différences partielles*, thèse de sciences mathématiques, Faculté des Sciences de Paris. Paris: Gauthier-Villars.

Poincaré, Henri. 1881. Sur les courbes définies par une équation différentielle. *Journal de mathématiques pures et appliquées* (III) 7(1881): 375–422.

Poincaré, Henri. 1882. Sur les courbes définies par une équation différentielle. *Journal de mathématiques pures et appliquées* (III) 8(1882): 251–296.

Poincaré, Henri. 1885a. Sur les courbes définies par une équation différentielle. *Journal de mathématiques pures et appliquées* (IV) 1(1885): 167–244.

Poincaré, Henri. 1885b. Sur l'équilibre d'une masse fluide animée d'un mouvement de rotation. *Acta Mathematica* 7(1885): 259–380.

Poincaré, Henri. 1886a. Sur les courbes définies par une équation différentielle. *Journal de mathématiques pures et appliquées* (IV) 2(1886): 151–217.

Poincaré, Henri. 1886b. *Notice sur les Travaux Scientifiques de Henri Poincaré*. Paris: Gauthier-Villars.

Poincaré, Henri. 1890. Sur le problème des trois corps et les équations de la dynamique. *Acta Mathematica* 13(1890): 1–270.

Poincaré, Henri. 1892. Les *Méthodes Nouvelles de la Mécanique Céleste*. Paris: Gauthier-Villars.

Poincaré, Henri. 1893. Les *Méthodes Nouvelles de la Mécanique Céleste*. Paris: Gauthier-Villars.

Poincaré, Henri. 1899. Les *Méthodes Nouvelles de la Mécanique Céleste*. Paris: Gauthier-Villars.

Poincaré, Henri. 1895. Analysis Situs. *Journal de l'École Polytechnique* 2e série, cahier 1: 1–121.

Poincaré, Henri. 1902a. Notice sur la télégraphie sans fil. *Annuaire du Bureau des longitudes* A1–A34.

Poincaré, Henri. 1902b. Sur la télégraphie sans fil. *Revue scientifique* 17(1902): 65–73.

Poincaré, Henri. 1905a. Sur la méthode horistique de Gylden. *Acta Mathematica* 29(1905): 235–271.

Poincaré, Henri. 1905b. *Leçons de Mécanique Céleste*, 3 vol. Paris: Gauthier-Villars.

Poincaré, Henri. 1907a. *Leçons de Mécanique Céleste*, 3 vol. Paris: Gauthier-Villars.

Poincaré, Henri. 1907b. *La théorie de Maxwell et les oscillations hertziennes: La télégraphie sans fil*, 3e éd. Paris: Gauthier-Villars.

Poincaré, Henri. 1908. Sur la télégraphie sans fil. *La Lumière Électrique* (II) 4, p. 259–266, 291–297, 323–327. *La Lumière Électrique* (II) 4: 355–359, 387–393.

Poincaré, Henri. 1909a. *Conférences sur la télégraphie sans fil. éd. La Lumière Électrique* Paris.

Poincaré, Henri. 1909b. La télégraphie sans fil. *Journal de l'Université des Annales* 3(1909): 541–552.

Poincaré, Henri. 1910. *Leçons de Mécanique Céleste*, 3 vol. Paris: Gauthier-Villars.

Poincaré, Henri. 1911. Sur diverses questions relatives à la télégraphie sans fil. *La Lumière Électrique* 13(1911): 7–12, 35–40, 67–72, 99–104.

Poincaré, Henri. 1912. Sur un théorème de Géométrie. *Rendiconti del Circolo Matématico di Palermo* 33(1912): 375–407.

Poincaré, Henri. 1921. Notice sur les Travaux Scientifiques de Henri Poincaré. *Acta Mathematica* 38(1921): 3–135.

Poincaré, Henri. 1947. О кривых, определяемых дифференциального уравнения [Sur les courbes définies par une équation différentielle], Gostekhizdat, 1947. Traduction russe éditée et annotée par A. A. Andronov, E. Leontovich et A. Majer.

Pomey, Jean-Baptiste. 1914. *Cours d'électricité théorique*. 3 vol. Paris: Gauthier-Villars.

Pomey, Jean-Baptiste. 1921. *Analogies Mécaniques de l'Électricité*. Paris: Gauthier-Villars.

Pomey, Jean-Baptiste. 1928. *Cours d'électricité théorique*. 3 vol. Paris: Gauthier-Villars.

Pomey, Jean-Baptiste. 1931. *Cours d'électricité théorique*. 3 vol. Paris: Gauthier-Villars.

Pontrjagin, Lev Semenoviéc. 1930. Sur une hypothèse fondamentale de la théorie de la dimension. *Comptes rendus hebdomadaires des séances de l'Académie des sciences* 190(23 avril 1930): 1105–1107.

Pontrjagin, Lev Semenoviéc. 1933. Les fonctions presque périodiques et l'Analysis situs. *Comptes rendus hebdomadaires des séances de l'Académie des sciences* 196(24 avril 1933): 1201–1203.

Pontrjagin, Lev Semenoviéc. 1934. On dynamical systems close to Hamiltonian systems. *Zhurnal Éksperimental' noï Teoreticheskoi Fiziki* 4(1934): 883–885.

Remaud, Patrice. 2004. *Une histoire de la genèse de l'automatique en France 1850–1950. De l'École de la régulation française au début du XXe siècle à l'émergence de l'automatique en France après la seconde guerre mondiale*. Thèse, C.D.H.T. du C.N.A.M. Paris. (http://laii.univ-poitiers.fr/THESES/remaud/index.php).

Richardson, L.F. 1937. The behaviour of an Osglim lamp. *Proceedings of the Royal Society of London. Series A, Mathematical and Physical Sciences* 162(910): 293–390.

Righi, Augusto. 1902. Sui fenomeni acustici dei condensatori. *Rendiconto dell'Accadémia della Scienza di Bologna* 6(1902): 188–231.

Robb, Alfred A. 1922. On a graphical solution of a class of differential equations occurring in wireless telegraphy. *Philosophical Magazine* 43(253): 206–215.

Rocard, Yves. 1932. Sur les oscillations à lampes réglées prés de la limite d'entretien. *Comptes rendus hebdomadaires des séances de l'Académie des sciences* 194(18 avril 1932): 1325–1327.

Rocard, Yves. 1935a. Sur certains types nouveaux d'oscillations mécaniques. In *Livre jubilaire de M. Brillouin*, 400–408. Paris: Gauthier-Villars.

Rocard, Yves. 1935b. *Sur la stabilité des locomotives*, Actualités scientifiques et industrielles, vol. 234. Paris: Hermann.

Rocard, Yves. 1935c. *Sur la stabilité des locomotives*, Actualités scientifiques et industrielles, vol. 279. Paris: Hermann.

Rocard, Yves. 1937a. Relaxation, Synchronisation et Démultiplication de Fréquence. *Onde Électrique* 16(1937): 396–412.

Rocard, Yves. 1937b. *Les phénomènes d'auto-oscillations dans les installations hydrauliques*, Actualités scientifiques et industrielles, vol. 544. Paris: Hermann.

Rocard, Yves. 1938. Oscillateurs à lampe stabilisés par leur propre circuit. *Journal de Physique et le Radium* 9(1938): 237–240.

Rocard, Yves. 1941. *Théorie des oscillateurs*. Paris: Édition de la Revue Scientifique.

Rocard, Yves. 1942. *Dualité des mécanismes d'autooscillation, Comptes rendus hebdomadaires des séances de l'Académie des sciences* 214: 601–602.

Rocard, Yves. 1943. *Dynamique générale des vibrations*. Paris: Masson.

Rocard, Yves. 1962. *Titres et Travaux Scientifiques*. Paris: Jouve Éditeur.

Rochot, Bernard. 1965. Léon Auger (1886–1964). *Revue d'histoire des sciences* 18(1965): 116–117.

Rodet, Julien. 1907. *Les lampes à incandescence électriques*. Paris: Gauthier-Villars.

Routh, Edward John. 1877. *A treatise on the stability of a given state of motion, particularly steady motion*. London: Macmillan and Co.

Ruelle, David, et Floris Takens. 1971. On the nature of turbulence. *Communications in Mathematical Physics* 20: 167–192.

Ruhmer, Ernst. 1908. *Wireless telephony*. New York: Van Nostrand Company.

Runge, Carl. 1912. *Graphical methods*. New York: Columbia University Press. Trad. allemande. *Graphische Methoden* Leipzig–Berlin: Teubner, 1915.

Rytov, Sergei Mikhailovich. 1957. Developments of the theory of nonlinear oscillations in the USSR. *Radiotekh. i elektronika* 2(1435): 169–192.

Rytov, Sergei Mikhailovich. 1979. Leonid Isanovich Mandel'shtam. *Soviet Physics Uspekhi* 22(1979): 826–832.

Sansone, Giovanni. 1949a. Valuatzioni asintotica degli integrali sulla equazione di Liénard che per $t \rightarrow -\infty$ tendono allo zero. *Atti della Accademia Nazionale dei Lincei, Rendiconti, Classe di Scienze Fisiche, Matematiche e Naturali* (VIII) 6(1949): 13–18.

Sansone, Giovanni. 1949b. Sopra una classa di equazioni di Liénard prive di integrali periodici. *Atti della Accademia Nazionale dei Lincei, Rendiconti, Classe di Scienze Fisiche, Matematiche e Naturali* (VIII) 6(1949): 159–160.

Schallreuter, Walter von. 1923. *Über Schwingungserscheinungen in Entladungsröhren*. Braunschweig: Vieweg.

Schlomiuk, Dana. 1993. Elementary first integrals of differential equations and invariant algebraic curves. *Expositiones Mathematicae* 11: 433–454.

Shaw, Robert S. 1981. Strange attractors, chaotic behavior, and information flow. *Zeitschrift für Naturforschung* 36a: 80–112.

Shohat, James Alexander. 1943. A new analytical method for solving Van der Pol's and Certain related types of non-linear differential equations homogeneous and non-homogeneous. *Journal of Applied Physics* 14(1943): 40–48.

Shohat, James Alexander. 1944. On Van der Pol's and related non-linear differential equations. *Journal of Applied Physics* 15(1944): 568–574.

Siegmund-Schultze, Reinhard. 2001. Rockefeller and the internationalization of mathematics between the two world wars: documents and studies for the social history of mathematics in the 20th century. Basel et al.: Birkhäuser Verlag.

Simon, Andrew L. 1998. *Made in Hungary: Hungarian contributions to universal culture*. Safety Harbor (Fla.): Simon Publications.

Simon, Hermann Theodor Simon. 1906. Zur Theorie der selbsttönenden Lichtbogens. *Physikalische Zeitschrift* 7(1906): 433–445.

Sire De Vilar, Henri. 1901. La dualité en électrotechnique. *L'Éclairage Électrique* 27(1901): 252–259, 278–285.

Stepanov, Vyacheslaw. 1925. Sur quelques généralisations des fonctions presque périodiques. *C.R.A.S.* 181: 90–92.

Stepanov, Vyacheslaw, and Andreï Tychonov. 1933. Sur les espaces des fonctions presque périodiques. *C.R.A.S.* 196: 1199–1201.

Stoker, James Johnston. 1950. *Nonlinear vibrations in mechanical and electrical systems*. New York: Interscience Publishers/John Wiley & Sons.

Stokes, Georges Gabriel. 1845. On the theories of the internal friction of fluids in motion, and of the equilibrium and motion of elastic solids. *Transactions of the Cambridge Philosophical Society* 8(1845): 287–305.

Strutt, John William, and Lord Rayleigh. 1877. *The theory of sound*, 2 vol. London: MacMillan and C°.

Strutt, John William, and Lord Rayleigh. 1883. On maintained vibrations. *Philosophical Magazine* (ser. 5) 15: 229–235.

Strutt, John William, and Lord Rayleigh. 1911. On the calculation of Chladni's figures for a square plate. *Philosophical Magazine* (ser. 6) 22(1911): 225–229.

Stumpers, F.L.H.M. 1960. Balth. Van der Pol's work on nonlinear circuits. *IRE Transactions on Circuit Theory* 7: 366–367.

Taton, René, dir. 1957–1964. Histoire Générale des Sciences, 4 tomes. Paris: P.U.F.

Teissié-Solier, Max, Louis Castagnetto, and Marcel Sabathe. 1937. Sur les battements qui accompagnent la formation des tourbillons alternés de Bénard Karman. *Comptes rendus hebdomadaires des séances de l'Académie des sciences* 205(28 juin 1937): 23–25.

Teissié-Solier, Max, and Louis Castagnetto. 1938. Sur la détermination du critérium des tourbillons alternés et les oscillations de relaxation. *Comptes rendus hebdomadaires des séances de l'Académie des sciences* 206(4 avril 1938): 1279–1281.

Thompson, Silvanus Phillips. 1892. On the physics of the voltaic arc. *The Electrician* XXIX(1892): 460.

Thompson, Silvanus Phillips. 1893. *Dynamo-electric machinery*, vol. I. New York: Home Book Company.

Thomson, William, and Lord Kelvin. 1853. On transient electric currents. *Philosophical Magazine* (Series 4) 5(1853): 393–405.

Thomson, William and Lord Kelvin. 1892. Graphical solution of dynamical problems. *Philosophical Magazine* 34: 443–448.

Timoshenko, Stephen Prokofievitch. 1928. Vibration problems in engineering. Trad. fr. *Théorie des Vibrations à l'usage des ingénieurs*. Paris: Librairie polytechnique Ch. Béranger, 1937.

Tissot, Camille. 1902. Observations sur l'arc chantant, (Congrès de l'Association française pour l'avancement des sciences à Montauban). *Éclairage Électrique* XXXII(1902): 354–355.

Tournès, Dominique. 2003a. L'intégration graphique des équations différentielles ordinaires. *Historia Mathematica* 30(2003): 457–493.

Tournès, Dominique. 2003b. Junius Massau et l'intégration graphique. *Revue d'histoire des mathématiques* 9(2003): 181–252.

Tournès, Dominique. 2003c. Du compas aux intégraphes: Les instruments du calcul graphique. *Revue Repères – IREM* 50(2003): 63–84.

Tricomi, Francesco. 1933. Integrazione di un'equazione differenziale presantatasi in electrotechnica. *Annali della R. Scuola Normale Superiore di Pisa* (Ser. 11) 2(1933).

Trotter, Alexander Pelham. 1924. Mrs Ayrton's work on the electric arc. *Nature* 113(1924): 48–49.

Trutat, B., and G. Bouttes. 1925. Les oscillations d'un moteur à excitation séparée alimenté par une génératrice série. *Revue générale de l'Electricité* 17(1925): 669–671.

Turpain, Albert. 1908. *La Télégraphie Sans Fil et les applications pratiques des ondes électriques*, 2e éd. Paris: Gauthier-Villars.

Vacquié, Serge. 1995. Arc électrique. *Techniques de l'ingénieur, Génie électrique* D4, N° D 2 870: 1–28.

Van Aalst, F. 1922. Sur l'entretien des oscillations électriques entretenues par une lampe à trois électrodes. *Comptes rendus hebdomadaires des séances de l'Académie des sciences* 175(26 décembre 1922): 1397–1399.

Van Bladel, Jean. 2006. The early history of URSI. In *Jubilee Svenska Nationalkommittén för Radiovetenskap* (SNRV 75 år) celebrating the 75th anniversary of the Swedish National Committee at the Royal Academy of Sciences in Stockholm, 1–12, 20 March 2006.

Van der Pol, Balthasar. 1920. A theory of the amplitude of free and forced triode vibrations. *Radio Review (London)* 1(1920): 701–710, 754–762.

Van der Pol, Balthasar. 1922. On oscillation hysteresis in a triode generator with two degrees of freedom. *The London, Edinburgh, and Dublin Philosophical Magazine and Journal of Science* (Ser. 6) 43: 700–719.

Van der Pol, Balthasar. 1924. Gedwongen trillingen in een systeem met nietlineairen weerstand (Ontvangst met teruggekoppelde triode). *Tijdschrift van het Nederlandsch Radiogenootschap* 2(1924): 57–73.

Van der Pol, Balthasar. 1925. Het onderling verband tusschen eenige moderne vorderingen in de draadlooze telegrafie en telefonie. *Polytechnish Weekblad* 19(1925): 791–794.

Van der Pol, Balthasar. 1926a. Over Relaxatietrillingen. *Physica* 6(1926): 154–157. Eindhoven

Van der Pol, Balthasar. 1926b. Over „Relaxatie-trillingen". *Tijdschrift van het Nederlandsch Radiogenootschap* 3: 25–40.

Van der Pol, Balthasar. 1926c. Über „Relaxationsschwingungen". *Jahrbuch der drahtlosen Telegraphie und Telephonie* 28(1926): 178–184.

Van der Pol, Balthasar. 1926d. On "Relaxation-oscillations". *The London, Edinburgh, and Dublin Philosophical Magazine and Journal of Science* (VII) 2(1926): 978–992.

Van der Pol, Balthasar. 1927a. Forced oscillations in a circuit with non-linear resistance (reception with reactive triode). *The London, Edinburgh, and Dublin Philosophical Magazine and Journal of Science* (VII) 3(1927): 65–80.

Van der Pol, Balthasar. 1927b. Over „Relaxatietrillingen" II. *Tijdschr. Ned. Radiogenoot* 3: 94–105.

Van der Pol, Balthasar. 1927c. Über „Relaxationsschwingungen" II. *Jahrbuch der drahtlosen Telegraphic und Telephonie* 29(1927): 114–118.

Van der Pol, Balthasar. 1930. Oscillations sinusoïdales et de relaxation. *Onde Électrique* 9(1930): 245–256, 293–312.

Van der Pol, Balthasar. 1931. Electrical and Mechanical Oscillations the period of which is proportional to a time constant. *Proceedings of the Third International Congress of Applied Mechanics* (Stockholm, 24–29 août 1930) 3: 178–180.

Van der Pol, Balthasar. 1934. The nonlinear theory of electric oscillations. *Proceedings of the Institute of Radio Engineers* 22: 1051–1086.

Van der Pol, Balthasar. 1938a. Oscillations de relaxation et démultiplication de fréquence. Paris: Hermann (Actualités scientifiques et industrielles, vol. 708): 69–80. (Réunion Internationale de Physique-Chimie-Biologie, Congrès du Palais de la Découverte. Paris, Octobre 1937).

Van der Pol, Balthasar. 1938b. Beyond radio. *Proceedings of world radio convention*, 3–15. I.R.E., Sydney, Avril.

Van der Pol, Balthasar. 1960. *Selected scientific papers*, vols. I & II, édité par H. Bremmer et C. J. Bouwkamp, introduction de H. B. G. Casimir. Amsterdam: North-Holland Publishing Co.

Van der Pol, B., and K. Posthumus. 1925. Au sujet du rendement des triodes. *Onde Électrique* 4(1925): 536–539.

Van der Pol, B., and Jan Van Der Mark. 1927a. Frequency Demultiplication. *Nature* 120(1927): 363–364.

Van der Pol, B., and Jan Van Der Mark. 1927b. Quelques expériences avec les triodes et les oscillations de relaxation. *Onde Électrique* 6(1927): 461–464.

Van der Pol, B., and Jan Van Der Mark. 1928a. Le battement du cœur considéré comme oscillation de relaxation et un modèle électrique du cœur. *Onde Électrique* 7(1928): 365–392.

Van der Pol, B., and Jan Van Der Mark. 1928b. The heartbeat considered as a relaxation oscillation, and an electrical model of the heart. *The London, Edinburgh, and Dublin Philosophical Magazine and Journal of Science* (VII) 6(1928): 763–775.

Van Dyke, Milton. 1982. *An Album of Fluid Motion*. Stanford: The Parabolic Press.

Vaquez, Henri, and Édouard Donzelot. 1926. *Les troubles du rythme cardiaque*. Paris: Librairie J. B. Baillière & Fils.

Vaquez, Henri. 1928. *Les maladies du cœur*, 2e éd. Paris: Librairie J. B. Baillière & Fils.
Verhulst, Ferdinand. 1996. *Nonlinear differential equations and dynamical systems*, 2e éd., New York: Springer.
Verhulst, Pierre-François. 1838. Notice sur la loi que suit la population dans son accroissement. *Correspondances Mathématiques et Physiques* X(1838): 113–121.
Villat, Henri. 1929. La théorie de l'expérience de M. Henri Bénard sur les tourbillons alternés dans une cuve limitée par deux parois fixes parallèles. *Annales scientifiques de l'École Normale Supérieure* (III) 46(1929): 259–281.
Vogel, Théodore. 1950. Sur certains mouvements périodiques à déferlements. *Comptes rendus hebdomadaires des séances de l'Académie des sciences* 231(17 juillet 1950): 207–209.
Vogel, Théodore. 1953. *Actes du Colloque international des vibrations non linéaires*, Île de Porquerolles, 18–21 septembre 1951, Publications Scientifiques et Techniques du Ministère de l'Air 281(1953): 1–17.
Vogel, Théodore. 1953a. Sur les systèmes déferlants. *Bulletin de la Société Mathématique de France* 81(1953): 63–75.
Vogel, Théodore. 1953b. Systèmes dynamiques héréditaires à déferlement. *Rendiconti del Seminario Matematico della Università di Padova* 22(1953): 64–80.
Volterra, Vito. 1926. Variazioni e fluttuazione del numero d'individui in specie animali conviventi. *Mem. Acad. Lincei.* (s. 6) 2(1926): 31–113.
Volterra, Vito. 1928. Variations and fluctuations of the number of individuals in animal species living Together. *Journal du Conseil international pour l'exploration de la mer* Copenhague III(1928): 3–51.
Volterra, Vito. 1931. *Leçons sur la Théorie Mathématique de la Lutte pour la Vie*, Rédigées par M. Brelot, Cahiers Scientifiques, vol. 7. Paris: Gauthier-Villars.
Von Karman, Théodore. 1940. The engineer grapples with nonlinear problems. *Bulletin of the American Mathematical Society* 46(1940): 615–683.
Von Karman, Théodore, and Hans Rubach. 1912. Über den mechanismus des flüssigkeits und luftwiderstandes. *Physikalische Zeitschrift* 13(1912): 49–59.
Walter, Scott, éd. 2007. *La correspondance entre Henri Poincaré et les physiciens, chimistes et ingénieurs*. Basel: Birkhäuser (coll. Publications des Archives Henri-Poincaré).
Weil, André. 1932. On systems of curves on a ring-shaped surface. *The Journal of the Indian, Mathematical Society* Part I, 19(1931–1932): 109–114.
Weil, André. 1936. Les familles de courbes sur le tore. *Matematicheskii Sbornik* 43(1936): 779–781.
Weil, André. 1948. L'avenir des mathématiques, dans F. Le Lionnais, dir., Les Grands courants de la pensée mathématique, *Cahiers du Sud*, 1948; rééd, 307–320; Paris: Hermann, 1998.
Weyl, Hermann. 1926. Integralgleichungen und fastperiodische Funktionen. *Mathematische Annalen* 97: 338–356.
Williame, Aimé. 1906. Note sur la théorie de l'arc chantant. *Annales de la Société scientifique de Bruxelles* 179–202.
Witkowski, Nicolas. 1995. La chasse à l'effet papillon. *Alliage* 22(1995): 46–53.
Witt, Aleksandr. 1932. Sur la stabilité du mouvement quasi-périodique. *Comptes rendus hebdomadaires des séances de l'Académie des sciences* 195(11 juillet 1932): 101–103.
Witt, Aleksandr. 1932. Sur l'amorçage des oscillations de très haute fréquence. *Comptes rendus hebdomadaires des séances de l'Académie des sciences* 195(28 novembre 1932): 1005–1007.
Witt, Aleksandr. 1937. Sur la théorie de la corde de violon. *Journal de Physique Technique de l'URSS* 4(1937): 261–288.
Witz, Aimé. 1889a. Des inversions de polarités dans les machines série-dynamos. *Comptes rendus hebdomadaires des séances de l'Académie des sciences* 108(juillet-décembre 1889): 1243–1246.
Witz, Aimé. 1889b. Recherches sur les inversions de polarité des série-dynamos. *Journal de Physique Théorique et Appliquée* (8e série) 1(1889): 581–586.

Witz, Aimé. 1890. Recherches sur les inversions de polarité des série-dynamos. *La Lumière Électrique* 4(1890): 193–196.
Zenneck, Jonathan. 1915. *Wireless telegraphy*. New York: McGraw-Hill.
Zenneck, Jonathan. 1929. The importance of radiotelegraphy in science. *Proceedings of the Institute of Radio Engineers* 17(1929): 89–114.

Index nominum

J.-M. Ginoux, *History of Nonlinear Oscillations Theory in France (1880–1940)*,
Archimedes 49, DOI 10.1007/978-3-319-55239-2

Index

J.-M. Ginoux, *History of Nonlinear Oscillations Theory in France (1880–1940)*, Archimedes 49, DOI 10.1007/978-3-319-55239-2

M

N

O

P

Q

R

S

T

W

MIX
Papier aus verantwortungsvollen Quellen
Paper from responsible sources
FSC® C105338

If you have any concerns about our products,
you can contact us on
ProductSafety@springernature.com

In case Publisher is established outside the EU,
the EU authorized representative is:
Springer Nature Customer Service Center GmbH
Europaplatz 3, 69115 Heidelberg, Germany

Printed by Libri Plureos GmbH
in Hamburg, Germany